Mathematical Engineering

Today, the development of high-tech systems is unthinkable without mathematical modeling and analysis of system behavior. As such, many fields in the modern engineering sciences (e.g. control engineering, communications engineering, mechanical engineering, and robotics) call for sophisticated mathematical methods in order to solve the tasks at hand.

The series Mathematical Engineering presents new or heretofore little-known methods to support engineers in finding suitable answers to their questions, presenting those methods in such manner as to make them ideally comprehensible and applicable in practice.

Therefore, the primary focus is—without neglecting mathematical accuracy—on comprehensibility and real-world applicability.

To submit a proposal or request further information, please use the PDF Proposal Form or contact directly: Dr. Thomas Ditzinger (thomas.ditzinger@springer.com)

Indexed by SCOPUS, zbMATH, SCImago.

Thach Ngoc Dinh · Andreas Rauh ·
Sze Zheng Yong · Zhenhua Wang
Editors

Set-Valued Approaches to Control and Estimation of Uncertain Systems

Theory and Applications

Springer

Editors
Thach Ngoc Dinh
Conservatoire National des Arts et Métiers
Paris, France

Andreas Rauh
Carl von Ossietzky Universität Oldenburg
Oldenburg, Niedersachsen, Germany

Sze Zheng Yong
Northeastern University
Boston, MA, USA

Zhenhua Wang
Harbin Institute of Technology
Harbin, China

ISSN 2192-4732 ISSN 2192-4740 (electronic)
Mathematical Engineering
ISBN 978-3-031-94238-9 ISBN 978-3-031-94239-6 (eBook)
https://doi.org/10.1007/978-3-031-94239-6

This Springer imprint is published by the registered company Springer Nature Switzerland AG
The registered company address is: Gewerbestrasse 11, 6330 Cham, Switzerland

Preface

This book brings together cutting-edge research and innovative methodologies in control and estimation under uncertainty. It is the outcome of two highly successful open invited sessions organized at the IFAC World Congress 2023 in Yokohama, Japan. The enthusiastic response and the quality of contributions inspired us to compile this volume, offering a cohesive and detailed exploration of set-valued techniques and their practical applications.

Uncertainty is an inherent aspect of real-world systems, whether arising from modeling inaccuracies, external disturbances, measurement noise, or unpredictable dynamics. Traditional point-based methods often fall short in providing robust and reliable solutions, especially when safety, performance, and guarantees are paramount. Set-valued approaches, by contrast, offer a powerful framework for handling uncertainty by working with sets of possible values for system states or inputs. This paradigm shift enables the design of control and estimation strategies that are inherently robust, capable of providing guaranteed bounds on system behavior, and adaptable to a wide range of applications—from cyber-physical systems and autonomous vehicles to energy systems and beyond.

The book is structured into 10 chapters, encompassing both theoretical and applications-oriented contributions. Nacim Meslem and Andreas Rauh present a secure state estimation algorithm for discrete-time linear systems affected by sensor anomalies. Leila Dadi, Thach Ngoc Dinh, Tarek Raïssi, Haifa Ethabet, and Mohamed Aoun propose an interval estimation method for switched systems, combining H_∞ observers with zonotopic analysis. Mohammad Khajenejad, Tarun Pati, and Sze Zheng Yong develop optimal interval observers for nonlinear systems with bounded Jacobians, while Hector Rios, Tarek Raïssi, and Denis Efimov introduce a sliding-mode-based robust interval predictive control strategy for constrained uncertain systems.

The volume is further enriched by several chapters presenting practical examples and applications. Ruochen Niu, Elikplim Gah, and Sze Zheng Yong address active model discrimination in switched piecewise affine systems with temporal logic constraints, offering three innovative algorithms for bilevel optimization problems. These algorithms are validated through applications on a lunar/planetary assisted drive system and a vehicle lane-changing scenario. Luc Jaulin investigates minimal contractors and separators for ellipse-based localization, applying his approach to robotic localization tasks. Jitao Li, Yuxi Liu, Yushi Zhang, Mingjun Zhang, and Zhenhua Wang propose a zonotopic fault detection method specifically tailored for autonomous underwater vehicles. Shuang Zhang, Sara Ifqir, and Vicenç Puig introduce a robust state-feedback control strategy for Linear Parameter Varying (LPV) systems utilizing a zonotopic Kalman filter, demonstrating its effectiveness in a path-tracking control application for vehicle lateral dynamics. Ekaterina Auer, Andreas Rauh, and Lorenz Gillner leverage GPU-based parameter identification for cooperative solid oxide fuel cell models. Lastly, Matthieu Borelle, Sylvain Bertrand, Cristina Stoica, Teodoro Alamo, and Eduardo F. Camacho propose a distributed moving horizon estimation technique designed for sensor networks with nonlinear measurements, illustrated through a simulation scenario of constrained localization for a non-cooperative target.

This volume stands as a testament to the growing importance of set-valued techniques in tackling the complexities of uncertain systems. We hope it will serve both as a valuable resource for researchers and practitioners and as a catalyst for future innovations in this highly dynamic and rapidly evolving field.

Paris, France — Thach Ngoc Dinh
Oldenburg, Germany — Andreas Rauh
Boston, USA — Sze Zheng Yong
Harbin, China — Zhenhua Wang
April 2025

Acknowledgements This edited book was supported in part by a fellowship by the *Hanse-Wissenschaftskolleg—Institute for Advanced Study (HWK).*

Contents

Secure State Estimation Algorithm for Discrete-Time Linear Systems: A Set-Valued Approach

Nacim Meslem and Andreas Rauh

Abstract Based on elimination by consistency techniques, a set-valued algorithm is proposed to estimate the actual state vector of discrete-time linear systems in the presence of both bounded uncertainties and sensor anomalies. This algorithm can be considered as an extension of a former predictor-corrector set-valued state estimator to the case where the feasible domains of the bounded uncertainties are time-varying. Moreover, the introduced algorithm comprises several set-membership tests that allow detecting the occurrence of sensor faults or output malicious attacks and then discarding them from the estimation process. Moreover, to defeat the attacker's strategy, only a sub-set of the available sensors is selected randomly at each time instant, to improve the efficiency of the estimation algorithm.

1 Introduction

Set-valued state estimation theory is closely related to elimination by consistency techniques (Jaulin and Walter 1993; Jaulin et al. 2001; Jaulin 2002; Chabert and Jaulin 2009). It consists in characterizing tight outer enclosures of the actual system state vector by taking into account the consistence between the predicted data obtained by simulations and the available data provided by sensors. In the unknown but bounded error paradigm, the problem has been investigated over the last two decades and numerous approaches have been proposed to solve it. For linear discrete-time systems, one can classify these approaches into three categories. The first one, is based on set-valued extensions of the classical Luenberger observer (Le et al. 2013; Mazenc et al. 2014; Cacace et al. 2015; Efimov et al. 2016; Loukkas et al. 2017; Meslem et al. 2018). Here, one can cite the interval observer design methods that rely on the

N. Meslem (✉)
GIPSA-lab, Univ. Grenoble Alpes, CNRS, Grenoble INP*, Grenoble, France
e-mail: nacim.meslem@grenoble-inp.fr

A. Rauh
Distributed Control in Interconnected Systems, Department of Computing Science, Carl von Ossietzky Universität Oldenburg, Oldenburg, Germany
e-mail: andreas.rauh@uni-oldenburg.de

T. N. Dinh et al. (eds.), *Set-Valued Approaches to Control and Estimation of Uncertain Systems*, Mathematical Engineering,
https://doi.org/10.1007/978-3-031-94239-6_1

positive system theory (Efimov et al. 2013; Mazenc et al. 2013; Cacace et al. 2015; Chambon et al. 2016; Chevet et al. 2021) and set-valued observer design methods that are based on the reachability analysis of the system state vector or of its estimation error (Xu et al. 2016; Wang et al. 2018; Meslem and Prieur 2021; Meslem et al. 2022; Meslem 2023). The second family of approaches extends the classical predictor-corrector principle to the case of set-valued computation. Many geometrical shapes (boxes, ellipsoids, zonotopes, ...) have been used to accomplish this objective (Chisci et al. 1996; Kieffer et al. 2002; Alamo et al. 2005; Ben Chabane et al. 2014a, b; Rauh et al. 2023). Note that the main drawback of these methods is the lack of a general convergence proof for the size of the estimated state enclosures. To overcome this issue, a third class of approaches has been emerged (Meslem and Ramdani 2018a, 2020; Meslem et al. 2020; Khan and Xie 2019; Khan et al. 2020), which relies on the observability matrix of the system to implement new and efficient correction stages as introduced in Meslem and Ramdani (2018b), Krebs et al. (2019).

Almost all the aforementioned set-valued state estimation approaches provide satisfactory estimation performance in the presence of additive state disturbances and measurement noise. However, the robustness and the resilience of these state estimation approaches against sensor faults and attacks are not yet studied in a comprehensive way. Only a few of the existing methods in the literature deal with these problems. However, the emergence of networked and cyber-physical systems renders robustness and resilience of the estimation algorithms to malicious attacks a major issue to be solved to guarantee the safety of this kind of systems, especially when being part of critical infrastructures. Notice that, a malicious attack can be viewed as a discrete action taken by a hacker that significantly disrupts the normal behavior of the cyber-physical system without being detected by the classical monitoring system devoted to face against faults. Indeed, a typical characteristic of a cyber attack is the concealment. That is, an attacker will tend to mask their intrusion by trying to mislead the supervision process of the cyber-physical system.

Motivated by the discussion above, the present work intends to introduce a novel set-valued state estimation methodology that ensures resilience against adversarial attacks on sensors and robustness in the presence of process and measurement noise, together with the ability to detect and isolate sensors faults. To achieve these objectives, first, the equations of the former set-valued state estimator (Meslem and Ramdani 2018b, 2020) are rewritten in a simpler and compact form and adapted to include the case of time-varying domains of uncertainties. Then, a random selection of sensors is proposed to deceive the attacker's strategy and getting around its vigilance. Moreover, this new set-valued state estimator is combined with two types of consistency tests. The first one is dedicated to detect and isolate sensor faults and the second type of tests is used to reveal malicious attacks that are not detectable from the sensor fault tests. Compared to related works, the main advantages of the proposed approaches can be highlighted by the following points:

- any restrictive assumption is required to apply the introduced state estimator. Indeed, only the classical observability assumption of linear systems is used.
- to cope with the propagation of pessimism, iterative set-valued computations are reduced as much as possible.

- a new prediction scheme, free from the system output, is introduced. The sensor anomalies detection strategy is built on top of it.
- based on the use of the observability matrix, a new set-inversion method that allows one to reconstruct a feasible domain of the whole state vector form the measurement is proposed.

However, in terms of computational complexity, the introduced secure estimation approach is more demanding than point-valued alternatives but stays numerically tractable. Indeed, all the involved matrices and boxes in the different procedures of the proposed set-valued state estimator are of constant sizes, which depend only on the system dimensions.

The remaining parts of this chapter are structured as follows. In Sect. 2, some needed definitions, notations and operations of interval analysis are briefly introduced. Then, based on interval analysis, a predictor-corrector set-valued state estimator is presented in Sect. 3 for discrete-time linear systems. Section 4 introduces the main contribution of this work, where the proposed state estimation algorithm is interfaced with a moving target strategy to randomly select sensors and set-membership tests to detect and isolate faulty sensors. In Sect. 5, a numerical example is studied to show the performance of the proposed secure state estimator and to support the theoretical results. Section 6 completes this paper with a conclusion and some perspectives.

2 Interval Computation

For sake of simplicity, interval analysis (Moore 1966; Alefeld and Mayer 2000; Jaulin et al. 2001) is applied to implement the proposed secure set-valued state estimator. However, to obtain more accurate estimation results, zonotopic set computations (Alamo et al. 2005; Combastel 2015; Scott et al. 2016) can be introduced in future work after a modification of some functions of the proposed algorithm.

In the sequel, an interval vector or a box of dimension n is denote by $[\mathbf{x}]$. More formally, by $[\mathbf{x}]$ one designates a subset of $\mathbb{R}^n$ that can be described in the form

$$\begin{aligned}[\mathbf{x}] &:= \{\mathbf{x} \in \mathbb{R}^n \mid \underline{\mathbf{x}} \leq \mathbf{x} \leq \overline{\mathbf{x}},\ \ \underline{\mathbf{x}},\ \overline{\mathbf{x}} \in \mathbb{R}^n\} \\ &:= [\underline{\mathbf{x}},\ \overline{\mathbf{x}}],\end{aligned} \tag{1}$$

where the real vectors $\underline{\mathbf{x}}$ and $\overline{\mathbf{x}}$ are its lower and upper bounds, respectively. By convention, the size or the width of an interval vector $[\mathbf{x}] \in \mathbb{R}^n$ is defined as the infinity norm $w([\mathbf{x}]) = \|\overline{\mathbf{x}} - \underline{\mathbf{x}}\|_\infty$ of the difference between its lower and upper bounds.

The sum of two boxes $[\mathbf{x}]$ and $[\mathbf{y}]$ in $\mathbb{R}^n$ is a Minkowski sum that we denote by $\oplus$. This sum returns an interval vector $[\mathbf{z}]$ in $\mathbb{R}^n$ according to

$$\begin{aligned}[\mathbf{z}] &:= [\mathbf{x}] \oplus [\mathbf{y}] \\ &:= \{\mathbf{x} + \mathbf{y} \mid \mathbf{x} \in [\mathbf{x}], \mathbf{y} \in [\mathbf{y}]\}.\end{aligned} \tag{2}$$

The upper and lower bound of $[\mathbf{z}]$ can be computed directly by

$$\overline{\mathbf{z}} = \overline{\mathbf{x}} + \overline{\mathbf{y}} \tag{3}$$

and

$$\underline{\mathbf{z}} = \underline{\mathbf{x}} + \underline{\mathbf{y}}. \tag{4}$$

Let $\mathbf{A}$ be a point-valued matrix of dimension $m \times n$ and $[\mathbf{x}]$ be an interval vector of dimension n. The linear image of the box $[\mathbf{x}]$ by the matrix $\mathbf{A}$ is enclosed by

$$\mathbf{A} \cdot [\mathbf{x}] := [\mathbf{A}^{+}\overline{\mathbf{x}} - \mathbf{A}^{-}\underline{\mathbf{x}},\ \mathbf{A}^{+}\underline{\mathbf{x}} - \mathbf{A}^{-}\overline{\mathbf{x}}], \tag{5}$$

where $\mathbf{A}^{+}$ and $\mathbf{A}^{-}$ are non-negative matrices deduced from $\mathbf{A}$ as

$$\mathbf{A}^{+} = \max(0, \mathbf{A}) \tag{6}$$

and

$$\mathbf{A}^{-} = \mathbf{A}^{+} - \mathbf{A}. \tag{7}$$

Here, the max operator is understood in a component-wise manner.

Lemma 1 *Let $\mathbf{A} \in \mathbb{R}^{m\times n}$ and $[\mathbf{x}] \subset \mathbb{R}^{n}$. An upper bound of the width of the box $[\mathbf{z}] = \mathbf{A} \cdot [\mathbf{x}]$ can be computed by*

$$w([\mathbf{z}]) \leq \|\mathbf{A}\|_{\infty} w([\mathbf{x}]), \tag{8}$$

where $\|\mathbf{A}\|_{\infty}$ is the maximum absolute row sum of the matrix $\mathbf{A}$.

Proof By a direct application of interval arithmetic, we have for all $i \in \{1, \ldots, n\}$ and $j \in \{1, \ldots, m\}$

$$\begin{aligned} w\big([z_i]\big) &= \sum_{j=1}^{m} \Big(a_{i,j}[x_j]\Big) \\ &\leq \Big(\sum_{j=1}^{m} |a_{i,j}|\Big) \max_j\{w\big([x_j]\big)\}. \end{aligned}$$

Thus, we can conclude that

$$\begin{aligned} \max_i\{w\big([z_i]\big)\} &= \|\overline{\mathbf{z}} - \underline{\mathbf{z}}\|_{\infty} = w([\mathbf{z}]) \\ &\leq \max_i\Big\{\Big(\sum_{j=1}^{m} |a_{i,j}|\Big)\Big\} \cdot \max_j\{w\big([x_j]\big)\} = \|\mathbf{A}\|_{\infty} \cdot \|\overline{\mathbf{x}} - \underline{\mathbf{x}}\|_{\infty} \\ &= \|\mathbf{A}\|_{\infty} \cdot w([\mathbf{x}]). \end{aligned}$$

This ends the proof. □

3 Set-Valued State Estimation

Consider the class of discrete-time systems whose dynamics can be described by the following difference inclusion

$$\begin{aligned} \mathbf{x}_{k+1} &\in \mathbf{A}\mathbf{x}_k + \mathbf{B}\mathbf{u}_k + \mathbf{E}[\mathbf{w}_k] \\ {}^i\mathbf{y}_k &\in {}^i\mathbf{C}\mathbf{x}_k + {}^i\mathbf{D}\mathbf{u}_k + {}^i\mathbf{F}[{}^i\mathbf{v}_k],\ i \in \{1, \ldots, N\} \\ \mathbf{x}_0 &\in [\mathbf{x}_0] \end{aligned} \tag{9}$$

where $\mathbf{x}_k \in \mathbb{R}^n$, $\mathbf{u}_k \in \mathbb{R}^m$, ${}^i\mathbf{y}_k \in \mathbb{R}^{n_i}$ stand for the state vector, input vector, output vectors of the system, respectively; N is the number of the considered groups of sensors and the matrices $\mathbf{A}$, $\mathbf{B}$, ${}^i\mathbf{C}$, ${}^i\mathbf{D}$, $\mathbf{E}$, and ${}^i\mathbf{F}$ are assumed to be constant with appropriate dimensions. The uncertain parts of (9) are modeled by boxes and defined as follows:

- $[\mathbf{w}_k]$: is the worst-case domain of the modeling error, which includes the state disturbances and process noise;
- $[{}^i\mathbf{v}_k]$: stands for the feasible domain of the output error, which includes measurement noise and sensor inaccuracy;
- $[\mathbf{x}_0]$: is the feasible set of the system initial state.

Moreover, in the case of cyber-physical and networked systems, sensor data could be subject to cyber-attacks or faults. That is, the available measurements ${}^{(m,i)}\mathbf{y}_k$ can be expressed as a sum of the actual system output ${}^i\mathbf{y}_k$ with a false value ${}^i\mathbf{a}_k$. More formally, one has

$${}^{(m,i)}\mathbf{y}_k = {}^i\mathbf{y}_k + {}^i\mathbf{a}_k,\ i \in \{1, \ldots, N\}. \tag{10}$$

Unlike classical point-valued state estimation approaches (Sorenson 1985; Luenberger 1971), set-valued methods (Polyak et al. 2004; Alamo et al. 2005; Wan et al. 2018) aim at estimating a guaranteed enclosure of the actual state vector of the system without requiring any statistical property about the uncertain parts of the system. This objective is introduced more formally in the next subsection.

3.1 Objective

From a mathematical point of view, set-valued state estimation consists in characterizing the reachable set $\mathcal{X}_k$ of the uncertain system (9) that is consistent with the available measurements. That is, at each time instant k, the aim is to compute an outer approximation of the set

$$\mathcal{X}_k = \left\{ \begin{array}{c} \mathbf{x}_k \mid \\ \text{Consistency with system dynamics} \\ \mathbf{x}_k = \mathbf{A}\mathbf{x}_{k-1} + \mathbf{B}\mathbf{u}_{k-1} + \mathbf{E}\mathbf{w}_{k-1}, \\ \mathbf{x}_{k-1} \in [\mathbf{x}_{k-1}] \\ \mathbf{w}_{k-1} \in [\mathbf{w}_{k-1}] \\ \wedge \\ \text{Consistency with measurements} \\ \mathbf{x}_k \in \textbf{Set-Inversion}\big([{}^i\mathbf{y}_k]\big) \end{array} \right\}. \tag{11}$$

In (11), by **Set-Inversion**(.) we denote a set-valued inversion function that returns an outer enclosure of the state vector $[{}^{\text{inv}}\mathbf{x}_k]$ from the feasible domain of system outputs

$$[{}^i\mathbf{y}_k] = {}^{(m,i)}\mathbf{y}_k \oplus (-{}^i\mathbf{F}[{}^i\mathbf{v}_k]),\ i \in \{1, \ldots N\} \tag{12}$$

Note that, in (12) ${}^{(m,i)}\mathbf{y}_k$ stands for the measured value of the system output at the time instant k by the ith sensor.

Thus, one can define a set-valued state estimator as a predictor-corrector algorithm that returns a sequence of outer approximations (boxes) of the set (11) with a converging width. All parts of the used set-valued algorithm in this contribution are derived in detail in the next subsections.

3.2 *Interval Predictor*

To avoid the propagation of pessimism in set-valued numerical algorithms, an interval extension of the general solution of discrete-time linear systems Krasnochtanova et al. (2010)

$$\mathbf{x}_k = \mathbf{A}^{k-s}\mathbf{x}_s + \sum_{j=0}^{k-s-1} \mathbf{A}^{k-s-j-1}\mathbf{B}\mathbf{u}_{s+j} + \sum_{j=0}^{k-s-1} \mathbf{A}^{k-s-j-1}\mathbf{E}\mathbf{w}_{s+j}, \tag{13}$$

where k is the current time instant and s stands for the initial time instant, is proposed. This extension can be expressed by the following compact interval formula,

$$[{}^p\mathbf{x}_k] = \mathbf{A}^{\sigma(k)}[\mathbf{x}_s] \oplus \mathbf{N}_{\sigma(k)}[\mathbf{P}_{s:s+\sigma(k)-1}] \oplus \mathbf{B}_{\sigma(k)}\mathbf{U}_{s:s+\sigma(k)-1}. \tag{14}$$

The parameters and matrices involved in (14) are defined by

$$
\begin{aligned}
\sigma(k) &= k - s \\
\mathbf{N}_{\sigma(k)} &= \left(\mathbf{A}^{\sigma(k)-1}\mathbf{E},\ \mathbf{AE},\ \ldots,\ \mathbf{E}\right) \\
\mathbf{B}_{\sigma(k)} &= \left(\mathbf{A}^{\sigma(k)-1}\mathbf{B},\ \mathbf{AB},\ \ldots,\ \mathbf{B}\right) \\
[\mathbf{P}_{s:s+\sigma(k)-1}] &= \left([\mathbf{w}_s]^T,\ [\mathbf{w}_{s+1}]^T,\ \ldots,\ [\mathbf{w}_{s+\sigma(k)-1}]^T]\right)^T \\
[\mathbf{U}_{s:s+\sigma(k)-1}] &= \left(\mathbf{u}_s^T,\ \mathbf{u}_{s+1}^T,\ \ldots,\ \mathbf{u}_{s+\sigma(k)-1}^T]\right)^T.
\end{aligned}
\tag{15}
$$

Note, in (14) the sequence of future state enclosures $[\mathbf{x}_k]$, $\forall k > s$ is directly computed from the initial box $[\mathbf{x}_s]$. Thus, the set-valued formula (14) is a means to avoid the *wrapping effect* and *dependency phenomenon*.

Remark 1 Compared to the former works (Meslem and Ramdani 2018b; Meslem et al. 2020), formulas (14)–(15) are more general. They take into account the case of time-varying domains of process and measurement noise. This fact allows one to use local bounds on these uncertain signals rather than their global bounds. Thus, by construction, tighter predictor enclosures are expected.

3.3 *Interval-Based State Correction at Past-Time Instants*

For sake of simplicity, we consider in this work the case where $n_i = 1$ for all $i \in \{1, \ldots N\}$. Thus, under the observability assumption on the pairs $(\mathbf{A}, {}^i\mathbf{C})$, it is always possible to characterize a guaranteed enclosure of the system state, at any past time instant $\sigma(k) = k - (n-1)$, from the knowledge of its $n-1$ successive outputs and inputs. The set-valued formula that allows one doing that is defined by

$$
[{}^{(\mathrm{inv},i)}\mathbf{x}_{\sigma(k)}] = {}^{(\mathrm{inv},i)}\hat{\mathbf{x}}_{\sigma(k)} \oplus {}^i\Xi_p[\mathbf{P}_{\sigma(k):\sigma(k)+n-2}] \oplus {}^i\Xi_z[{}^i\mathbf{Z}_{\sigma(k):\sigma(k)+n-1}], \tag{16}
$$

where ${}^{(\mathrm{inv},i)}\hat{\mathbf{x}}_{\sigma(k)}$ stands for a point-valued estimate of the state vector at the past time instant $\sigma(k)$,

$$
{}^{(\mathrm{inv},i)}\hat{\mathbf{x}}_k = ({}^iO)^{-1}\left({}^i\mathbf{Y}_{\sigma(k):\sigma(k)+n-1} - {}^iO_u\mathbf{U}_{\sigma(k):\sigma(k)+n-1}\right), \tag{17}
$$

while the boxes ${}^i\Xi_p[\mathbf{P}_{\sigma(k):\sigma(k)+n-2}]$ and ${}^i\Xi_z[{}^i\mathbf{Z}_{\sigma(k):\sigma(k)+n-1}]$ represent guaranteed enclosures of the estimation error, which are related to modeling errors and measurement noise, respectively, with

$$
\begin{aligned}
{}^i\Xi_p &= -({}^iO)^{-1}\,{}^iO_d \text{ and} \\
{}^i\Xi_z &= -({}^iO)^{-1}\,{}^i\mathbf{H}.
\end{aligned}
\tag{18}
$$

All vectors, matrices and boxes used in (16)-(18) are defined as follows:

- $\mathbf{U}_{\sigma(k):\sigma(k)+n-1}$ and ${}^{i}\mathbf{Y}_{\sigma(k):\sigma(k)+n-1}$ represent the inputs and outputs of the systems collected over a time-window of size equal to n

$$\mathbf{U}_{\sigma(k):\sigma(k)+n-1} = \begin{pmatrix} \mathbf{u}_{\sigma(k)} \\ \mathbf{u}_{\sigma(k)+1} \\ \vdots \\ \mathbf{u}_{\sigma(k)+n-1} \end{pmatrix} \tag{19}$$

$${}^{i}\mathbf{Y}_{\sigma(k):\sigma(k)+n-1} = \begin{pmatrix} {}^{(m,i)}\mathbf{y}_{\sigma(k)} \\ {}^{(m,i)}\mathbf{y}_{\sigma(k)+1} \\ \vdots \\ {}^{(m,i)}\mathbf{y}_{\sigma(k)+n-1} \end{pmatrix}; \tag{20}$$

- ${}^{i}\mathbf{H} = \mathrm{diag}(n, {}^{i}\mathbf{F})$ is an n blocks diagonal matrix, where each of its diagonal blocks i corresponds to ${}^{i}\mathbf{F}$;
- ${}^{(m,i)}\mathbf{y}_{\sigma(k)}$ stands for the measured output at the time instant $\sigma(k)$;
- $[{}^{i}\mathbf{Z}_{\sigma(k):\sigma(k)+n-1}] = \left([{}^{i}\mathbf{v}_{\sigma(k)}]^T, [{}^{i}\mathbf{v}_{\sigma(k)+1}]^T, \ldots, [{}^{i}\mathbf{v}_{\sigma(k)+n-1}]^T]\right)^T$ stands for a vertical aggregation of the feasible domains of the output error;
- $[\mathbf{P}_{\sigma(k):\sigma(k)+n-2}] = \left([\mathbf{w}_{\sigma(k)}]^T, [\mathbf{w}_{\sigma(k)+1}]^T, \ldots, [\mathbf{w}_{\sigma(k)+n-2}]^T]\right)^T$ stands for a vertical aggregation of the feasible domains of state disturbances;
- the expressions of the matrices ${}^{i}O$, ${}^{i}O_u$, and ${}^{i}O_d$ are given for all $i \in \{1, \ldots, N\}$ as follows:

$${}^{i}O = \begin{pmatrix} {}^{i}\mathbf{C} \\ {}^{i}\mathbf{CA} \\ \vdots \\ {}^{i}\mathbf{CA}^{n-1} \end{pmatrix}, \tag{21}$$

$${}^{i}O_d = \begin{pmatrix} \mathbf{0} & \mathbf{0} & \ldots & \mathbf{0} \\ {}^{i}\mathbf{CE} & \mathbf{0} & \ldots & \vdots \\ \vdots & \vdots & \ldots & \vdots \\ {}^{i}\mathbf{CA}^{n-2}\mathbf{E} & {}^{i}\mathbf{CA}^{n-3}\mathbf{E} & \ldots & {}^{i}\mathbf{CE} \end{pmatrix}, \tag{22}$$

$${}^{i}O_u = \begin{pmatrix} {}^{i}\mathbf{D} & \mathbf{0} & \ldots & \mathbf{0} & \mathbf{0} \\ {}^{i}\mathbf{CB} & {}^{i}\mathbf{D} & \ldots & \vdots & \vdots \\ \vdots & \vdots & \ldots & \vdots & \vdots \\ {}^{i}\mathbf{CA}^{n-2}\mathbf{B} & {}^{i}\mathbf{CA}^{n-3}\mathbf{B} & \ldots & {}^{i}\mathbf{CB} & {}^{i}\mathbf{D} \end{pmatrix}. \tag{23}$$

Based on (16), for all $k > n-1$, the predicted state enclosure can be corrected at the past time instant $\sigma(k)$ by

$$[{}^c\mathbf{x}_{\sigma(k)}] := [{}^{(\text{inv},i)}\mathbf{x}_{\sigma(k)}] \cap [{}^p\mathbf{x}_{\sigma(k)}]. \tag{24}$$

Then, from $[{}^c\mathbf{x}_{\sigma(k)}]$ one can predict the future enclosure of the state vector at time instant $k+1$ by

$$[{}^p\mathbf{x}_{k+1}] := \mathbf{A}^n[{}^c\mathbf{x}_{\sigma(k)}] \oplus \mathbf{N}_n[\mathbf{P}_{\sigma(k):k}] \oplus \mathbf{B}_n\mathbf{U}_{\sigma(k):k}, \tag{25}$$

where

$$\begin{aligned} \mathbf{N}_n &= \left(\mathbf{A}^{n-1}\mathbf{E},\ \mathbf{AE},\ \ldots\ \mathbf{E}\right) \\ \mathbf{B}_n &= \left(\mathbf{A}^{n-1}\mathbf{B},\ \mathbf{AB},\ \ldots\ \mathbf{B}\right). \end{aligned} \tag{26}$$

3.4 Interval State Estimator

The previous set-valued operators, namely, set-inversion (16), set-intersection (24), and set-propagation (25), are gathered in the same algorithm to design a predictor-corrector set-valued state estimator that provides an outer enclosure of the reachable set defined in (11). This algorithm is constituted by two main phases detailed as follows.

Phase 1: Interval-based predictor

- **For** $k := 1$ to $k := n-1$
  1. $[{}^p\mathbf{x}_k] := \mathbf{A}^k[\mathbf{x}_0] \oplus \mathbf{N}_k[\mathbf{P}_{0:k-1}] \oplus \mathbf{B}_k\mathbf{U}_{0:k-1}$
  2. $[\mathbf{x}_k] := [{}^p\mathbf{x}_k]$
- **end**

Phase 2: Interval-based predictor-corrector

- **For** $k \geq n-1$ to ∞
  3. $\sigma(k) := k - (n-1)$
  4. **Set-inversion: For** $i = 1$ **to** $i = N$

$$[{}^{(\text{inv},i)}\mathbf{x}_{\sigma(k)}] = {}^{(\text{inv},i)}\hat{\mathbf{x}}_{\sigma(k)} \oplus^i \Xi_p[\mathbf{P}_{\sigma(k):\sigma(k)+n-2}] \oplus^i \Xi_z[{}^i\mathbf{Z}_{\sigma(k):\sigma(k)+n-1}]$$

  5. **Set-intersection**

$$[{}^c\mathbf{x}_{\sigma(k)}] := \cap_{i=1}^N[{}^{(\text{inv},i)}\mathbf{x}_{\sigma(k)}] \cap [{}^p\mathbf{x}_{\sigma(k)}]$$

6. **Set-propagation**

$$[{}^p\mathbf{x}_{k+1}] := \mathbf{A}^n[{}^c\mathbf{x}_{\sigma(k)}] \oplus \mathbf{N}_n[\mathbf{P}_{\sigma(k):k}] \oplus \mathbf{B}_n\mathbf{U}_{\sigma(k):k}$$

7. $[\mathbf{x}_{k+1}] := [{}^p\mathbf{x}_{k+1}]$

- **end**
- **Return** $[\mathbf{x}_k],\ k \in \{1, 2, \ldots, \infty)$

It is worth pointing out that the first phase of the proposed algorithm does not need measurements to estimate an outer enclosure of the state vector. Thus, it is safe against sensor anomalies (faults or cyber-attacks). From the time instant $k = n - 1$, enough measurements are available to perform the set-inversion function. Thus, over a receding horizon time-window, the second stage of this algorithm uses the available data to carry out the correction procedure.

Remark 2 Compared to related works (Meslem and Ramdani 2018b; Meslem et al. 2020), the proposed algorithm merges all available data to obtain tighter state enclosures by discarding all inconsistent parts between them.

Remark 3 Compared to the classical step-by-step iterative set-valued estimation methods, the proposed interval state estimator is more demanding in terms of computational complexity. However, all the matrices and boxes involved in both prediction and correction phases are of bounded sizes. This means that the complexity of the proposed approach is not growing in time and depends only of the system dimensions. Thus, we can claim the numerical tractability of the proposed estimation algorithm.

Remark 4 Notice that the considered multi-sensors strategy is necessary to be able to detect, isolate and ensure guaranteed state estimation in the presence of anomalies. However, the price to pay is the increase of the computational complexity.

Proposition 1 *Under the observability assumption of the pairs*

$$(\mathbf{A}, {}^i\mathbf{C}),\ i \in \{1, \ldots N\}, \tag{27}$$

the proposed algorithm provides an interval sequence $[\mathbf{x}_k],\ k \in \{1, 2, \ldots\}$, *such that:*

- *For all* $k \geq (n-1)$, *the width of the state enclosure* $[\mathbf{x}_k]$ *is lower than,*

$$w\big([\mathbf{x}_k]\big) \leq \beta_v \min_{i \in \{1,\ldots N\}} \{w\big([{}^i\mathbf{Z}_{\sigma(k):\sigma(k)+n-1}]\big)\} + \beta_d w\big([\mathbf{P}_{\sigma(k):\sigma(k)+n-1}]\big), \tag{28}$$

where

- $\beta_v = \left\|\mathbf{A}^n\right\|_\infty \min_{i\in\{1,\ldots N\}}\{\left\|{}^i\Xi_z\right\|_\infty\}$ *and*
- $\beta_d = \left\|\mathbf{N}_n\right\|_\infty + \left\|\mathbf{A}^n\right\|_\infty \min_{i\in\{1,\ldots N\}}\{\left\|{}^i\Xi_p\right\|_\infty\}$.

Proof The proof is omitted because it is quite similar to that provided in Meslem and Ramdani (2018b). □

Remark 5 Based on (28), we can claim that in the case where the considered system is free from uncertainties ($[\mathbf{w}_k] = \mathbf{0}$ and $[\mathbf{v}_k] = \mathbf{0}$) the proposed algorithm is able to reconstruct exactly the actual state vector of the system for all $k \geq n - 1$.

4 Main Results

Now, consider the case where the system outputs are subject to additive anomalies represented by the unknown signal ${}^{i}\mathbf{a}_k$. For instance, ${}^{i}\mathbf{a}_k$ can be a senor fault, failure, or malicious attack. Note, an attack is a discrete action taken by an agent to significantly disrupt the normal behavior of a cyber-physical system without being detected by its monitoring system. Recall that, in these situations, the received measurements from sensors can be decomposed in the form

$$ {}^{(m,i)}\mathbf{y}_k = {}^{i}\mathbf{y}_k + {}^{i}\mathbf{a}_k. \tag{29} $$

Thus, to be resilient in this hostile context, the objective of the proposed set-valued state estimator remains the same. That is, tight outer approximations of the reachable set (11) have to be computed despite the presence of malicious signals ${}^{i}\mathbf{a}_k$. To achieve this goal, various prevention, detection, and isolation strategies are proposed in the remaining parts of this section.

4.1 Set-Membership Filtering

First, to be able to manage efficiently the impact of the output anomalies on the accuracy and the correctness of the estimated state enclosures, we propose in this work the use of a bundle of sensors for which the system is observable. Let us denote by $p < N$ the number of the used senors and define the following set of integers

$$ \mathcal{P} = \{1, \ldots, p\}. \tag{30} $$

Note that these sensors either measure the same or different system outputs. In case where the system is free from anomalies (*i.e.*, ${}^{i}\mathbf{a}_k = 0$), we have the inclusions

$$ {}^{(m,i)}\mathbf{y}_k \in [{}^{(p,i)}\mathbf{y}_k],\ i \in \mathcal{P} \tag{31} $$

with

$$ [{}^{(p,i)}\mathbf{y}_k] = {}^{i}\mathbf{C}[{}^{p}\mathbf{x}_k] + {}^{i}\mathbf{D}\mathbf{u}_k + {}^{i}\mathbf{F}[{}^{i}\mathbf{v}_k],\ i \in \mathcal{P}. \tag{32} $$

Thus, the idea is to design for each sensor (or group of sensors in the more general case) a correction stage that can be evaluated independently of the conveyed information by the other sensors. This allows one

- to increase the tightness of the estimated state enclosures by intersecting all valid inverted state enclosures,
- to detect and isolate the faulty sensors by simple comparisons with the predicted feasible domains of system outputs, and
- to increase the resilience against malicious attacks by selecting randomly the used sensors in the correction stage.

In what follows, these points will be detailed in a more formal way. We start by introducing a fault detection test and then we address the case of sensor attacks.

4.1.1 Fault Detection and Isolation

For each sensor $i \in \mathcal{P}$, we propose to check the following inclusion test to detect an anomaly occurrence:

$$^{(m,i)}\mathbf{y}_k \in [^{(p,i)}\mathbf{y}_k], \begin{cases} \text{True} \Rightarrow s_i = 1 \text{ (Healthy sensor)} \\ \text{False} \Rightarrow s_i = 0 \text{ (Faulty sensor).} \end{cases} \tag{33}$$

Then, from the results of the set-membership test (33), all sensors with $s_i = 0$ are discarded and those with $s_i = 1$ are retained and form the set of valid sensors

$$\mathcal{S} = \{i \in \mathcal{P} \mid s_i = 1\}. \tag{34}$$

Remark 6 All unknown signals $^{i}\mathbf{a}_k$ that induce the violation of the inclusion test (33) are considered as sensor faults.

4.1.2 Prevention of Faulty Estimates Due to Attacks and Resilience

To be able to deceive a possible malicious attack that attempts to corrupt the available measurements without activating the detection test (33), we propose the following moving target deference strategy that is based on a random selection of valid sensors:

- From $\mathcal{S}$ select randomly a subset $\mathcal{S}^{\star}$ of valid sensors to perform several set-inversion stages. That is, based on the interval formula (16) we obtain

$$\forall l \in \mathcal{S}^{\star},\ [^{(\text{inv},l)}\mathbf{x}_{\sigma(k)}]. \tag{35}$$

- Discard all inconsistent boxes $[^{(\text{inv},l)}\mathbf{x}_{\sigma(k)}]$ that satisfy

$$[^{(\mathrm{inv},l)}\mathbf{x}_{\sigma(k)}] \cap [^{p}\mathbf{x}_{\sigma(k)}] = \emptyset \tag{36}$$

and form a new subset

$$\mathcal{S}^{\star\star} = \{l \in \mathcal{S}^{\star} \mid (36) \text{ is false}\}. \tag{37}$$

- Correct the predicted state enclosure at the past time instant $\sigma(k)$ by intersecting all valid inverted state enclosures. That is,

$$[^{c}\mathbf{x}_{\sigma(k)}] = \left(\cap_{l\in\mathcal{S}^{\star\star}} [^{(\mathrm{inv},l)}\mathbf{x}_{\sigma(k)}]\right) \cap [^{p}\mathbf{x}_{\sigma(k)}] \tag{38}$$

Remark 7 It is worth point out that all unknown signals ${}^{i}\mathbf{a}_k$ that induce the violation of the intersection test (36) are considered as malicious attacks while that violating the membership test (33) are considered as sensor faults.

In the next subsection, we endow the second phase of the predictor-corrector state estimation algorithm by the above selection, detection, and filtering tests to design an anomaly resilient set-valued state estimator.

4.2 Secure State Estimation Algorithm

The first phase of the introduced state estimation algorithm, in Sect. 2, does not use measurements. Thus, it is not subject to sensor anomalies and it is not considered in this subsection. The steps of the new secure predictor-corrector estimation procedure are gathered in the following algorithm.

Phase 2: Secure predictor-corrector estimator

3. **Get** $\mathcal{S}$ from Phase 1 (based on the set-membership test (33))
4. **Select** randomly a subset $\mathcal{S}^{\star}$ from $\mathcal{S}$

- **For** $k \geq n-1$ to ∞

5. $\sigma(k) := k - (n-1)$

6. **Set-inversion**: $\forall l \in \mathcal{S}^{\star}$, **compute**

$$[^{(\mathrm{inv},l)}\mathbf{x}_{\sigma(k)}] = {}^{(\mathrm{inv},l)}\hat{\mathbf{x}}_{\sigma(k)} \oplus^{l} \Xi_p[\mathbf{P}_{\sigma(k):\sigma(k)+n-2}] \oplus^{l} \Xi_z[^{l}\mathbf{Z}_{\sigma(k):\sigma(k)+n-1}]$$

7. **Form** the subset $\mathcal{S}^{\star\star}$ (based on the set-membership tests (36)-(37))
8. **Correction step**

$$[^{c}\mathbf{x}_{\sigma(k)}] = \left(\cap_{l\in\mathcal{S}^{\star\star}} [^{(\mathrm{inv},l)}\mathbf{x}_{\sigma(k)}]\right) \cap [^{p}\mathbf{x}_{\sigma(k)}]$$

9. **Set-propagation**

$$[{}^p\mathbf{x}_{k+1}] := \mathbf{A}^n[{}^c\mathbf{x}_{\sigma(k)}] \oplus \mathbf{N}_n[\mathbf{P}_{\sigma(k):k}] \oplus \mathbf{B}_n\mathbf{U}_{\sigma(k):k}$$

10. $[\mathbf{x}_{k+1}] := [{}^p\mathbf{x}_{k+1}]$
11. **Form** a new valid sensors set $\mathcal{S}$
12. **Select** randomly a subset $\mathcal{S}^\star$ from $\mathcal{S}$

- **end**
- **Return** $[\mathbf{x}_k],\ k \in \{1, 2, \ldots, \infty)$

The fault detection test is carried out for the first time in Line 3 and then in Line 11. To cope with possible malicious attacks, a moving target defense strategy is applied in Lines 4 and 12 by a random selection of sensors. Line 6 computes, from the data provided by the selected sensors, inverted state enclosures at the past time instant $\sigma(k)$. Then, Line 7 discards all inverted state enclosures that are impacted by malicious attacks. Line 8, performs a set-filtering operation to increase the tightness of the computed state enclosure at the past time instant $\sigma(k)$. Then, a future state enclosure at the time instant $k + 1$ is computed in Line 9.

5 Illustrative Example

To illustrate the performance of the proposed secure set-valued state estimator, let us consider the following numerical example borrowed from Chang (2006). The state and output equations of this system are defined by the matrices

$$\mathbf{A} = \begin{pmatrix} 0.9630 & 0.0181 & 0.0187 \\ 0.1808 & 0.8195 & -0.0514 \\ -0.1116 & 0.0344 & 0.95861 \end{pmatrix}, \ \mathbf{B} = \begin{pmatrix} 0 \\ 0 \\ 1 \end{pmatrix}, \tag{39}$$

$$\mathbf{E} = \begin{pmatrix} 0.0996 & 0.0213 \\ 0.0050 & 0.1277 \\ 0.1510 & 0.0406 \end{pmatrix}, \tag{40}$$

$${}^1\mathbf{C} = \begin{pmatrix} 1 & 0 & -1 \end{pmatrix}, \ {}^2\mathbf{C} = \begin{pmatrix} -1 & 1 & 1 \end{pmatrix}, \tag{41}$$

and

$${}^1\mathbf{F} = {}^2\mathbf{F} = 1, \ {}^1\mathbf{D} = {}^2\mathbf{D} = 0. \tag{42}$$

The feasible domains of measurement noise, bounded state disturbances and initial states of this system are

$${}^i\mathbf{v}_k \in [-0.01, 0.01], \ i \in \{1, 2\}, \ \forall k \geq 0, \tag{43}$$

$$\mathbf{w}_k \in ([-0.1, 0.1], [-0.1, 0.1])^T, \ \forall k \geq 0, \tag{44}$$

and

$$\mathbf{x}_0 \in ([-10, 10], [-3, 3], [-10, 10])^T. \tag{45}$$

In the sequel, the considered system is simulated from the initial condition

$$\mathbf{x}_0 = (5, 0, 5)^T \tag{46}$$

and under uniformly distributed random state disturbance and measurement noise. In addition, the considered system input is chosen as

$$\mathbf{u}_k = 5\sin(100k). \tag{47}$$

For this example, both pairs $(\mathbf{A}, {}^1\mathbf{C})$ and $(\mathbf{A}, {}^2\mathbf{C})$ are observable. Then, the observability matrices ${}^1\mathcal{O}$ and ${}^2\mathcal{O}$ are non-singular. So far, we have introduced all integers, vectors and matrices needed to apply the proposed secure set-valued state estimator. Notice that in the different simulation tests we have set $p = N = 2$.

5.1 Test 1: Sensors Free From Anomalies

In this first simulation test, we consider the case where the data provided by the two sensors are not subject to any fault or attack.

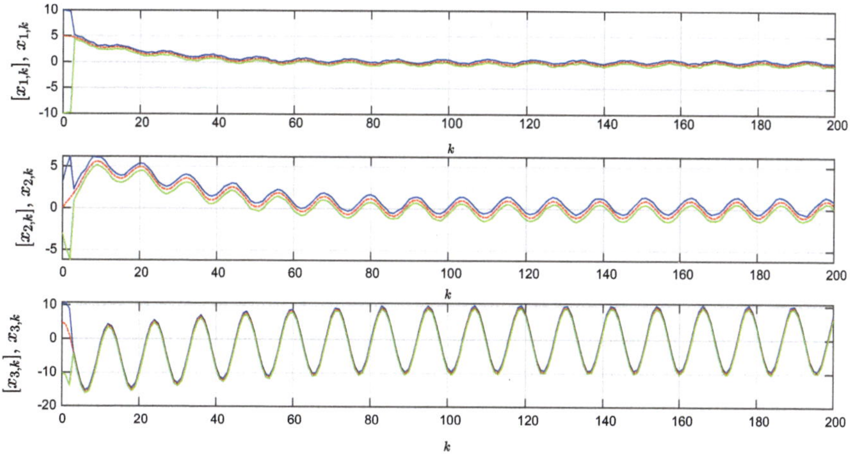

Fig. 1 Test 1, dashed lines represent the actual state variables of the system, while the solid lines correspond to the estimated intervals of these state variables

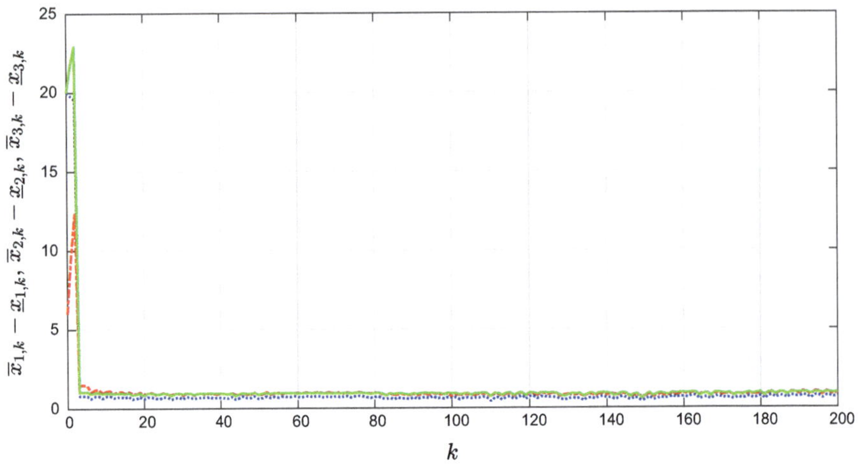

Fig. 2 Test 1, the widths of the estimated intervals of the state variables $[x_{j,k}],\ j \in \{1, 2, 3\}$

Figure 1 shows the estimated state enclosures of each state variable of the system. After the initial prediction stage, which lasts two iterations, the correction stage succeeds to improve the tightness of the estimated state enclosures. In fact at the time instant $k = 2$, we have enough output measurements to carry out the set-version procedure and to correct the state enclosure at the past-time instant $k = 0$. Then, from the corrected state enclosure, the prediction step is performed to obtain a tight state enclosure at the future time instant $k = 3$. The impact of the correction stage on the accuracy of the estimated state enclosure is clearly illustrated in Fig. 2, where the width of each estimated interval $[x_{j,k}]$, $j \in \{1, 2, 3\}$, is plotted. In fact, the width of the predicted state enclosure undergoes a significant decrease at the time instant $k = 3$, where the correction stage is applied for the first time. Thus, from $k = 2$, the sequence of operations (set-inversion, set-correction, and set-prediction) is repeated at each time instant. It is worth recalling that the applied sensor data in the set-inversion operation are collected over a receding horizon time-window. Also, based on the curves in Fig. 2, we can determine graphically that at the steady state the width of the estimated state enclosures $w\big([\mathbf{x}_\infty]\big)$ is lower than 1 and its upper bound given by (15) is equal to 8.11.

Remark 8 It is worth pointing out that compared to interval observer design approaches for LTI systems, the proposed set-valued state estimator has the advantage of not requiring the restrictive positivity assumption on the dynamics of the estimation error. Moreover, on some numerical examples, a performance comparison study, between a former version of the proposed estimation approach and an interval observer method, has been carried out in Meslem and Ramdani (2020) where the superiority of this new scheme to perform set-valued state estimation is shown.

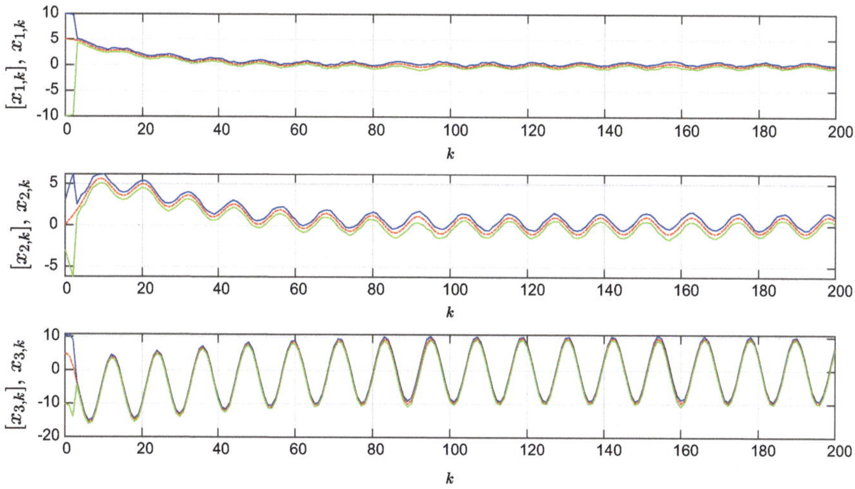

Fig. 3 Test 2, estimated state enclosure in the presence of sensor faults. Dashed lines represent the actual state variables of the system, while the solid lines correspond to the estimated intervals of these state variables

5.2 *Test 2: Sensors Subject to Faults*

In this second test, we assume that the two sensors are affected by faults defined as follows

$$^{1}a_k = \begin{cases} 3 \text{ if } 79 \leq k \leq 88 \\ 0 \quad \text{otherwise} \end{cases} \tag{48}$$

and

$$^{2}a_k = \begin{cases} 3 \text{ if } 149 \leq k \leq 158 \\ 0 \quad \text{otherwise.} \end{cases} \tag{49}$$

The obtained simulation results in this scenario are depicted in Fig. 3. Despite the presence of sensor faults, the proposed set-valued state estimator manages to provide guaranteed enclosures of the actual state vector of the system. The impact of the apparition of the faults, defined in (48)-(49), on the accuracy of the estimation error is shown in Fig. 4. We observe, in this figure, an increase of the width of the estimated state enclosures over the time sequences $k \in \{79, \ldots, 88\}$ and $k \in \{149, \ldots, 158\}$. In fact, during these time intervals only the valid sensor is used to perform the correction stage of the proposed set-valued estimator.

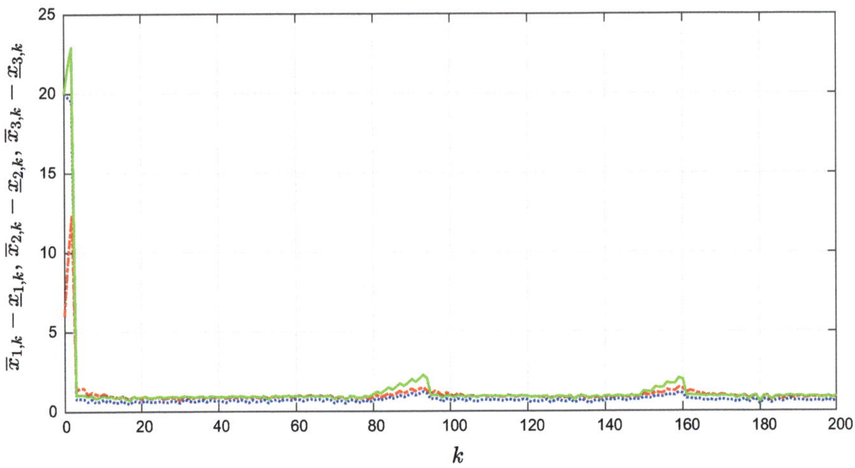

Fig. 4 Test 2, the widths of the estimated intervals of the state variables $[x_{j,k}]$, $j \in \{1, 2, 3\}$ in the presence of sensor faults

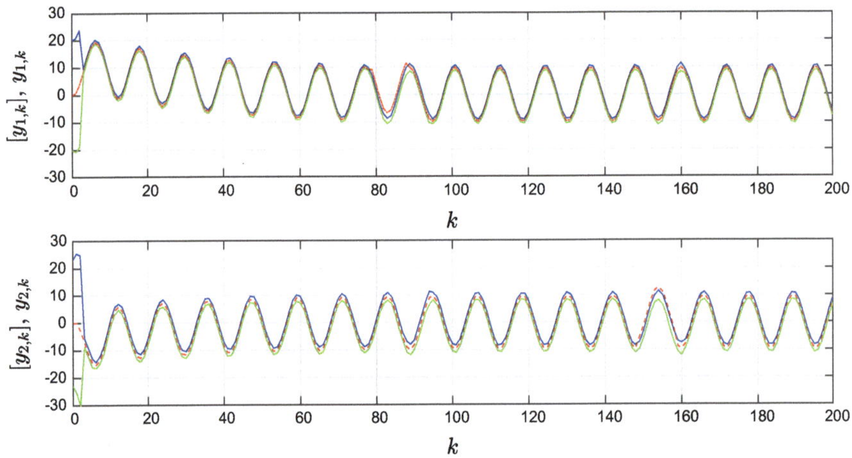

Fig. 5 Test 2, dashed lines represent the measured system outputs and the solid lines correspond to the predicted enclosure of the system outputs

The ability of the introduced state estimator to detect the considered sensor faults are illustrated in Figs. 5 and 6. Figure 5 shows that over the time sequences, where the faults appear, the predicted enclosures of the system outputs do not include the measured system output. This fact is highlighted in Fig. 6 where zooms around these time sequences are performed to show clearly that the proposed senor validation test (33) is not satisfied.

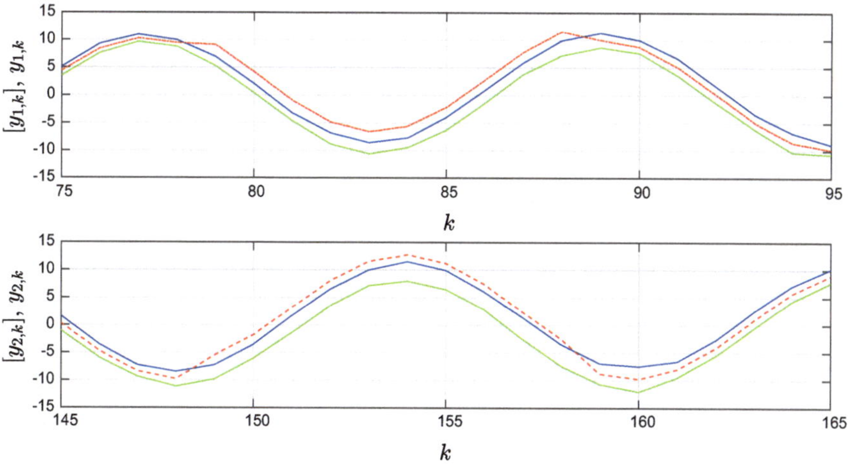

Fig. 6 Test 2, zooms around the time sequences $k \in \{79, \ldots, 88\}$ for the first output and $k \in \{149, \ldots, 158\}$ for the second one. Dashed lines represent the measured system outputs and the solid lines correspond to the predicted enclosure of the system outputs

5.3 *Test 3: Sensors Subject to Malicious Attacks*

Now, let us consider the case of malicious sensor attacks that are defined as

$$ {}^{1}a_k = \begin{cases} 0.1 & \text{if } 79 \leq k \leq 80 \\ 0 & \text{otherwise} \end{cases} \tag{50} $$

and

$$ {}^{2}a_k = \begin{cases} 0.1 & \text{if } 149 \leq k \leq 150 \\ 0 & \text{otherwise.} \end{cases} \tag{51} $$

Note, in this case the sensor validation test (33) fails to detect the presence of these attacks as illustrated in Fig. 7. Indeed, as observed in this figure, even though the measurements are corrupted by malicious attacks the predicted output set still encloses the measured system output. However, thanks to the state validation test (36), these kinds of attacks are detected. This fact is illustrated in Figs. 8 and 9, where the predicted and inverted state enclosures around the time sequences $k \in \{79, 80\}$ and $k \in \{149, 150\}$ are presented. We observe in these figures that when false data are used there are no intersections between the prediction and inverted state enclosures. Figure 8 corresponds to the case, where the first system output is attacked, while Fig. 9 illustrates the case, where the second system output is subject to the attack.

Figure 10 shows that the proposed set-valued state estimator is able to provide guaranteed enclosures of the actual state vector of the system in the presence of malicious attacks that are not detectable by the sensor validation test (33). Finally, the impact of the appearance of the attacks on the estimation performance is illustrated

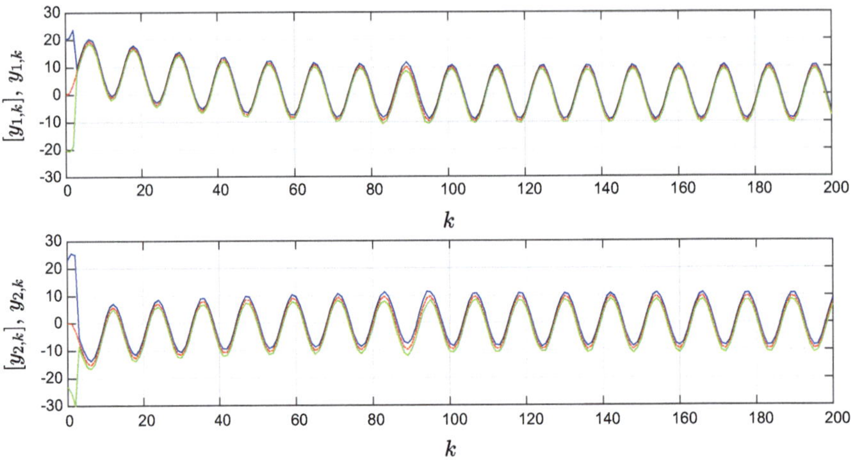

Fig. 7 Test 3, case where the system outputs are affected by malicious attacks. Dashed lines represent the measured system outputs and the solid lines correspond to the predicted enclosures of the system outputs

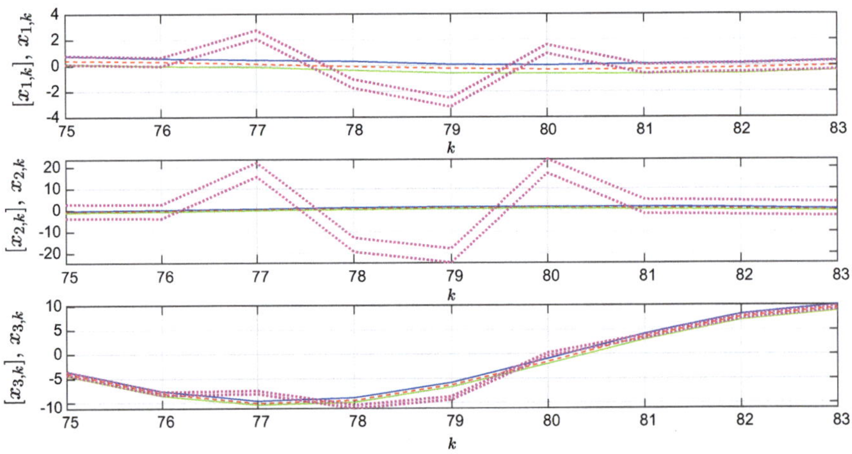

Fig. 8 Test 3, solid lines correspond to the predicted state enclosure while the dotted lines represent the inverted state enclosure from the data generated by the first sensor

in Fig. 11, where we observe an inflation of the width of the estimated state enclosure at each occurrence of an attack. This is related to the use of only the validated data to perform the correction stage, in these situations.

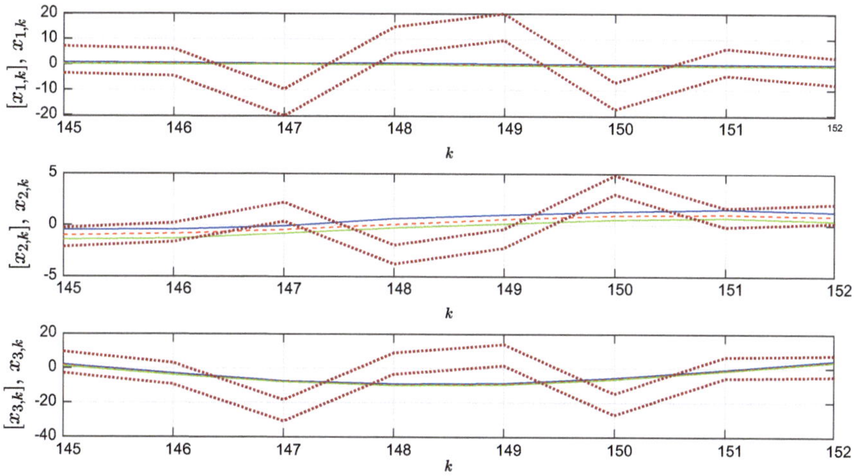

Fig. 9 Test 3, solid lines correspond to the predicted state enclosure while the dotted lines represent the inverted state enclosure from the data generated by the second sensor

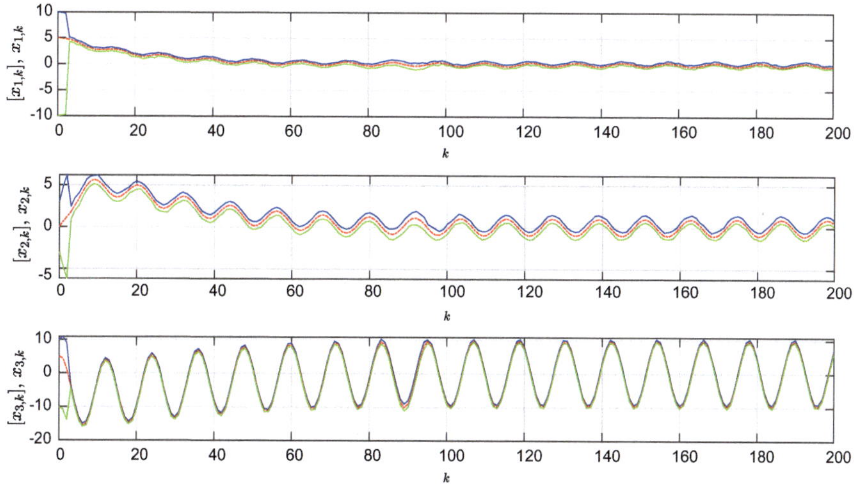

Fig. 10 Test 3, estimated state enclosure in the presence of malicious attacks. Dashed lines represent the actual state variables of the system, while the solid lines correspond to the estimated intervals of these state variables

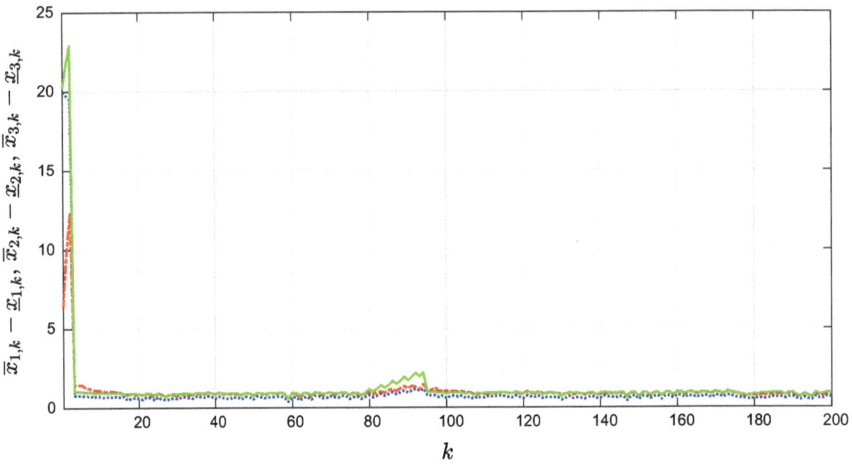

Fig. 11 Test 3, the widths of the estimated intervals of the state variables $[x_{j,k}]$, $j \in \{1, 2, 3\}$ in the presence of malicious attacks

6 Conclusions and Outlook

A set-valued state estimator able to manage sensor faults and attacks has been proposed in this work. This estimator is based on a new predictor-corrector principle, where the observability matrices of the system plays a fundamental role. The implementation of the proposed estimation algorithm is based on interval computation. Moreover, this algorithm is endowed by some set-membership tests that allows one to cope with sensor faults and attacks. Simulation results are presented, under different scenarios, to show the effectiveness of the proposed set-valued state estimation strategy.

In future works, the proposed secure estimation methodology will be extended to more complex classes of cyber-physical and networked systems. Also, specific types of malicious attacks and nonlinear systems will be considered.

References

Alamo T, Bravo JM, Camacho EF (2005) Guaranteed state estimation by zonotopes. Automatica 41:1035–1043

Alefeld G, Mayer G (2000) Interval analysis: theory and applications. J Comput Appl Math 121:421–464

Ben Chabane S, Stoica Maniu C, Alamo T, Camacho EF, Dumur D (2014) Improved set-membership estimation approach based on zonotopes and ellipsoids. In: Proceedings of the European control conference, Strasbourg, France, pp 993–998

Ben Chabane S, Stoica Maniu C, Alamo T, Camacho EF, Dumur D (2014) A new approach for guaranteed state estimation. In: Proceedings on IFAC world congress, Cape town, South Africa, pp 6533–6537

Cacace F, Germani A, Manes C (2015) A new approach to design interval observers for linear systems. IEEE Trans Autom Control 60:1665–1670

Chabert G, Jaulin L (2009) Contractor programming. Artif Intell 173(11):1079–1100

Chambon E, Burlion L, Apkarian P (2016) Overview of linear time-invariant interval observer design: towards a non-smooth optimisation-based approach. IET Control Theory Appl 10:1258–1268

Chang J-L (2006) Applying discrete-time proportional integral observers for state and disturbance estimations. IEEE Trans Autom Control 51:814–818

Chevet T, Rauh A, Dinh TN, Marzat J, Raïssi T (2021) Robust interval observer for systems described by the Fornasini–Marchesini second model. IEEE Control Syst Lett 6:1940–1945

Chisci L, Garulli A, Zappa G (1996) Recursive state bounding by parallelotopes. Automatica 32:1049–1055

Combastel C (2015) Zonotopes and kalman observers: gain optimality under distinct uncertainty paradigms and robust convergence. Automatica 55:265–273

Efimov D, Perruquetti W, Raïssi T, Zolghadri A (2013) On interval observer design for time invariant discrete time systems. In: Proceedings of the European control conference, Zurich, Switzerland, pp 266–269

Efimov D, Perruquetti W, Raïssi T, Zolghadri A (2016) Design of interval observers for estimation and stabilization of discrete-time LPV systems. IMA J Math Control Inf 33(4):1051–1066

Jaulin L (2002) Global optimization and constraint satisfaction. Springer, Berlin, Heidelberg

Jaulin L, Kieffer M, Didrit O, Walter E (2001) Applied interval analysis: with examples in parameter and state estimation, robust control and robotics. Springer, London

Jaulin L, Walter E (1993) Set inversion via interval analysis for nonlinear bounded-error estimation. Automatica 29(4):1053–1064

Khan A, Xie W (2019) Interval state estimator design using the observability matrix for multiple input input multiple output linear time-varying discrete-time systems. IEEE Access 7:167566–167576

Khan A, Xie W, Liu L (2020) Set-membership interval state estimator design using the observability matrix for discrete-time switched linear systems. IEEE Sens J 20(11):6121–6129

Kieffer M, Jaulin L, Walter E (2002) Guaranteed recursive nonlinear state estimation using interval analysis. Int J Adapt Control Signal Proc 16:193–218

Krasnochtanova I, Rauh A, Kletting M, Aschemann H, Hofer EP, Schoop K-M (2010) Interval methods as a simulation tool for the dynamics of biological wastewater treatment processes with parameter uncertainties. Appl Math Modell 34:744–762

Krebs S, Meslem N, Hohmann S (2019) Dynamic set-inversion procedure to design interval-based state estimators for discrete-time lpv systems. In: IEEE 58th conference on decision and control (CDC), Nice, France, pp 3190–3195

Le VTH, Stoica C, Alamo T, Camacho EF, Dumur D (2013) Zonotopic guaranteed state estimation for uncertain systems. Automatica 49(1):3418–3424

Loukkas N, Martinez JJ, Meslem N (2017) Set-membership observer design based on ellipsoidal invariant sets. In: Proceedings of the IFAC 2017 world congress, Toulouse, France, pp 6471–6476

Luenberger DG (1971) An introduction to observers. IEEE Trans Autom Control 16(6):596–602

Mazenc F, Dinh TN, Niculescu SI (2013) Robust interval observers for discrete time systems of luenberger type. In: Proceedings of the American Control Conference, Washington, United States, pp 2484–2489

Mazenc F, Dinh TN, Niculescu SI (2014) Interval observers for discrete-time systems. Int J Robust Nonlinear Control 24:2867–2890

Meslem N (2023) Tight set-valued state estimation by combining reachability analysis and set-filtering approaches. Transactions of the Institute of Measurement and Control. https://doi.org/10.1177/01423312231190193

Meslem N, Loukkas N, Martinez JJ (2018) Using set invariance to design robust interval observers for discrete-time linear systems. Int J Robust Nonlinear Control 28(11):3623–3639

Meslem N, Martinez JJ, Ramdani N, Besançon G (2020) An $\mathcal{H}_\infty$ interval observer for uncertain continuous-time linear systems. Int J Robust Nonlinear Control 30(5):1886–1902

Meslem N, Prieur C (2021) Reachability-based approach to design interval state observers. In: 24th international symposium on mathematical theory of networks and systems MTNS 2020, Cambridge, United Kingdom, pp 125–130

Meslem N, Ramdani N (2018) A comparison of two methods for state estimation: a statistical kalman filter, and a deterministic interval-based approach. In: Proceedings of the 26th mediterranean conference on control and automation (MED 2018), pp 127–132, Zadar, Croatia

Meslem N, Ramdani N (2018) Forward-backward set-membership state estimator based on interval analysis. In: IEEE, editor, Proceedings of the American control conference, Milwaukee, Wisconsin, USA, pp 5161–5166

Meslem N, Ramdani N (2020) A new approach to design set-membership state estimators for discrete-time linear systems based on the observability matrix. Int J Control 93(11):2541–2550

Meslem N, Raïssi T, Besançon G (2022) Further results on the design of a class of discrete-time set-valued state estimators. Int J Robust Nonlinear Control 32(2):649–668

Moore RE (1966) Interval analysis. Prentice-Hall, Englewood Cliff, NJ

Polyak BT, Nazin SA, Durieu C, Walter E (2004) Ellipsoidal parameter or state estimation under model uncertainty. Automatica 40:1171–1179

Rauh A, Soueidan J, Rohou S, Jaulin L (2023) Experimental validation of an ellipsoidal state estimation procedure for a magnetic levitation system. In: 22nd IFAC world congress, Yokohama, Japan, pp 8494–8499

Scott JK, Raimondo DM, Marseglia GR, Braatz RD (2016) Constrained zonotopes: a new tool for set-based estimation and fault detection. Automatica 69:126–136

Sorenson HW (1985) Kalman filtering: theory and application. IEEE Press selected reprint series

Wan J, Sharma S, Sutton R (2018) Guaranteed state estimation for nonlinear discrete-time systems via indirectly implemented polytopic set computation. IEEE Trans Autom Control 63(12):4317–4322

Wang Z, Lim CC, Shen Y (2018) Interval observer design for uncertain discrete-time linear systems. Syst Control Lett 116:41–46

Xu F, Tan J, Wang X, Puig V, Liang B, Yuan B (2016) A novel design of unknown input observers using set-theoretic methods for robust fault detection. In: 2016 American control conference (ACC), Boston, MA, USA, pp 5957–5961

Interval Estimation for Linear Switched Systems Using H_∞ Observer and Zonotopic Analysis

Leila Dadi, Thach Ngoc Dinh, Tarek Raïssi, Haifa Ethabet, and Mohamed Aoun

Abstract In this chapter, we address the challenge of designing an interval state estimation for discrete-time linear switched systems in the Unknown But Bounded Error (UBBE) context. The considered systems are subject to unknown but bounded disturbances and measurement noise in predefined zonotopes. The main contribution lies in the development of a two-step interval state estimation approach which combines a robust observer design providing more degrees of design freedom with a zonotopic analysis technique. First, we design a switched observer based on the H_∞ formalism, which reduces the impact of unknown uncertainties and provides a point estimate of the system state. The observer gains are computed by solving a Linear Matrix Inequality (LMI) obtained using multiple Lyapunov functions, while adhering to the Average Dwell Time (ADT) concept. Second, we combine the robust point estimation with zonotopic analysis techniques to achieve an interval estimation. The proposed approach is evaluated through simulation results on an academic example, demonstrating its effectiveness and advantages over existing methods.

Keywords Interval estimation · Zonotope · Switched linear systems · H_∞ design · LMI

L. Dadi · H. Ethabet · M. Aoun
Research Laboratory Modeling, Analysis and Control of Systems (MACS) LR16ES22, National Engineering School of Gabes, University of Gabes, Gabes, Tunisia
e-mail: mohamed.aoun@macs.tn

T. N. Dinh (✉) · T. Raïssi
Conservatoire National des Arts et Metiers (CNAM), Cedric-Lab, Paris, France
e-mail: ngoc-thach.dinh@lecnam.net

T. Raïssi
e-mail: tarek.raissi@cnam.fr

T. N. Dinh
Hanse-Wissenschaftskolleg—Institute for Advanced Study (HWK), Delmenhorst, Germany

T. N. Dinh et al. (eds.), *Set-Valued Approaches to Control and Estimation of Uncertain Systems*, Mathematical Engineering,
https://doi.org/10.1007/978-3-031-94239-6_2

1 Introduction

In recent years, switched systems have garnered significant attention in the scientific community due to their ability to represent complex and nonlinear behaviors and their real-life applications. Switched systems (Liberzon and Stephen Morse 1999), as a subclass of hybrid dynamical systems, consist of multiple discrete or continuous-time subsystems and a switching rule that controls the switching between them, specifying the active subsystem at each time instant. These systems are widely used to model engineering and physical systems such as switched circuits (Huang et al. 2020), flight control systems (Linh and Morgansen 2010), and chemical processes (Mhaskar et al. 2005), among others.

State estimation for switched systems remains a challenging and active area of research. The problem of state estimation for switched systems was initially studied in 1970 (Ackerson and Fu 1970), and many results have been developed since then, e.g., (Arichi et al. 2015; Birouche et al. 2006; Li et al. 2019) and the references therein. Note that the aforementioned observers provide point estimates of the system state, so their applicability is strongly dependent on system uncertainties.

In the presence of uncertainties, disturbances, and measurement noise, researchers have proposed the use of unknown-but-bounded error descriptions, where modeling errors, disturbances, and measurement noise are unknown but belong to known sets. To handle state bounds, researchers have developed interval observers (Gouzé et al. 2000; Ethabet et al. 2017; Marouani et al. 2018; Dinh et al. 2020; Marouani et al. 2022) and set-valued approaches. While interval observers calculate all admissible values and provide two bounds that enclose the real system trajectories, set-valued estimators aim to provide a compact set containing possible state values using geometrical sets such as ellipsoids (Liu et al. 2016), parallelotopes (Chisci et al. 1996), zonotopes (Combastel 2003) etc.

Zonotopes are linear images of hypercubes, and they have been used in various studies to guarantee the state estimation of different families of dynamical systems. For instance, zonotope-based estimation has been designed for discrete-time linear delayed and time-invariant systems in Sehli et al. (2020) and Tang et al. (2019), respectively. A new zonotopic Kalman filter-based interval observer (Chevet et al. 2021) is proposed for a discrete-time linear time-invariant system subject to unknown inputs and unknown but bounded perturbation and measurement noise. Ifqir et al. proposed a set-membership state estimation approach for uncertain discrete-time switched systems based on zonotopes using W_i-Radius Minimization in Ifqir et al. (2020). In addition, Zhang et al. (2020), Fei et al. (2021), Ifqir et al. (2022) addressed zonotope-based interval estimation and zonotopic set-membership state estimation approaches for switched systems. Recently, Fei et al. (2022) investigated interval estimation based on zonotopic techniques for asynchronously switched positive systems.

Compared to interval observers, zonotopic methods do not require coordinate changes to ensure the non-negativity of the estimation errors. Therefore, the estimation pessimism can be reduced. However, using purely zonotopic methods is a trade-off between accuracy and computation burden. The present paper proposes a new approach for guaranteed state estimation of discrete-time linear switched systems affected by unknown but bounded disturbances. The proposed approach combines the TNL observer design (where T, N and L denote the weighting matrices and gain used in this strategy) with the zonotopic analysis to achieve a good balance between computation complexity and estimation accuracy. In particular, the paper proposes the use of an H_∞ observer to reduce the impact of unknown uncertainties, followed by the calculation of a zonotopic set for the estimation error to generate two bounds for the state. Compared to existing approaches, the proposed method does not require the non-negativity properties of the estimation error, leading to improved estimation accuracy. The main goal of this paper is to achieve a set-membership approach to estimate the state for discrete-time switched linear systems subject to bounded uncertainties. The main contributions of this paper lie in the following aspects:

- A new observer-based zonotopic technique is designed to achieve the state estimation for switched systems subject to uncertainties. The state estimation method provides not only a point estimation but also computes an outer approximation set. Compared with Marouani et al. (2022), the proposed method gives more accurate interval state estimation.
- By integrating weighted matrices, the proposed observer structure provides additional degrees of design freedom.
- The designed method can attenuate the effect of disturbances by introducing an H_∞ performance in order to guarantee accurate interval estimation results.

The present chapter is organized as follows: Sect. 2 gives some preliminary results. Section 3 presents the problem statement. The main results are detailed in Sect. 4. In Sect. 5, simulations with a comparative example are shown to illustrate the efficiency of the proposed approach. Section 6 concludes the chapter and gives some perspectives.

2 Preliminaries

In this section, some preliminary results including some set definitions and properties that will be used along this chapter are introduced.

2.1 Notations, Basic Results

The following standard notations and definitions are used in this paper. $\mathbb{N}$, $\mathbb{Z}$ and $\mathbb{R}$ denote respectively the sets of natural numbers, integers and real numbers. The sets of positive real numbers and positive integers are denoted by $\mathbb{R}_+ = \{\tau \in \mathbb{R} : \tau \geq 0\}$ and $\mathbb{Z}_+ = \mathbb{Z} \cap \mathbb{R}_+$, respectively. $|x|$ denotes the elementwise absolute value of a vector $x \in \mathbb{R}^n$. $\|x\|_2 = x^T x$ represents the Euclidean norm of a signal x. The L_∞ norm of x is defined as $\|x\|_\infty = \sup_{k\in\mathbb{N}} \|x_k\|_2$. I denotes the identity matrix with any dimension. $\underline{x}$ and $\overline{x}$ are the lower and upper bounds of a variable x such that $\underline{x} \leq x \leq \overline{x}$. The matrices A^T and A^+, with $A \in \mathbb{R}^{n\times m}$, denote respectively the transpose and the Moore-Penrose pseudo-inverse of matrix A. A function $\gamma : \mathbb{R}_+ \to \mathbb{R}_+$ is a $\mathcal{K}$ function if it is continuous, strictly increasing and satisfying $\gamma(0) = 0$, it is $\mathcal{K}_\infty$ function if it is a $\mathcal{K}$ function and $\lim_{s\to+\infty} \gamma(s) = +\infty$. The relation $P \prec 0$ ($P \succ 0$) indicates that $P = P^T$ is strictly negative (positive) definite. The comparison operators $\leq$ and $\geq$ should be understood elementwise for vectors and matrices. The operators $\oplus$ and $\odot$ represent the Minkowski sum and the linear product, respectively. The asterisk $*$ denotes the symmetric term in a symmetric matrix. The following properties and definitions are used in this paper.

Lemma 1 (Radhakrishna and Mitra 1972) *Given three matrices $A \in \mathbb{R}^{a\times b}$, $B \in \mathbb{R}^{b\times c}$ and $C \in \mathbb{R}^{a\times c}$. If* $\text{rank}(B) = c$, *then the general solution of the equation $AB = C$ is defined by*

$$A = CB^\dagger + S(I - BB^\dagger), \tag{1}$$

where $S \in \mathbb{R}^{a\times b}$ is an arbitrary matrix.

Lemma 2 (Schur Complement Zhang (2006)) *Given the matrices $R = R^T$, $Q = Q^T$ and R with appropriate dimensions. The following LMIs are equivalent:*

(i) $$\begin{bmatrix} Q & S \\ S^T & R \end{bmatrix} \succ 0.$$

(ii) $R \succ 0;\ Q - SR^{-1}S^T \succ 0.$

(iii) $Q \succ 0;\ R - S^T Q^{-1} S \succ 0.$

2.2 Zonotopes

Since zonotope technique represents an interesting way for the estimation of the system state, therefore, it is important to recall some properties and definitions of zonotopes.

Definition 1 (Combastel 2003) An s-order zonotope $\mathrm{Z} \subset \mathbb{R}^n$ is an affine transformation of a hypercube $\mathrm{B}^s = \left[-1,\, 1\right]^s$ in $\mathbb{R}^n$ which can be expressed as follows:

$$\mathrm{Z} = \langle p, H\rangle \;= p + H\mathrm{B}^s \;=\; \left\{z \in \mathbb{R}^n : z = p + Hb\right\}, \tag{2}$$

where $p \in \mathbb{R}^n$ denotes the center of Z and $H \in \mathbb{R}^{n\times s}$ is called the generator matrix of Z.

Definition 2 (Zhang et al. 2020) For a zonotope $\mathrm{Z} \subset \mathbb{R}^n$, its interval hull $\mathrm{Box}(Z)$ is defined as the smallest interval vector containing it, which is denoted as follows:

$$\mathrm{Z} \subseteq \mathrm{Box}(\mathrm{Z}) = \left[\underline{z}, \overline{z}\right], \tag{3}$$

where $\left[\underline{z}, \overline{z}\right] = \left\{z \in \mathrm{Z},\ \underline{z} < z < \overline{z}\right\}$ is an interval vector, $\underline{z}$ and $\overline{z}$ denote the lower and upper bounds of z.

Property 1 (Combastel 2003) *Consider two zonotopes* $\mathrm{Z}_1 = \langle p_1, H_1\rangle$ *and* $\mathrm{Z}_2 = \langle p_2, H_2\rangle$. *The Minkowski sum of* Z_1 *and* Z_2 *is given by:*

$$\mathrm{Z}_1 \oplus \mathrm{Z}_2 = \langle p_1 + p_2\,,\ [H_1\quad H_2]\rangle\,, \tag{4}$$

where $p_1, p_2 \in \mathbb{R}^n$ *are known vectors and* $H_1, H_2 \in \mathbb{R}^{n\times s}$ *are determined matrices.*

Property 2 (Combastel 2003) *The linear product of a zonotope* $\mathrm{Z} = \langle p, H\rangle$ *and a matrix* $K \in \mathbb{R}^{m\times n}$ *is denoted as* $\odot$ *and defined as:*

$$K \odot \mathrm{Z} = \langle Kp\,,\ KH\rangle\,. \tag{5}$$

Property 3 (Zhang et al. 2020) *For a zonotope* $\mathrm{Z} = \langle p, H\rangle \subset \mathbb{R}^n$, *its interval hull* $\mathrm{Box}(\mathrm{Z}) = [\underline{z}, \overline{z}]$ *can be obtained by*

$$\begin{cases} \underline{z}_i = p_i - \sum\limits_{j=1}^{s} \left|H_{i,j}\right|, & i = 1, \ldots, n \\ \overline{z}_i = p_i + \sum\limits_{j=1}^{s} \left|H_{i,j}\right|, & i = 1, \ldots, n \end{cases}, \tag{6}$$

with s *is the column number of* H. *According to Definitions 1 and 2, the interval hull of* $\mathrm{Z} = \langle p, H\rangle$ *can also be computed by*

$$\mathrm{Z} = p \oplus H\mathrm{B}^s \subseteq \mathrm{Box}(\mathrm{Z}) = p \oplus \overline{H}\mathrm{B}^n, \tag{7}$$

where $\overline{H} \in \mathbb{R}^{n\times n}$ *is a diagonal matrix as follows:*

$$\overline{H} = \mathrm{Diag}\left(\left[\sum_{j=1}^{s}\left|H_{1,j}\right|, \ldots, \sum_{j=1}^{s}\left|H_{n,j}\right|\right]\right).$$

In the application of zonotopic analysis, the order of the zonotope linearly increases without the reduction operator. Therefore, a useful reduction method has been introduced in Combastel (2003) in order to bound the order of the zonotope. The following property is proposed to describe this reduction operator.

Property 4 (Combastel 2003) *A high-dimensional zonotope* $\mathrm{Z} \subset \mathbb{R}^n$ *can be bounded by a lower dimensional one via a reduction operator*

$$\mathrm{Z} = \langle p, H \rangle \subseteq \langle p, \downarrow_l (H) \rangle \subseteq \mathrm{Box}(\mathrm{Z}), \quad n < l < s, \tag{8}$$

where $p \in \mathbb{R}^n$ *and* $H \in \mathbb{R}^{n \times s}$ *are the center and the generation matrix of Z.* l *is the maximum number of columns of the generated matrix after reduction.* $\downarrow_l (H)$ *represents the complexity reduction operator with* $n < l < s$.

2.3 Average Dwell Time

Definition 3 (Hespanha and Morse 1999) For each switching signal σ and each $k_b \geq k_a \geq 0$, let $N_\sigma(k_b, k_a)$ denote the number of discontinuities of σ in the open interval (k_a, k_b). If there exist a positive integer $N_0 \geq 0$ and a scalar $\tau_a > 0$ such that

$$N_\sigma(k_b, k_a) \leq N_0 + \frac{k_b - k_a}{\tau_a} \tag{9}$$

holds for all k_a and k_b, then τ_a is called the average dwell time (ADT) and N_0 the chatter bound. In this chapter, we suppose that $N_0 = 0$ for simplicity as commonly used in the literature.

2.4 Input to State Stability

Input-to-State Stability (ISS) is a useful stability notion to study the robustness of control systems affected by exogenous signals such as disturbances and measurement noises.

Definition 4 (Jiang and Wang 2001; Sontag 2008) A function $\gamma : \mathbb{R}_+ \rightarrow \mathbb{R}_+$ is a $\mathcal{K}$ function if it is continuous, strictly increasing and $\gamma(0) = 0$, it is $\mathcal{K}_\infty$ function if it is a $\mathcal{K}$ function and $\lim_{s \rightarrow +\infty} \gamma(s) = +\infty$.

A function $\beta : \mathbb{R}_+ \times \mathbb{R}_+ \rightarrow \mathbb{R}_+$ is called a $\mathcal{KL}$ function if, for each fixed values $t \geq 0$, the function $\beta(., t)$ is a $\mathcal{K}$ function and for each fixed values $r \geq 0$, the function $\beta(r, .)$ is decreasing and $\lim_{t \rightarrow +\infty} \beta(r, t) = 0$.

Input-to-State Stability for discrete-time switched systems can be achieved by using multiple Lyapunov functions under a constrained switching signal under the necessary conditions listed in the following Lemma.

Lemma 3 (Zhu et al. 2018) *Consider the discrete-time switched linear system*

$$\begin{cases} x_{k+1} = A_{\sigma_k} x_k + B_{\sigma_k} u_k + w_k \\ y_k = C_{\sigma_k} x_k + v_k \end{cases}, \sigma \in \overline{1, N}. \tag{10}$$

Let $0 < \lambda < 1$ *and* $\mu > 1$. *Suppose that there exist* $V_{\sigma(k)} : \mathbb{R}^n \to \mathbb{R}_+$ *and two* $\mathcal{K}_\infty$ *functions* c_1 *and* c_2 *such that for each* $\sigma_k = q, q \neq l$, *the following conditions hold:*

$$c_1(\|x_k\|) \leq V_q(x_k) \leq c_2(\|x_k\|), \tag{11}$$

$$\Delta V_q(x_k) \leq -\lambda V_q(x_k), \tag{12}$$

$$V_q(x_k) \leq \mu V_l(x_k). \tag{13}$$

Then, the system (10) *is Input to State Stable (ISS) for any switching signal with an average dwell time satisfying*

$$\tau_a \geq \tau_a^* = -\frac{\ln(\mu)}{\ln(1-\lambda)}. \tag{14}$$

3 Problem Statement

Consider the following switched linear system described by

$$\begin{cases} x_{k+1} = A_q x_k + B_q u_k + w_k \\ y_k = C x_k + v_k \end{cases}, q \in \overline{1, N} \tag{15}$$

where $x \in \mathbb{R}^{n_x}$, $y \in \mathbb{R}^{n_y}$ and $u \in \mathbb{R}^{n_u}$ are the state, the control input and the output vector. $w \in \mathbb{R}^{n_w}$ and $v \in \mathbb{R}^{n_v}$ are the disturbances and the measurement noise assumed to be bounded. A_q, B_q and C are known matrices with appropriate dimensions. q is the index of the active subsystem and N is the number of linear subsystems. Let $\{t_q\}$ denote the switching time instants. In this paper, it is supposed that the active subsystem and the switching instants are known. The following assumption is considered.

Assumption 1 The initial state vector x_0, the disturbances w_k and the measurement noise v_k are assumed to be unknown but bounded by known zonotopes:

$$\begin{aligned} x_0 &\in \mathcal{X}_0 = \langle p_0 \, , \, H_0 \rangle, \\ w_k &\in \mathcal{W} = \langle 0 \, , \, H_w \rangle, \\ v_k &\in \mathcal{V} = \langle 0 \, , \, H_v \rangle, \end{aligned} \tag{16}$$

where p_0 is a known vector, H_0, H_w and H_v are known diagonal matrices.

The objective of this work is to derive an interval vector $[\underline{x}_k, \overline{x}_k]$ containing the real state x_k such that

$$\underline{x}_k \leq x_k \leq \overline{x}_k\,, \quad \forall k \in \mathbb{Z}_+.$$

holds despite the disturbances and the measurement noise.

In the sequel, the objective is to develop a robust state interval estimation for the switched linear system (15). To achieve this goal, we construct an H_∞ switched observer using the T-N-L structure (where T, N and L denote the weighting matrices and gain used in this structure) to estimate the system state accurately. Next, we use zonotopic analysis techniques to obtain an interval state estimation.

4 Main Results

This section introduces a novel method for designing interval state estimation using the T-N-L structure (Wang et al. 2018) and zonotopic analysis. Stability analysis is conducted by employing multiple Lyapunov functions under an ADT switching signal. Thanks to T-N-L structure, the proposed approach provides more design degrees of freedom.

Consider the following observer for the system (15)

$$\begin{cases} \xi_{k+1} = T_q A_q \hat{x}_k + T_q B_q u_k + L_q (y_k - C\hat{x}_k), \\ \hat{x}_k = \xi_k + N_q y_k, \end{cases} \tag{17}$$

where ξ_k is an intermediate variable and $\hat{x}_k$ denotes the point state estimation. The matrices $L_q \in \mathbb{R}^{n_x \times n_y}$ are the observer gains to be computed later. The weighted matrices T_q and N_q are constant matrices that should be calculated while satisfying the following condition:

$$T_q + N_q C = I. \tag{18}$$

The relation (18) can be rewritten as follows

$$\begin{bmatrix} T_q & N_q \end{bmatrix} \begin{bmatrix} I \\ C \end{bmatrix} = I. \tag{19}$$

Based on Lemma 1, the matrices T_q , N_q are given as:

$$\begin{bmatrix} T_q & N_q \end{bmatrix} = \begin{bmatrix} I \\ C \end{bmatrix}^{\dagger} + S_q \left(I - \begin{bmatrix} I \\ C \end{bmatrix} \begin{bmatrix} I \\ C \end{bmatrix}^{\dagger} \right). \tag{20}$$

where $S_q, S_q \in \mathbb{R}^{n_x \times (n_x+n_y)}$ $\forall q \in \overline{1,N}$ are arbitrary matrices which will be designed such that all matrices T_q are of full rank. $\begin{bmatrix} I \\ C \end{bmatrix}^\dagger$ is the pseudo-inverse matrix of $\begin{bmatrix} I \\ C \end{bmatrix}$.

Define the estimation error as

$$e_k = x_k - \hat{x}_k. \tag{21}$$

Under the new structure of the proposed observer (17), the main objective is to compute the observer gains L_q that minimize the effect of state disturbances and measurement noise on the estimation error (21).

Bearing in mind (18). From (15), the vector x_k satisfies the following

$$x_{k+1} = T_q A_q x_k + T_q B_q u_k + T_q w_k + N_q C x_{k+1}. \tag{22}$$

It follows that

$$e_{k+1} = (T_q A_q - L_q C) e_k + T_q w_k - L_q v_k - N_q v_{k+1}, \tag{23}$$

which can be rewritten as

$$e_{k+1} = \mathcal{A}_q e_k + \mathcal{D}_q d_k, \tag{24}$$

where

$$\mathcal{A}_q = T_q A_q - L_q C, \;\; \mathcal{D}_q = \begin{bmatrix} I \; L_q \; N_q \end{bmatrix}, \tag{25}$$

$$d_k = \begin{bmatrix} T_q w_k \\ -v_k \\ -v_{k+1} \end{bmatrix}. \tag{26}$$

4.1 Robust State Observer Design Based H_∞ Approach

This part proposes an observer design for the switched system defined by (15) that is robust against uncertainties. We introduce the following theorem to determine the weighting matrices T_q, N_q, and L_q such that the estimation error in (24) is stable and satisfies the H_∞ performance requirement:

$$\|e\|_2 \leq \gamma^2 \|d\|_2, \tag{27}$$

where γ is a positive real number. The observer gains L_q reduce the impact of the disturbances term d on the estimation error.

Theorem 1 *Consider the system* (15). *Suppose that there exists piecewise Lyapunov functions defined by $V_q(e_k) = e_k^T P_q e_k$ and $V_l(e_k) = e_k^T P_l e_k$. The error dynamics in*

(24) *is input-to-state stable and satisfies the* H_∞ *performances* (27) *if for given* $0 < \lambda < 1$ *and* $0 < \beta < 1$*, there exist symmetric and positive definite matrices* $P_q \in \mathbb{R}^{n_x \times n_x}$*, constant matrices* $M_l \in \mathbb{R}^{n_x \times n_x}$*,* $W_q \in \mathbb{R}^{n_x \times n_y}$*,* $Y_q \in \mathbb{R}^{n_x \times (n_x+n_y)}$*, and positive scalars* $0 < c_1 < c_2$*,* $\gamma > 0$ *such that the following conditions hold for all* $l, q \in \overline{1..N}$ *,* $q \neq l$*,*

$$c_1 I \preceq P_q \preceq c_2 I, \tag{28}$$

$$\begin{bmatrix} (\lambda-1)P_q & * & * & * & * \\ 0 & -\gamma^2 I & * & * & * \\ 0 & 0 & -\gamma^2 I & * & * \\ 0 & 0 & 0 & -\gamma^2 I & * \\ K_{1q} & P_q & W_q & K_{2q} & -P_q \end{bmatrix} \preceq 0, \tag{29}$$

$$\begin{bmatrix} M_l & P_q \\ P_q & P_q \end{bmatrix} \succeq 0, \tag{30}$$

$$\min_{P_q, M_l} \beta\mu + (1-\beta)\gamma \quad q \in \overline{1..N}, \tag{31}$$

with

$$\begin{aligned} M_l &= \mu P_l, \\ K_{1q} &= P_q \Theta^+ \alpha_1 A_q + Y_q \psi \alpha_1 A_q - W_q C, \\ K_{2q} &= P_q \Theta^+ \alpha_2 + Y_q \psi \alpha_2, \end{aligned}$$

$$\alpha_1 = \begin{bmatrix} I \\ 0 \end{bmatrix}, \ \alpha_2 = \begin{bmatrix} 0 \\ I \end{bmatrix}, \ \Theta = \begin{bmatrix} I \\ C \end{bmatrix}, \ \psi = I - \Theta\Theta^+.$$

Also, if the state vector x *is bounded, then* $\hat{x}$ *is also bounded under a switching signal with an average dwell time* τ_a *satisfying*

$$\tau_a \geq \tau_a^* = -\frac{\ln(\mu)}{\ln(1-\lambda)}, \tag{32}$$

where $\mu = \frac{c_2}{c_1}$ *and the matrices* L_q*,* T_q *and* N_q *are given by*

$$\begin{cases} L_q = P_q^{-1} W_q \\ T_q = \Theta^\dagger \alpha_1 + P_q^{-1} Y_q \psi \alpha_1 \\ N_q = \Theta^\dagger \alpha_2 + P_q^{-1} Y_q \psi \alpha_2 \end{cases}. \tag{33}$$

In addition, the estimation error (24) *is bounded by*

$$\lim_{k\to\infty} \|e_k\| < \frac{\gamma^2}{c_1 \lambda} \|d_k\|_\infty^2. \tag{34}$$

Proof 1 To show the stability of (24) and the H_∞ performances, sufficient conditions are developed in terms of LMIs based on multiple Lyapunov functions defined by:

$$V_q(e_k) = e_k^T P_q e_k, \quad P_q = P_q^T > 0. \tag{35}$$

The increment of the Lyapunov function (35) is given by:

$$\begin{aligned} \Delta V_q(e_k) &= V_q(e_{k+1}) - V_q(e_k) \\ &= e_k^T \mathcal{A}_q{}^T P_q \mathcal{A}_q e_k - e_k^T P_q e_k + d_k^T \mathcal{D}_q{}^T P_q \mathcal{A}_q e_k \\ &\quad + e_k^T \mathcal{A}_q{}^T P_q \mathcal{D}_q d_k + d_k^T \mathcal{D}_q{}^T P_q \mathcal{D}_q d_k. \end{aligned} \tag{36}$$

To satisfy the H_∞ performances, we are interested in computing observer gains L_q which minimise the following cost function (Dinh et al. 2020):

$$\begin{aligned} &\min_{L_q} \ \gamma^2, \qquad \forall q = 1, \ldots, N, \\ &\text{subject to } \ \|e\|_2 \leq \gamma^2 \|d\|_2. \end{aligned} \tag{37}$$

By adding and subtracting the terms $\lambda e_k^T P_q e_k$ and $\gamma^2 d_k^T d_k$, the above equality then implies

$$\Delta V_q(e_k) = \begin{bmatrix} e_k \\ d_k \end{bmatrix}^T \Lambda_q \begin{bmatrix} e_k \\ d_k \end{bmatrix} - \lambda e_k^T P_q e_k + \gamma^2 d_k^T d_k, \tag{38}$$

where

$$\Lambda_q = \begin{bmatrix} \mathcal{A}_q{}^T P_q \mathcal{A}_q - (1-\lambda) P_q & \mathcal{A}_q{}^T P_q \mathcal{D}_q \\ \mathcal{D}_q{}^T P_q \mathcal{A}_q & \mathcal{D}_q^T P_q \mathcal{D}_q - \gamma^2 I \end{bmatrix}. \tag{39}$$

Note that (39) can be seen as

$$\Lambda_q = \begin{bmatrix} (\lambda - 1) P_q & 0 \\ 0 & -\gamma^2 I \end{bmatrix} + \begin{bmatrix} \mathcal{A}_q^T P_q \\ \mathcal{D}_q^T P_q \end{bmatrix} P_q^{-1} \begin{bmatrix} \mathcal{A}_q^T P_q \\ \mathcal{D}_q^T P_q \end{bmatrix}^T \tag{40}$$

Then, by applying Schur Complement Lemma, $\Lambda_q \preceq 0$ is equivalent to

$$\begin{bmatrix} (\lambda - 1) P_q & * & * \\ 0 & -\gamma^2 I & * \\ P_q \mathcal{A}_q & P_q \mathcal{D}_q & -P_q \end{bmatrix} \preceq 0. \tag{41}$$

Bearing in mind (25) and (33), (41) is equivalent to (29).

In addition, $\Lambda_q \preceq 0$ can be expressed as

$$\begin{bmatrix} \mathcal{A}_q^T P_q \mathcal{A}_q - P_q & * \\ \mathcal{D}_q{}^T P_q \mathcal{A}_q & \mathcal{D}_q^T P_q \mathcal{D}_q \end{bmatrix} + \begin{bmatrix} \lambda P & 0 \\ 0 & -\gamma^2 I \end{bmatrix} \prec 0. \tag{42}$$

Pre- and post- multiplying (42) with $\begin{bmatrix} e_k^T & d_k^T \end{bmatrix}$ and its transpose, it follows that:

$$\begin{bmatrix} e_k \\ d_k \end{bmatrix}^T \begin{bmatrix} \mathcal{A}_q^T P_q \mathcal{A}_q - P_q & \mathcal{A}_q^T P_q \mathcal{D}_q \\ * & \mathcal{D}_q^T P_q \mathcal{D}_q \end{bmatrix} \begin{bmatrix} e_k \\ d_k \end{bmatrix} + \begin{bmatrix} e_k \\ d_k \end{bmatrix}^T \begin{bmatrix} \lambda P_q & 0 \\ 0 & -\gamma^2 I \end{bmatrix} \begin{bmatrix} e_k \\ d_k \end{bmatrix} \prec 0.$$

Therefore, we obtain

$$\Delta V_q(e_k) \leq -\lambda e_k^T P_q e_k + \gamma^2 \|d_k\|^2 . \tag{43}$$

Thus,

$$\Delta V_q(e_k) < -\lambda V_q(e_k) + \gamma^2 d_k^T d_k. \tag{44}$$

Note that when $w_k = 0$ and $v_k = 0$, then $d_k = 0$ and (44) is equivalent to

$$\Delta V_q(e_k) < -\lambda V_q(e_k) < 0. \tag{45}$$

Consequently, (12) is satisfied.

From (44), one can obtain

$$\begin{aligned} V_q(e_k) &\leq (1-\lambda)^k V_q(e_0) + \gamma^2 \sum_{m=0}^{k-1} (1-\lambda)^m \|d_k\|_\infty^2 \\ &\leq (1-\lambda)^k V_q(e_0) + \gamma^2 \tfrac{(1-\lambda^k)}{\lambda} \|d_k\|_\infty^2 \\ &\leq (1-\lambda)^k V_q(e_0) + \tfrac{\gamma^2}{\lambda} \|d_k\|_\infty^2 . \end{aligned} \tag{46}$$

From (28), we arrive at

$$c_1 \|e_k\| \leq V_q(e_k), \tag{47}$$

which allows to deduce that

$$\|e_k\| \leq \frac{1}{c_1}((1-\lambda)^k V_q(e_0) + \frac{\gamma^2}{\lambda} \|d_k\|_\infty^2). \tag{48}$$

Hence, when $k \to \infty$,

$$\lim_{k\to\infty} \|e_k\| < \frac{\gamma^2}{c_1\lambda} \|d_k\|_\infty^2, \tag{49}$$

which shows that the estimation error is bounded by a known term $\frac{\gamma^2}{c_1\lambda}\|d_k\|_\infty^2$. Note that for given c_1 and λ, the latter depends only on γ. Accordingly, the minimization of γ allows us to improve the estimation accuracy. Furthermore, we aim to minimize μ in order to obtain an optimum dwell time. The resolution of such a problem leads to solving a problem of linear optimization, which consists of searching a minimization function. For these purposes, we add the following objective function to the LMIs conditions:

$$\beta\mu + (1-\beta)\gamma, \quad \beta \in [0\ \ 1].$$

Now, let us focus on the stability at the switching instants. By defining $V_l(e_k) = e_k^T P_l e_k$, $P_l = P_l^T > 0$, the inequality defined in (13) can be rewritten as follows

$$\mu P_l - P_q \succeq 0, \tag{50}$$

where $q, l \in \overline{1..N}$, $q \neq l$, q is the current mode.

By using the Schur Complement Lemma, inequality (50) implies

$$\begin{bmatrix} \mu P_l & I \\ I & P_q^{-1} \end{bmatrix} \succeq 0. \tag{51}$$

Let us multiply the both sides of (51) by $\begin{bmatrix} I & 0 \\ 0 & P_q \end{bmatrix}$, one can deduce that (51) is equivalent to (30) with $M_l = \mu P_l$.

Integrating the inequality (12) over the interval $[k_q, k)$ leads to:

$$V_q(e_k) \leq (1-\lambda)^{k-k_q} V_q(e_{k_q}), \forall q \in \overline{1..N}. \tag{52}$$

For all $q, l \in \overline{1..N}$, $q \neq l$,

$$\begin{aligned} V_q(e_k) &\leq (1-\lambda)^{k-k_q} V_q(e_{k_q}) \frac{V_l(e_{k_l})}{V_l(e_{k_l})} \\ &\leq (1-\lambda)^{k-k_q} \frac{e_{k_q}^T P_q e_{k_q}}{e_{k_l}^T P_l e_{k_l}} V_l(e_{k_l}). \end{aligned} \tag{53}$$

Using (11), we obtain:

$$\frac{e_{k_q}^T P_q e_{k_q}}{e_{k_l}^T P_l e_{k_l}} \leq \frac{c_2}{c_1}. \tag{54}$$

The inequality in (53) is equivalent to:

$$V_q(e_k) \leq \frac{c_2}{c_1}(1-\lambda)^{k-k_q} V_l(e_{k_l}). \tag{55}$$

Thus, at the switching instant $k = k_q$, the inequality (55) can be rendered as follows:

$$\begin{aligned} V_q(e_k) &\leq \tfrac{c_2}{c_1} V_l(e_{k_l}) \\ &\leq \mu V_l(e_{k_l}), \end{aligned} \tag{56}$$

with $\mu = \frac{c_2}{c_1} > 1$ because $c_2 > c_1$.

Consequently, the input-to-state stability introduced in Lemma 3 is satisfied, which proves that the estimation error dynamics in (24) is input-to-state stable and satisfies the H_∞ performance.

4.2 Interval State Estimation

Note that $\hat{x}$ represents a single-point approximation of x_k, and thus cannot provide any information about the state bounds. The goal of this section is to achieve interval estimation for (15) using zonotopic techniques based on the designed observer (17). To begin, we compute an approximation set of the error e_k given in (23) by leveraging the known zonotopic bounds of disturbances and measurement noise presented in Assumption 1. From this, we can deduce the bounds of the state x_k. Specifically, we use (21) to obtain the following expression

$$x_k = \hat{x}_k + e_k. \tag{57}$$

The vector $\hat{x}_k$ can be directly calculated by (17). So, if one can find $\overline{e}_k$ and $\underline{e}_k$ such that $\underline{e}_k \leq e_k \leq \overline{e}_k$, then from (57), the interval estimation of x_k can be derived as follows

$$\begin{cases} \overline{x}_k = \hat{x}_k + \overline{e}_k \\ \underline{x}_k = \hat{x}_k + \underline{e}_k \end{cases}, \tag{58}$$

where $\overline{x}_k$ and $\underline{x}_k$ are respectively the upper and the lower bounds of the state x_k.

We introduce the following theorem in order to compute the interval estimation of e_k.

Theorem 2 *If Assumption 1 is satisfied and there exists a zonotope Z^e_{k-1} such that $e_{k-1} \in Z^e_{k-1} = \langle 0,\ H^e_{k-1} \rangle$, then the interval estimation of e_k can be obtained as follows*

$$\begin{cases} \underline{e}_k(i) = -\sum_{j=1}^{n_h} \left| H^e_k(i,j) \right|, & i = 1, \ldots, n_x \\ \overline{e}_k(i) = \sum_{j=1}^{n_h} \left| H^e_k(i,j) \right|, & i = 1, \ldots, n_x \end{cases}, \tag{59}$$

where $\underline{e}_k(i)$ and $\overline{e}_k(i)$ represent the ith component of $\underline{e}_k$ and $\overline{e}_k$. n_h is the column number of H^e_k which satisfies

$$H^e_k = \left[(T_q A_q - L_q C) {\downarrow_l} (H^e_{k-1}) \ T_q H_w \ -L_q H_v \ -N_q H_v \right] \tag{60}$$

Proof 2 From Assumption 1, the initial state satisfies $x_0 \in \mathcal{X}_0 = \langle p_0,\ H_0 \rangle$. Considering $\hat{x}_0 = p_0$ deduces that the initial error $e_0 \in \mathcal{X}_0 \oplus (-\hat{x}_0) = \langle 0,\ H_0 \rangle \triangleq \mathcal{Z}^e_0$. Assume that $e_{k-1} \in Z^e_{k-1} = \langle 0,\ H^e_{k-1} \rangle$.

Based on (23) and (16), at the step k, $e_k \in Z_k^e$ where

$$\begin{aligned} Z_k^e =&(T_q A_q - L_q C) Z_{k-1}^e \oplus T_q \mathcal{W} \oplus (-L_q)\mathcal{V} \oplus (-N_q)\mathcal{V} \\ =&(T_q A_q - L_q C)\left\langle 0,\, H_{k-1}^e \right\rangle \oplus T_q \left\langle 0,\, H_w \right\rangle \\ &\oplus (-L_q)\left\langle 0,\, H_v \right\rangle \oplus (-N_q)\left\langle 0,\, H_v \right\rangle . \end{aligned} \tag{61}$$

Using the properties of zonotopes, we can easily obtain $e_k \in Z_k^e = \left\langle 0,\, H_k^e \right\rangle$ with

$$H_k^e = \left[(T_q A_q - L_q C)(H_{k-1}^e)\ T_q H_w \ \ - L_q H_v \ \ - N_q H_v \right]. \tag{62}$$

As mentioned in Sect. 2, the column number of the matrix H_{k-1}^e linearly increases, which can result in computational complexity. To address this issue, we use the reduction operator $\downarrow_l(.)$ to limit the column number to a fixed value. By applying this operator, we ensure that the inclusion property $\left\langle 0,\, H_{k-1}^e \right\rangle \subseteq \left\langle 0, \downarrow l(H_{k-1}^e) \right\rangle$ is satisfied. Consequently, by replacing the matrix H_{k-1}^e with its reduced form $\downarrow l(H_{k-1}^e)$ in equation (62), we obtain equation (60).

By using Property 3, the interval estimation of e_k can be calculated as follows

$$\begin{cases} \underline{e}_k(i) = -\sum\limits_{j=1}^{n_h} \left| H_k^e(i,j) \right| \ , \ \ i = 1, ..., n_x \\ \overline{e}_k(i) = \ \ \sum\limits_{j=1}^{n_h} \left| H_k^e(i,j) \right| \ , \ \ i = 1, ..., n_x \end{cases}, \tag{63}$$

which ends the proof.

5 Numerical Example

In this section, a numerical example is provided to show the effectiveness of the proposed approach. Consider the discrete-time switched linear system (15) where

$$A_1 = \begin{bmatrix} 0.6 & 0.5 & 1 \\ 0.2 & 0 & 1 \\ 0 & 0.2 & 0 \end{bmatrix}, A_2 = \begin{bmatrix} 0.9 & -0.8 & 0 \\ 0.1 & 0 & 1 \\ 0 & 0.5 & 0.1 \end{bmatrix}, A_3 = \begin{bmatrix} 0.5 & 0.5 & 0.1 \\ 0.4 & 0 & 0.5 \\ 0.1 & 0.2 & 0 \end{bmatrix}$$

$$B_1 = \begin{bmatrix} 1 \\ 0.1 \\ 1.3 \end{bmatrix}, \quad B_2 = \begin{bmatrix} 0.1 \\ 1 \\ 1 \end{bmatrix}, \quad B_3 = \begin{bmatrix} 1.2 \\ 1 \\ 0.5 \end{bmatrix}, \quad C = \begin{bmatrix} 1.2 & 0.01 & 0 \\ 0.1 & 1.1 & 0.1 \end{bmatrix}.$$

In this example, w_k and v_k are uniformly distributed bounded signals such that $w_k \in \mathcal{W} = \langle 0, H_w \rangle$ and $v_k \in \mathcal{V} = \langle 0, H_v \rangle$ with $H_w = 0.2$ and $H_v = 0.1$. According to Assumption 1, we select $p_0 = \begin{bmatrix} 0 & 0 & 0 \end{bmatrix}^T$ and $H_0 = I_3$. Besides, we set the reduction

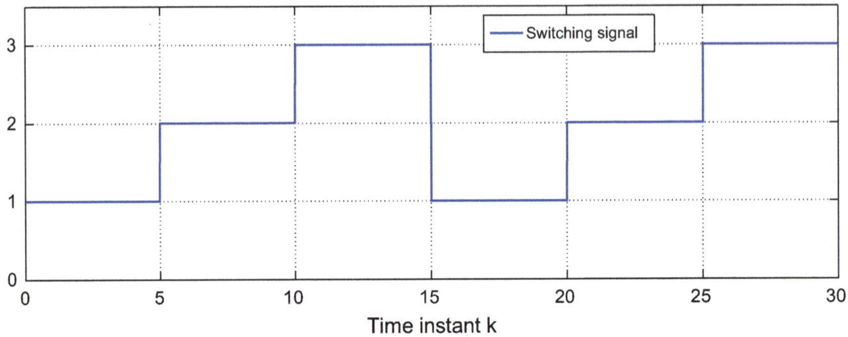

Fig. 1 Switching signal

order of $\downarrow_l (H_k^e)$ as $l = 8$ to limit the column number of the generator matrix. The switching signal, which indicates the active subsystem is plotted in Fig. 1 .

Robust state estimation results are given in the sequel using multiple quadratic Lyapunov functions with an ADT switching signal constraint. The simulation was carried out by using Matlab optimization tools (Yalmip/Sedumi). The conservatism of gain matrices is reduced by integrating weighted matrices T_q and N_q. The existence of a solution for LMIs proposed in Theorem 1 allows one to improve the accuracy of state estimation and to obtain an optimum dwell time.

The optimization problem in (31) can then be solved and the Lyapunov matrices obtained:

$$P_1 = \begin{bmatrix} 0.5228 & -0.0092 & 0.0002 \\ -0.0092 & 0.5211 & 0.0004 \\ 0.0002 & 0.0004 & 0.5326 \end{bmatrix}, \quad P_2 = \begin{bmatrix} 0.5271 & 0.0021 & -0.0044 \\ 0.0021 & 0.5011 & -0.0100 \\ -0.0044 & -0.0100 & 0.5987 \end{bmatrix},$$

$$P_3 = \begin{bmatrix} 0.5305 & -0.0015 & -0.0003 \\ -0.0015 & 0.5127 & 0.0008 \\ -0.0003 & 0.0008 & 0.5348 \end{bmatrix}.$$

The set of observer gains $L_q = P_q{}^{-1} W_q$ are given by:

$$L_1 = \begin{bmatrix} 0.0968 & 0.1354 \\ -0.1001 & -0.1627 \\ -0.0101 & 0.1760 \end{bmatrix}, \quad L_2 = \begin{bmatrix} 0.2876 & -0.2589 \\ -0.0519 & 0.0227 \\ 0.0546 & 0.3332 \end{bmatrix},$$

$$L_3 = \begin{bmatrix} 0.1856 & 0.2443 \\ 0.0072 & -0.0620 \\ 0.0576 & 0.1574 \end{bmatrix}.$$

The weighted matrices T_q and N_q can be obtained as follows:

$$T_1 = \begin{bmatrix} 0.3196 & -0.2903 & -0.0259 \\ -0.3584 & 0.3690 & -0.0571 \\ 0.0054 & 0.0076 & 1.0007 \end{bmatrix}, \quad N_1 = \begin{bmatrix} 0.5454 & 0.2590 \\ 0.2511 & 0.5714 \\ -0.0040 & -0.0069 \end{bmatrix},$$

$$T_2 = \begin{bmatrix} 0.3711 & -0.0272 & -0.0020 \\ -0.0779 & 0.0633 & -0.0852 \\ 0.1279 & -0.0655 & 0.9939 \end{bmatrix}, \quad N_2 = \begin{bmatrix} 0.5224 & 0.0200 \\ -0.0061 & 0.8516 \\ -0.1116 & 0.0606 \end{bmatrix},$$

$$T_3 = \begin{bmatrix} 0.5668 & -0.0680 & -0.0059 \\ -0.1281 & 0.1854 & -0.0740 \\ -0.0361 & 0.0132 & 1.0012 \end{bmatrix}, \quad N_3 = \begin{bmatrix} 0.3561 & 0.0585 \\ 0.0451 & 0.7401 \\ 0.0311 & -0.0123 \end{bmatrix}.$$

Note that T_q and N_q are designed to satisfy the condition (18). In the simulation study, $\mu = 1.2$ which leads to an ADT $\tau_a > 0.0792$ and $\gamma = 0.9423$. The switching signal in Fig. 1 verifies the ADT constraint.

Next, interval state estimation can be calculated using the analysis detailed in Theorem 2. Simulation results of the proposed approach as well as the comparison with Marouani et al. (2022) are shown in Figs. 2 , 3 and 4 where solid lines correspond

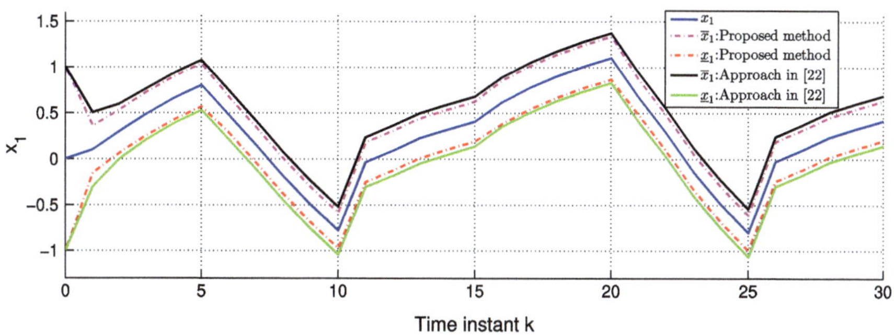

Fig. 2 x_1 and its interval estimation

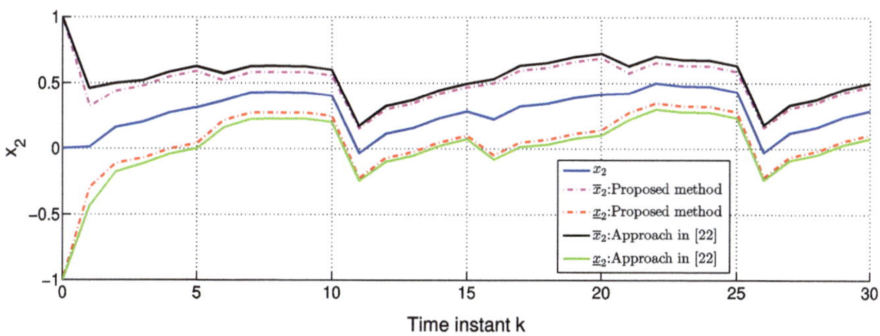

Fig. 3 x_2 and its interval estimation

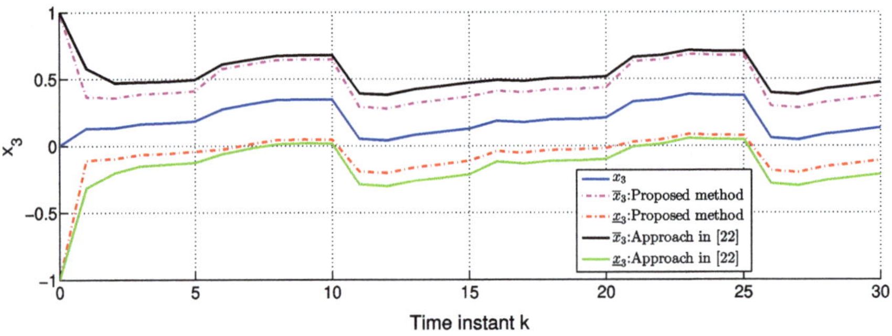

Fig. 4 x_3 and its interval estimation

to the real state and the approach in Marouani et al. (2022), while dotted lines correspond to the proposed method.

The approach in Marouani et al. (2022) suffers from ensuring the non-negativity property, which causes additional conservatism while the two-step interval estimation proposed in this letter integrates robust observer design with zonotopic analysis and hence, gets rid of the non-negativity requirement. The simulations show that the proposed zonotopic interval estimation gives more accurate results than the approach in Marouani et al. (2022).

6 Conclusion

This paper presents a new approach to design zonotope-based state estimator for discrete-time switched linear systems subject to uncertainties. We combine a robust observer design with zonotopic analysis. Stability conditions of the proposed observer are ensured in terms of LMIs using multiple Lyapunov functions. A comparative simulation shows the efficiency of the proposed contribution. For future works, extensions of these results to switched systems with unknown switching rules will be considered. Also, to guarantee more accurate estimation results, another promising perspective is to introduce the L_∞ performance in the observer design.

Acknowledgements This research was supported in part by a fellowship from the Hanse-Wissenschaftskolleg—Institute for Advanced Study (HWK).

References

Ackerson GA, Fu KS (1970) On state estimation in switching environments. IEEE Trans Autom Control 15(1):10–17

Arichi F, Djemai M, Cherki B, Manamanni N (2015) Continuous and discrete state estimation for a class of nonlinear switched systems. IEEE Trans Circuits Syst II 62(7):691–695

Birouche A, Daafouz J, Iung C (2006) Observer design for a class of discrete time piecewise-linear systems. IFAC Proc Vol 39(5):12–17

Boyd S, Ghaoui LE, Feron E, Balakrishnan V (1994) Linear matrix inequalities in system and control theory. SIAM

Chevet T, Ngoc Dinh T, Marzat J, Wang Z, Raïssi T (2021) Zonotopic Kalman filter-based interval estimation for discrete-time linear systems with unknown inputs. IEEE Control Syst Lett 6:806–811

Chisci L, Garulli A, Zappa G (1996) Recursive state bounding by parallelotopes. Automatica 32(7):1049–1055

Combastel C (2003) A state bounding observer based on zonotopes. In: IEEE European control conference (ECC), pp 2589–2594

Ethabet H, Raïssi T, Amairi M, Aoun M (2017) Interval observers design for continuous-time linear switched systems. IFAC-PapersOnLine 50(1):6259–6264

Fei Z, Yang L, Sun X-M, Ren S (2021) Zonotopic set-membership state estimation for switched systems with restricted switching. IEEE Trans Autom Control 67(11):6127–6134

Fei Z, Chen W, Zhao X (2022) Interval estimation for asynchronously switched positive systems. Automatica 143:110427

Gouzé J-L, Rapaport A, Hadj-Sadok MZ (2000) Interval observers for uncertain biological systems. Ecolog Model 133(1–2):45–56

Huang J, Ma X, Zhao X, Che H, Chen L (2020) Interval observer design method for asynchronous switched systems. IET Control Theory Appl 14(8):1082–1090

Ifqir S, Puig V, Ichalal D, Ait-Oufroukh N, Mammar S (2020) Zonotopic set-membership estimation for switched systems based on wi-radius minimization: vehicle application. IFAC-PapersOnLine 53(2):7446–7451

Ifqir S, Puig V, Ichalal D, Ait-Oufroukh N, Mammar S (2022) Zonotopic set-membership state estimation for switched systems. J Frank Inst 359(16):9241–9270

Jiang Z-P, Wang Y (2001) Input-to-state stability for discrete-time nonlinear systems. Automatica 37(6):857–869

Joao P. Hespanha and A. Stephen Morse. Stability of switched systems with average dwell-time. In: Proceedings of the 38th IEEE conference on decision and control, vol 3, pp 2655–2660

Li J, Pan K, Zhang D, Qingyu S (2019) Robust fault detection and estimation observer design for switched systems. Nonlinear Anal Hybrid Syst 34:30–42

Liberzon D, Stephen Morse A (1999) Basic problems in stability and design of switched systems. IEEE Control Syst Mag 19(5):59–70

Linh V, Morgansen KA (2010) Stability of time-delay feedback switched linear systems. IEEE Trans Autom Control 55(10):2385–2390

Liu Y, Zhao Y, Falin W (2016) Ellipsoidal state-bounding-based set-membership estimation for linear system with unknown-but-bounded disturbances. IET Control Theory Appl 10(4):431–442

Marouani G, Dinh TN, Raïssi T, Messaoud H (2018) Interval observers design for discrete-time linear switched systems. In: IEEE European control conference (ECC), pp 801–806

Marouani G, Dinh TN, Wang Z, Raïssi T, Messaoud H (2022) Interval observers design for discrete-time linear switched systems. In: Complex systems: spanning control and computational cybernetics: foundations, pp 245–268

Mhaskar P, El-Farra NH, Christofides PD (2005) Predictive control of switched nonlinear systems with scheduled mode transitions. IEEE Trans Autom Control 50(11):1670–1680

Ngoc Dinh T, Marouani G, Raïssi T, Wang Z, Messaoud H (2020) Optimal interval observers for discrete-time linear switched systems. Int J Control 93(11):2613–2621

Radhakrishna RC, Mitra SK (1972) Generalized inverse of a matrix and its applications. In: Proceedings of the sixth berkeley symposium on mathematical statistics and probability, vol 1, pp 601–620

Sehli N, Wang Z, Taarit KI, Raïssi T, Ksouri M (2020) Interval estimation for linear discrete-time delay systems. IFAC-PapersOnLine vol 53, no 2, pp 4798–4803

Sontag ED (2008) Input to state stability: basic concepts and results. Nonlinear and optimal control theory, pp 163–220

Tang W, Wang Z, Wang Y, Raïssi T, Shen Y (2019) Interval estimation methods for discrete-time linear time-invariant systems. IEEE Trans Autom Control 64(11):4717–4724

Wang Z, Lim C-C, Shen Y (2018) Interval observer design for uncertain discrete-time linear systems. Syst Control Lett 116:41–46

Zhang F (2006) The Schur complement and its applications. Springer Sci Bus Media 4

Zhang W, Wang Z, Raïssi T, Ngoc Dinh T, Dimirovski G (2020) Zonotope-based interval estimation for discrete-time linear switched systems. IFAC World Cong 53(2):4707–4712

Zhu K, Song Y, Ding D, Wei G, Liu H (2018) Robust MPC under event-triggered mechanism and round-robin protocol: an average dwell-time approach. Inf Sci 457:126–140

Optimal Interval Observers for Bounded Jacobian Nonlinear Dynamical Systems

Mohammad Khajenejad, Tarun Pati, and Sze Zheng Yong

Abstract In this chapter, we introduce two interval observer designs for discrete-time (DT) and continuous-time (CT) nonlinear systems with bounded Jacobians that are affected by bounded uncertainties. Our proposed methods utilize the concepts of mixed-monotone decomposition and embedding systems to design correct-by-construction interval framers, i.e., the interval framers inherently bound the true state of the system without needing any additional constraints. Further, our methods leverage techniques for positive/cooperative systems to guarantee global uniform ultimate boundedness of the framer error, i.e., the proposed interval observer is input-to-state stable. Specifically, our two interval observer designs minimize the $\mathcal{H}_\infty$ and L_1 gains, respectively, of the associated linear comparison system of the framer error dynamics. Moreover, our designs adopt a multiple-gain observer structure, which offers additional degrees of freedom, along with coordinate transformations that may improve the feasibility of the resulting optimization programs. We will also discuss and propose computationally tractable optimization formulations to compute the observer gains. Finally, we compare the efficacy of the proposed designs against existing DT and CT interval observers.

1 Introduction

State estimation is of great importance across a spectrum of cyber-physical systems and engineering domains, from autonomous vehicles to power systems, serving as the bedrock for monitoring, decision-making, and control processes. Among the diverse

M. Khajenejad (✉)
The University of Tulsa, Tulsa, OK, USA
e-mail: mohammad-khajenejad@utulsa.edu

T. Pati · S. Z. Yong
Northeastern University, Boston, MA, USA
e-mail: pati.ta@northeastern.edu

S. Z. Yong
e-mail: s.yong@northeastern.edu

T. N. Dinh et al. (eds.), *Set-Valued Approaches to Control and Estimation of Uncertain Systems*, Mathematical Engineering,
https://doi.org/10.1007/978-3-031-94239-6_3

array of state estimation methods, interval observers emerge as a compelling solution for systems affected by uncertainty. Interval observers furnish interval-valued state estimates, i.e., upper and lower bounds of the states, offering a unique advantage in scenarios where uncertainties manifest as sets rather than point values. This is particularly pertinent when faced with non-stochastic uncertainties or when the distributions governing these uncertainties are unknown. Thus, this chapter considers the interval observer design problem for nonlinear bounded-error continuous-time (CT) and discrete-time (DT) dynamical systems using a unified framework.

1.1 Literature Review

A vast body of scholarly literature delves into the intricacies of designing set-valued/interval observers for a wide spectrum of systems, including linear, nonlinear, cooperative/monotone, mixed-monotone, distributed, hybrid and Metzler dynamics (Chebotarev et al. 2015; Wang et al. 2015; Tahir and Açıkmeşe 2021; Khajenejad et al. 2021; Khajenejad and Yong 2020; Farina and Rinaldi 2000; Khajenejad and Yong 2022; Pati et al. 2023; Khajenejad et al. 2022, 2025). Across this literature, a prevailing approach involves synthesizing observer gains to simultaneously ensure the Schur/Hurwitz stability and cooperativity of observer error dynamics, which, while conceptually sound, often leads to significant theoretical and computational challenges. To address these complexities, researchers have explored diverse strategies, such as harnessing interval arithmetic-based methods (Kieffer and Walter 2006), transforming systems into positive forms (Cacace et al. 2014), and employing time-varying/invariant state transformations, e.g., Tahir and Açıkmeşe (2021). However, for general classes of nonlinear systems, the task remains formidable, necessitating the utilization of bounding mappings/decomposition functions (Khajenejad and Yong 2023) to reformulate observer design problems as semidefinite programs (SDPs) or optimizations with linear matrix inequality (LMI) constraints (Wang et al. 2012, 2015; Efimov and Raïssi 2016; Briat and Khammash 2016; Tahir and Açıkmeşe 2021; Khajenejad et al. 2022a; Khajenejad and Yong 2022b; Khajenejad et al. 2023a, b, c). Nevertheless, the obtained matrix inequalities can still impose significant restrictions, potentially rendering solutions infeasible for certain systems due to the imposition of various conditions and upper bounds. In response to this challenge, numerous coordinate transformations have been suggested to alleviate design constraints and enable the computation of feasible observer gains (Dinh et al. 2014; Mazenc and Bernard 2011). However, unfortunately, existing approaches are unable to concurrently synthesize or design both the coordinate transformation and observer gains.

Moreover, recent advancements in uncertain systems have prompted considerations of additional design criteria, such as robustness against noise and uncertainty. This has led to the translation of design problems into SDPs encompassing a combination of framer satisfaction (i.e., the correctness property), observer stability, and noise attenuation constraints, albeit at the cost of potential conservatism and increased computational burden (Tahir and Açıkmeşe 2021; Cacace et al. 2014; Efimov and Raïssi 2016; Khajenejad and Yong 2021). In prior research endeavors, an $\mathcal{H}_\infty$-optimal observer was proposed by the authors in Khajenejad and Yong (2022) to mitigate these issues by minimizing the L_2-gain of the observer error system without coordinate transformation for both CT and DT systems, which involves solving of an SDP. On the other hand, an interval observer design with additional degrees of freedom (i.e., observer gains) was introduced in Pati et al. (2022) for both CT and DT systems to minimize the L_1-gain of the error system, which instead involves solving of a mixed-integer linear program (MILP). Building upon these foundations, the present chapter introduces a unified framework for designing $\mathcal{H}_\infty$ or L_1 robust interval observers for both CT and DT bounded-error nonlinear systems with bounded Jacobians.

2 Preliminaries

Notation. $\mathbb{R}^n$, $\mathbb{R}^n_{>0}$, $\mathbb{R}^{n\times p}$, $\mathbb{N}_n$ and $\mathbb{N}$ denote the n-dimensional Euclidean space, positive vectors of size n, matrices of size n by p, natural numbers up to n and natural numbers, respectively. For a vector $v \in \mathbb{R}^n$, its vector p-norm is given by $\|v\|_p \triangleq (\sum_{i=1}^n |v_i|^p)^{\frac{1}{p}}$ and $\mathrm{diag}(v)$ represents a diagonal matrix whose diagonal elements are elements of v, while for a matrix $M \in \mathbb{R}^{n\times p}$, M_{ij} represents its j-th column and i-th row entry, $\mathrm{sgn}(M)$ represents its element-wise signum function, $M^\oplus \triangleq \max(M, \mathbf{0}_{n\times p})$, $M^\ominus \triangleq M^\oplus - M$, and $|M| \triangleq M^\oplus + M^\ominus$ is its element-wise absolute value. Moreover, M^{d} is a diagonal matrix with (only) the diagonal elements of the square matrix $M \in \mathbb{R}^{n\times n}$, $M^{\mathrm{nd}} \triangleq M - M^{\mathrm{d}}$ is the matrix with only its off-diagonal elements, and $M^{\mathrm{m}} \triangleq M^{\mathrm{d}} + |M^{\mathrm{nd}}|$ is the "Metzlerized" matrix.[1] Further, $M \succ 0$ and $M \prec 0$ (or $M \succeq 0$ and $M \preceq 0$) denote that M is positive and negative (semi-)definite, respectively, while all vector and matrix inequalities are element-wise inequalities, and the matrices of zeros and ones of dimension $n \times p$ are denoted as $\mathbf{0}_{n\times p}$ and $\mathbf{1}_{n\times p}$, respectively. In addition, a continuous function $\alpha : [0, a) \to \mathbb{R}_+$ is said to belong to class $\mathcal{K}$ if it is strictly increasing and $\alpha(x) = 0$, (and consequently it is positive definite, i.e., $\alpha(x) = 0$ for $x = 0$ and $\alpha(x) > 0$ otherwise). Moreover, the function α belongs to class $\mathcal{K}_\infty$ if it belongs to class $\mathcal{K}$, $a = \infty$, and $\lim_{r\to\infty} \alpha(r) = \infty$, i.e., α is unbounded. Finally, a continuous function $\lambda : [0, a) \times [0, \infty) \to \mathbb{R}_+$ is said to belong to class $\mathcal{KL}$ if for each fixed $t \geq 0$, the function $\lambda(s, t)$ belongs to class $\mathcal{K}$, for each fixed $s \geq 0$, $\lambda(s, t)$ is decreasing in t and $\lim_{t\to\infty} \lambda(s, t) = 0$.

[1] A Metzler matrix is a square matrix in which all the off-diagonal components are nonnegative (equal to or greater than zero).

Definition 1 (*Interval*) An (n-dimensional) interval denoted as $\mathcal{I} \triangleq [\underline{z}, \overline{z}] \subset \mathbb{R}^n$ is the set of vectors $z \in \mathbb{R}^{n_z}$ that satisfy $\underline{z} \leq z \leq \overline{z}$. A matrix interval follows a similar definition.

Definition 2 (*Jacobian Sign-Stability*) A vector-valued function $g : \mathcal{Z} \subset \mathbb{R}^{n_z} \to \mathbb{R}^p$ is Jacobian sign-stable (JSS), if in its domain $\mathcal{Z}$, the entries of its Jacobian matrix do not change signs, i.e., if one of the following hold:

$$J^g_{ij}(z) \geq 0 \text{ or } J^g_{ij}(z) \leq 0$$

for all $z \in \mathcal{Z}, \forall i \in \mathbb{N}_p, \forall j \in \mathbb{N}_{n_z}$, where $J^g(z)$ represents the Jacobian matrix of the mapping g evaluated at $z \in \mathcal{Z}$.

Proposition 1 (Jacobian Sign-Stable Decomposition Khajenejad et al. 2022a, Proposition 2) *For a mapping $g : \mathcal{Z} \subset \mathbb{R}^{n_z} \to \mathbb{R}^p$, if $J^g(z) \in [\underline{J}^g, \overline{J}^g]$ for all $z \in \mathcal{Z}$, where $\underline{J}^g, \overline{J}^g \in \mathbb{R}^{p \times n_z}$ are known matrices, then the function g can be decomposed as the sum of a JSS mapping μ and a remainder affine mapping Hz (that is also JSS), in an additive* remainder-form*:*

$$\forall z \in \mathcal{Z}, g(z) = Hz + \mu(z), \tag{1}$$

where the matrix $H \in \mathbb{R}^{p \times n_z}$, satisfies

$$\forall (i, j) \in \mathbb{N}_p \times \mathbb{N}_{n_z}, H_{ij} = \underline{J}^g_{ij} \vee H_{ij} = \overline{J}^g_{i,j}. \tag{2}$$

Remark 1 As can be seen from (1), for an n-dimensional vector field (i.e., $n_z = p = n$), we have $2^{(n^2)}$ choices for H, all of which lead to correct JSS decompositions and can be applied. The investigation of the best choice of H for interval observer designs is a subject of future work. In this chapter and to implement the observer on the simulation examples, we selected H similar to Yang et al. (2019), which we showed in our previous work Khajenejad and Yong (2023) that it belongs to the set of choices provided in Proposition 1.

Definition 3 (*Mixed-Monotonicity and Decomposition Functions* (Abate et al., 2020, Definition 1),(Yang et al., 2019, Definition 4)) Consider the uncertain discrete-time (DT) or continuous-time (CT) dynamical system with initial state $x_0 \in \mathcal{X}_0 \triangleq [\underline{x}_0, \overline{x}_0] \subset \mathbb{R}^n$ and an augmentation of known inputs and process noise $w_t \in \mathcal{W} \triangleq [\underline{w}, \overline{w}] \subset \mathbb{R}^{n_w}$:

$$x_t^+ = g(z_t) \triangleq g(x_t, w_t), \quad z_t \triangleq [x_t^\top \ w_t^\top]^\top, \tag{3}$$

Moreover, $g : \mathcal{Z} \subset \mathbb{R}^{n_z} \to \mathbb{R}^n$ is the vector field with augmented state $z_t \in \mathcal{Z} \triangleq \mathcal{X} \times \mathcal{W} \subset \mathbb{R}^{n_z}$ as its domain, where $\mathcal{X}$ is the entire state space and $n_z = n + n_w$.

- If (3) is a DT system, i.e., $x_t^+ \triangleq x_{t+1}$, then, a mapping $g_d : \mathcal{Z} \times \mathcal{Z} \to \mathbb{R}^n$ is a DT mixed-monotone decomposition function with respect to g, if

(i) $g_d(z, z) = g(z)$,
(ii) g_d is monotone increasing in its first argument, i.e.,

$$\hat{z} \geq z \Rightarrow g_d(\hat{z}, z') \geq g_d(z, z'),$$

(iii) g_d is monotone decreasing in its second argument, i.e.,

$$\hat{z} \geq z \Rightarrow g_d(z', \hat{z}) \leq g_d(z', z).$$

- On the other hand, if (3) is a CT system, i.e., if $x_t^+ \triangleq \dot{x}_t$, then a mapping $g_d : \mathcal{Z} \times \mathcal{Z} \to \mathbb{R}^n$ is a CT mixed-monotone decomposition function with respect to g, if

(i) $g_d(z, z) = g(z)$,
(ii) g_d is monotone increasing in its first argument only with respect to "off-diagonal" arguments, i.e.,

$$\forall (i, j) \in \mathbb{N}_n \times \mathbb{N}_{n_z} \wedge i \neq j, \hat{z}_j \geq z_j, \hat{z}_i = z_i \Rightarrow g_{d,i}(\hat{z}, z') \geq g_{d,i}(z, z'),$$

(iii) g_d is monotone decreasing in its second argument, i.e.,

$$\hat{z} \geq z \Rightarrow g_d(z', \hat{z}) \leq g_d(z', z).$$

Proposition 2 (Tight Decomposition Functions for Linear Systems) *Suppose the system in* (3) *is linear time-invariant, i.e.,*

$$x_t^+ = g(x_t, w_t) = A_g x_t + B_g w_t. \tag{4}$$

Then, the mapping g admits the following decomposition function:

$$g_d(x_1, w_1, x_2, w_2) = A_g^{\uparrow} x_1 - A_g^{\downarrow} x_2 + B_g^{\oplus} w_1 - B_g^{\ominus} w_2, \tag{5}$$

where

- $A_g^{\uparrow} \triangleq A_g^{\oplus}$ *and* $A_g^{\downarrow} \triangleq A_g^{\ominus}$ *if* (4) *is a DT system, and*
- $A_g^{\uparrow} \triangleq A_g^{\mathrm{nd},\oplus} + A_g^{\mathrm{d}}$ *and* $A_g^{\downarrow} \triangleq A_g^{\mathrm{nd},\ominus}$ *if* (4) *is a CT system.*

Proof Defining $z \triangleq (x, w)$, it is straightforward to see from (5) that

$$\begin{aligned} g_d(z,z) &= (A_g^{\uparrow} - A_g^{\downarrow})x + (B_g^{\oplus} - B_g^{\ominus})w \\ &= \begin{cases} (A_g^{\oplus} - A_g^{\ominus})x + B_g w, & \text{(DT)} \\ (A_g^{nd,\oplus} - A_g^{nd,\ominus} + A_g^d)x + B_g w = (A_g^{nd} + A_g^d)x + B_g w, & \text{(CT)} \end{cases} \\ &= A_g x + B_g w = g(z). \end{aligned}$$

This proves condition (i) in Definition 3 for both DT and CT cases.

Moreover, in the DT case, let $\hat{z} = (\hat{x}, \hat{w}) \geq z = (x, w) \Leftrightarrow \hat{x} \geq x, \hat{w} \geq w$. Then, for any $z' = (x', w') \in \mathcal{Z}$, the following holds

$$g_d(\hat{z}, z') = A_g^{\oplus}\hat{x} - A_g^{\ominus}x' + B_g^{\oplus}\hat{w} - B_g^{\ominus}x' \geq A_g^{\oplus}x - A_g^{\ominus}x' + B_g^{\oplus}w - B_g^{\ominus}x' = g_d(z, z'),$$

where the inequality follows from the non-negativity of $A_g^{\oplus}$ and $B_g^{\oplus}$, which preserves the order of any two ordered multiplicands, and in particular, $\hat{x}, x$ and $\hat{w}, w$. By similar arguments, $g_d(z', \hat{z}) \leq g_d(z', z)$. In other words, conditions (ii) and (iii) for the DT case in Definition 3 hold.

Furthermore, to prove condition (ii) for the CT case, we consider an arbitrary dimension $i \in \mathbb{N}_n$ and an ordered $\hat{z}, z \in \mathcal{Z}$ such that $\hat{z}_j \geq z_j, \forall j \in \mathbb{N}_{n_z}, j \neq i$ and $\hat{z}_i = \hat{z}_i$. Then, for any $z' \in \mathcal{Z}$ we have

$$\begin{aligned} g_{d,i}(\hat{z}, z') &= (A_{g,i}^{nd,\oplus} + A_{g,i}^d)\hat{x} - A_{g,i}^{nd,\ominus}x' + B_{g,i}^{\oplus}\hat{w} - B_{g,i}^{\ominus}w' \\ &\geq (A_{g,i}^{nd,\oplus} + A_{g,i}^d)x - A_{g,i}^{nd,\ominus}x' + B_{g,i}^{\oplus}w - B_{g,i}^{\ominus}w' g_{d,i}(z, z'), \end{aligned}$$

where the inequality follows from the non-negativity of $A_{g,i}^{nd,\oplus}$ and $B_{g,i}^{\oplus}$, and the facts that $\hat{x}_j \geq x_j, \forall j \neq i, \hat{x}_i = x_i$, and $\hat{w} \geq w$. This proves condition (ii) for the CT case in Definition 3. Finally, the proof of (iii) for the CT case is similar to its counterpart for the DT case.

Proposition 3 (Tight and Tractable DT Decomposition Functions for Nonlinear JSS Mappings Khajenejad et al. 2022a, Proposition 4 and Khajenejad et al. 2022a, Lemma 3) *Let $\mu : \mathcal{Z} \subset \mathbb{R}^{n_z} \to \mathbb{R}^p$ be a JSS mapping on its domain, where $J^{\mu} \in [\underline{J}^{\mu}, \overline{J}^{\mu}]$. Then, it admits a tight DT decomposition function for each μ_i, $i \in \mathbb{N}_p$, as follows:*

$$\mu_{d,i}(z_1, z_2) = \mu_i(D^i z_1 + (I_{n_z} - D^i)z_2), \tag{6}$$

for any ordered $z_1, z_2 \in \mathcal{Z}$, where D^i is a binary diagonal matrix determined by which vertex of the interval $[z_2, z_1]$ or $[z_1, z_2]$ that maximizes (if $z_2 \leq z_1$) or minimizes (if $z_2 > z_1$) the function μ_i, and can be found in closed-form as:

$$D^i = \mathrm{diag}(\max(\mathrm{sgn}(\overline{J}_i^{\mu}), \mathbf{0}_{1,n_z})), \tag{7}$$

where $\mathrm{diag}(v)$ *is a diagonal matrix whose diagonal elements are elements of* v. *Moreover, for any interval domain* $\underline{z} \leq z \leq \overline{z}$, *with* $\underline{z}, z, \overline{z} \in \mathcal{Z}$, *the following inequality holds:*

$$\delta_z^{\mu} \triangleq \mu_d(\overline{z}, \underline{z}) - \mu_d(\underline{z}, \overline{z}) \leq F_{\mu}(\overline{z} - \underline{z}), \tag{8}$$

where $F_{\mu} = (\overline{J}^{\mu})^{\oplus} + (\underline{J}^{\mu})^{\ominus}$.

Definition 4 (*One-Sided Decomposition Functions* Khajenejad and Yong 2023, Definition 2) Consider system (3) and suppose there exist mixed-monotone mappings $\overline{g}_d, \underline{g}_d : \mathcal{Z} \times \mathcal{Z} \to \mathcal{R}^n$ such that for any $z, \underline{z}, \overline{z} \in \mathcal{Z}$ the following statement holds:

$$\underline{z} \leq z \leq \overline{z} \Rightarrow \underline{g}_d(\underline{z}, \overline{z}) \leq g(z) \leq \overline{g}_d(\overline{z}, \underline{z}),$$

where with a slight abuse of notation, we overload the notation of $\underline{g}_d(\underline{z}, \overline{z})$ and $\overline{g}_d(\overline{z}, \underline{z})$ when (3) is a CT system to represent the case with a fixed z_i for the ith function f_i. Then, $\overline{g}_d$ and $\underline{g}_d$ are called upper and lower one-sided decomposition functions for g over $[\underline{z}, \overline{z}]$, respectively.

Definition 5 (*Generalized Embedding System* Khajenejad and Yong 2023, Definition 3) For an n-dimensional uncertain system (3) with any pair of one-sided decomposition function $\overline{g}_d, \underline{g}_d$, its generalized embedding system is defined as the following $2n$-dimensional certain dynamical system with initial condition $\left[\overline{x}_0^{\top} \; \underline{x}_0^{\top}\right]^{\top}$:

$$\begin{bmatrix} \underline{x}_t^{+} \\ \overline{x}_t^{+} \end{bmatrix} = \begin{bmatrix} \underline{g}_d(\left[(\underline{x}_t)^{\top} \; \underline{w}^{\top}\right]^{\top}, \left[(\overline{x}_t)^{\top} \; \overline{w}^{\top}\right]^{\top}) \\ \overline{g}_d(\left[(\overline{x}_t)^{\top} \; \overline{w}^{\top}\right]^{\top}, \left[(\underline{x}_t)^{\top} \; \underline{w}^{\top}\right]^{\top}) \end{bmatrix}. \tag{9}$$

Proposition 4 (Framer Property Khajenejad and Yong 2023, Proposition 3) *The solution to a generalized embedding system* (9) *with one-sided decomposition functions* $\overline{g}_d, \underline{g}_d$ *corresponding to the system dynamics in* (3) *has a* state framer property, *i.e., it is guaranteed to frame the unknown state trajectory* x_t *of* (3)*:*

$$\underline{x}_t \leq x_t \leq \overline{x}_t, \forall t \in \mathbb{T}.$$

3 Problem Formulation

System Dynamics. We consider a class of uncertain/noisy discrete-time (DT) or continuous-time (CT) nonlinear systems given as:

$$\mathcal{G}: \begin{cases} x_t^{+} = f(x_t) + Bu_t + Ww_t, \\ y_t = h(x_t) + Du_t + Vv_t, \end{cases} \quad \text{for all } t \in \mathbb{T}, \tag{10}$$

where

- $x_t^+ = x_{t+1}$ and $\mathbb{T} = \{0\} \cup \mathbb{N}$, if $\mathcal{G}$ is a DT system, and
- $x_t^+ = \dot{x}_t$ and $\mathbb{T} = \mathbb{R}_{\geq 0}$, if $\mathcal{G}$ is a CT system.

Moreover, $x_t \in \mathcal{X} \subset \mathbb{R}^n$, $w_t \in \mathcal{W} \triangleq [\underline{w}, \overline{w}] \subset \mathbb{R}^{n_w}$, $v_t \in \mathcal{V} \triangleq [\underline{v}, \overline{v}] \subset \mathbb{R}^{n_v}$, $u_t \in \mathcal{U} \triangleq [\underline{u}, \overline{u}] \subset \mathbb{R}^m$ and $y_t \in \mathbb{R}^l$ are state, process noise, measurement noise, known control input and output measurement signals, respectively. Further, $f : \mathbb{R}^n \to \mathbb{R}^n$ denotes the nonlinear state vector field, while $h : \mathbb{R}^n \to \mathbb{R}^l$ is a mapping that can represent system observation/measurements and/or state constraints. Additionally, $W \in \mathbb{R}^{n \times n_w}$, $B \in \mathbb{R}^{n \times m}$, $V \in \mathbb{R}^{l \times n_v}$ and $D \in \mathbb{R}^{l \times m}$ are known matrices. Furthermore, we assume the following:

Assumption 1 (*Known Disturbance Bounds, Measurements, and Inputs*) The disturbance/noise and input bounds $\underline{w}, \overline{w}, \underline{v}, \overline{v}$ and $\underline{u}, \overline{u}$, as well as the signals y_t (output) and u_t (input, if any) are known at all times. Moreover, the initial state x_0 is such that $x_0 \in \mathcal{X}_0 = [\underline{x}_0, \overline{x}_0]$ with known bounds $\underline{x}_0$ and $\overline{x}_0$.

Assumption 2 (*Bounded Jacobian Nonlinearities*) The mappings/functions f and h are known and differentiable, and have bounded Jacobians in their domain.[2] Moreover, the lower and upper bounds of their Jacobian matrices over the entire state space $\mathcal{X}$, i.e., $\underline{J}^f, \overline{J}^f \in \mathbb{R}^{n \times n}$ and $\underline{J}^h, \overline{J}^h \in \mathbb{R}^{l \times n}$ are known.

We aim to estimate the state trajectories of the plant $\mathcal{G}$ in (10). To pose this problem formally, we first define the notions of correctness, input-to-state stability (ISS), and optimal interval observer, as follows.

Definition 6 (Correct Interval Framers and Framer Errors) The signals/sequences $\overline{x}, \underline{x} : \mathbb{T} \to \mathbb{R}^n$ are called upper and lower framers for the system states x_t of the nonlinear plant $\mathcal{G}$ in (10) if

$$\underline{x}_t \leq x_t \leq \overline{x}_t, \ \forall t \in \mathbb{T}, \forall w_t \in \mathcal{W}, \forall v_t \in \mathcal{V}. \tag{11}$$

Further, $e_t \triangleq \overline{x}_t - \underline{x}_t$ is called the *framer error* at time t. Any dynamical system whose states are correct framers for the system states of the plant $\mathcal{G}$, i.e., with $e_t \geq 0, \forall t \in \mathbb{T}$, is called a *correct* interval framer for system (10).

Definition 7 (*Input-to-State Stability*) An interval framer is input-to-state stable (ISS), if the framer error (cf. Definition 6) is bounded as follows:

$$\|e_t\|_2 \leq \beta(\|e_0\|_2, t) + \rho(\|\delta\|_{\ell_\infty}), \forall t \in \mathbb{T}, \tag{12}$$

where β and ρ are functions of classes $\mathcal{KL}$ and $\mathcal{K}_\infty$, respectively, with the ℓ_∞ signal norm $\|\delta\|_{\ell_\infty} \triangleq \sup_{t \in [0,\infty)} \|\delta_t\|_2 = \|\delta\|_2$. Moreover, ε_t and $\delta_t = \delta \triangleq \left[\delta_w^\top \ \delta_v^\top\right]^\top$ are the framer error and combined noise signals, with $\delta_w \triangleq \overline{w} - \underline{w}$ and $\delta_v \triangleq \overline{v} - \underline{v}$.

---

[2] The differentiability assumption is primarily for ease of exposition and can be relaxed to a weaker assumption (cf. Khajenejad and Yong (2023) for more details).

Definition 8 *(L_1 or $\mathcal{H}_\infty$-Robust and Optimal Interval Observer)* An interval framer $\hat{\mathcal{G}}$ is L_1-robust or $\mathcal{H}_\infty$-robust and optimal, if the L_1-gain (with $s = \ell_1$) or $\mathcal{H}_\infty$-gain (with $s = \ell_2$), respectively, of the framer error system $\tilde{\mathcal{G}}$, defined below, is minimized:

$$\|\tilde{\mathcal{G}}\|_s \triangleq \sup_{\|\delta\|_s=1} \|e\|_s, \; s \in \{\ell_1, \ell_2\}, \tag{13}$$

where $\|\nu\|_{\ell_1} \triangleq \int_0^\infty \|\nu_t\|_1 dt$, $\|\nu\|_{\ell_2} \triangleq \sqrt{\int_0^\infty \|\nu_t\|_2^2 dt}$ (for CT systems) and $\|\nu\|_{\ell_1} \triangleq \sum_0^\infty \|\nu_t\|_1$, $\|\nu\|_{\ell_2} \triangleq \sum_0^\infty \|\nu_t\|_2^2$ are the ℓ_1 and ℓ_2 signal norms for $\nu \in \{e, \delta\}$, respectively.

The optimal observer design problem can be stated as follows:

Problem 1 (ISS & L_1 or $\mathcal{H}_\infty$-Optimal Interval Observer Design) Given the nonlinear system in (10), design a correct and L_1 or $\mathcal{H}_\infty$-robust optimal interval observer (cf. Definitions 8) whose framer error (cf. Definition 6) is input-to-state stable (ISS).[3]

4 Proposed Interval Observer

4.1 Interval Observer Structure and Framer Property (Correctness)

We begin by deriving an equivalent representation of the system dynamics for system $\mathcal{G}$ in (10). Specifically, we first leverage Assumption 2 to decompose the functions f and h into two components based on the results in Proposition 1:

$$\begin{aligned} f(x) &= Ax + \phi(x), \\ h(x) &= Cx + \psi(x), \end{aligned} \tag{14}$$

where $A \in \mathbb{R}^{n \times n}$ and $C \in \mathbb{R}^{l \times n}$ are chosen such that ϕ and ψ are JSS mappings (cf. Definition 2). Then, we propose an equivalent system representation for plant $\mathcal{G}$ in (10) with additional degrees of freedom (that will be our to-be-designed observer gains) through the following lemma.

Lemma 1 (Equivalent System Representation) *Consider plant $\mathcal{G}$ in* (10) *and suppose that Assumptions 1–2 hold. Let $L, N \in \mathbb{R}^{n \times l}$ and $T \in \mathbb{R}^{n \times n}$ be arbitrary matrices that satisfy the following:*

$$T + NC = I_n. \tag{15}$$

---

[3] If desired, we can replace the 2-norm in the original ISS definition with any norm that is more aligned with the norm used for the robustness, and the satisfaction of the former would also imply the latter by norm equivalence.

Then, the system dynamics for plant $\mathcal{G}$ in (10) *can be equivalently written as*

$$\begin{cases} \xi_t^+ = M_x \xi_t + T\phi(x_t) - L\psi(x_t) - N\rho(x_t, w_t, u_t) \\ \qquad + M_w w_t - M_v v_t + M_u u_t + (M_x N + L)y_t, \\ x_t = \xi_t + Ny_t - NVv_t - NDu_t, \end{cases} \tag{16}$$

with initial condition $\xi_0 = x_0 - Ny_0 + NDu_0 + NVv_0$, *where*

$$\begin{aligned} &M_x \triangleq TA - LC - NA_2, \ M_w \triangleq TW - NW_2, \\ &M_u \triangleq TB - NB_2 - (M_x N + L)D, M_v \triangleq (M_x N + L)V. \end{aligned} \tag{17}$$

Moreover, A and C are computed through the JSS decompositions in (14), *while* $A_2 \in \mathbb{R}^{l \times n}$, $W_2 \in \mathbb{R}^{l \times n_w}$, *and* $B_2 \in \mathbb{R}^{l \times m}$ *are chosen such that the following decomposition hold* $\forall (x, w, u) \in \mathcal{X} \times \mathcal{W} \times \mathcal{U}$ *(cf. Definition 2 and Proposition 1):*

$$\psi^+(x, w, u) = A_2 x + W_2 w + B_2 u + \rho(x, w, u), \tag{18}$$

such that ρ *is a JSS mapping, with*

- $\psi^+(x, w, u) \triangleq \dot{\psi}(x, w, u) = \frac{\partial \psi}{\partial x}(f(x) + Ww + Bu)$ *if* $\mathcal{G}$ *is a CT system, and*
- $\psi^+(x, w, u) \triangleq \psi(x^+, w, u) = \psi(f(x) + Ww + Bu)$ *if* $\mathcal{G}$ *is a DT system.*

Proof We begin by defining an auxiliary state

$$\xi_t \triangleq x_t - N(y_t - Vv_t - Du_t). \tag{19}$$

Then, from the second equation in (14) and from (19), as well as by choosing N to satisfy (15), we have:

$$\begin{aligned} \xi_t &= x_t - N(y_t - Vv_t - Du_t) = x_t - N(Cx_t + \psi(x_t)) = \underbrace{(I - NC)}_{T} x_t - N\psi(x_t) \\ &= Tx_t - N\psi(x_t). \end{aligned} \tag{20}$$

Now, we are ready to derive the dynamics of the auxiliary state ξ_t starting from (20), and using (10), (14) and (18):

$$\begin{aligned} \xi_t^+ &= Tx_t^+ - N\psi^+(x_t, w_t, u_t) \\ &= T(Ax_t + \phi(x_t) + Ww_t + Bu_t) - N(A_2 x_t + W_2 w_t + B_2 u_t + \rho(x_t, w_t, u_t)), \end{aligned} \tag{21}$$

with $x_t^+ = Ax_t + \phi(x_t) + Ww_t + Bu_t$ from (10) and the second equation in (14), as well as $\psi^+(x_t, w_t, u_t) = A_2 x_t + W_2 w_t + B_2 u_t + \rho(x_t, w_t, u_t)$ from (18). Next, from the second equations in (10) and (14), we have

$$L(y_t - Cx_t - \psi(x_t) - Vv_t - Du_u) = 0,$$

for any matrix L with appropriate dimensions. Adding this 'zero term' to the right-hand side of (21) yields:

$$\begin{aligned}\xi_t^+ &= M_x x_t + T\phi(x_t) - L\psi(x_t) - N\rho(x_t, w_t, u_t) + M_w w_t - LVv_t \\ &+ (TB - NB_2)u_t + Ly_t.\end{aligned} \tag{22}$$

Finally, plugging x_t from (19) into the first term on the right-hand side of (22) results in the first equation in (16), while the second equation in (16) is a reorganization of the first equality in (20).

4.1.1 Construction of Interval Framers

From the equivalent system in (16), we can construct an embedding system by formulating the corresponding decomposition functions for both CT and DT cases, respectively, as follows.

First, note that by Proposition 2, the component of the dynamics in (16) that is affine in ξ, w, and v (with known y and u), i.e.,

$$g^\ell(\xi, w, v) \triangleq M_x\xi + M_w w - M_v v + M_u u + (M_x N + L)y, \tag{23}$$

admits a tight decomposition function (cf. Proposition 2 for details):

$$\begin{aligned}g_d^\ell(\xi_1, w_1, v_1, \xi_2, w_2, v_2) &= M_x^\uparrow \xi_1 - M_x^\downarrow \xi_2 + M_w^\oplus w_1 - M_w^\ominus w_2 + M_v^\ominus v_1 - M_v^\oplus v_2 \\ &+ M_u u + (M_x N + L)y,\end{aligned} \tag{24}$$

where

- $M_x^\uparrow \triangleq M_x^\oplus$ and $M_x^\downarrow \triangleq M_x^\ominus$ if (4) is a DT system, and
- $M_x^\uparrow \triangleq M_x^{nd,\oplus} + M_x^d$ and $M_x^\downarrow \triangleq M_x^{nd,\ominus}$ if (4) is a CT system.

Moreover, the nonlinear component of the dynamics in (16) consists of a matrix-weighted summation of the JSS mappings ϕ, ψ and ρ:

$$g_d^\nu(x, w, u) \triangleq T\phi(x) - L\psi(x) - N\rho(x, w, u), \tag{25}$$

for which we construct a DT decomposition function using Proposition 3 (to compute tight decomposition functions for each of the JSS mappings ϕ, ψ, and ρ), and Proposition 2 (to compute a decomposition function for the linear combination/weighted sum of the JSS functions):

$$
\begin{aligned}
&g_d^v(x_1, w_1, u_1, x_2, w_2, u_2) = \\
&T^{\oplus}\phi_d(x_1, x_2) - T^{\ominus}\phi_d(x_2, x_1) + L^{\ominus}\psi_d(x_1, x_2) - L^{\oplus}\psi_d(x_2, x_1) \\
&+N^{\ominus}\rho_d(x_1, w_1, v_1, x_2, w_2, v_2) - N^{\oplus}\rho_d(x_2, w_2, v_2, x_1, w_1, v_1).
\end{aligned} \tag{26}
$$

Next, based on the fact that the summation of the decomposition functions construct a decomposition function for the summation of functions, we compute a decomposition function for the dynamics in (16) by adding the two decomposition functions in (24) and (26):

$$
\begin{aligned}
&g_d(\xi_1, x_1, w_1, v_1, u_1, \xi_2, x_2, w_2, v_2, u_2) = \\
&g_d^{\ell}(\xi_1, w_1, v_1, \xi_2, w_2, v_2) + g_d^v(x_1, w_1, u_1, x_2, w_2, u_2).
\end{aligned} \tag{27}
$$

Finally, by leveraging g_d, we compute an embedding system for the dynamics in (16) whose states frame the states of (16), and hence, equivalently the states of the original system (10), by Khajenejad and Yong (2023, Proposition 3). This process is summarized in the following Theorem.

Theorem 1 (Correctness) *Consider system* (10), *and let L, T and N be arbitrary matrices with appropriate dimensions that satisfy* (15). *Then, the following dynamical system initialized at $[\underline{x}_0^T \ \overline{x}_0^T]^\top$, constructs a framer for* (16)*:*

$$
\begin{cases}
\underline{\xi}_t^+ = M_x^{\uparrow}\underline{\xi}_t - M_x^{\downarrow}\overline{\xi}_t + M_w^{\oplus}\underline{w} - M_w^{\ominus}\overline{w} + M_v^{\ominus}\underline{v} - M_v^{\oplus}\overline{v} + M_u u_t + (M_x N + L)y_t \\
\qquad +T^{\oplus}\phi_d(\underline{x}_t, \overline{x}_t) - T^{\ominus}\phi_d(\overline{x}_t, \underline{x}_t) + L^{\ominus}\psi_d(\underline{x}_t, \overline{x}_t) - L^{\oplus}\psi_d(\overline{x}_t, \underline{x}_t) \\
\qquad +N^{\ominus}\rho_d(\underline{x}_t, \underline{w}, \underline{u}, \overline{x}_t, \overline{w}, \overline{u}) - N^{\oplus}\rho_d(\overline{x}_t, \overline{w}, \overline{u}, \underline{x}_t, \underline{w}, \underline{u}), \\
\overline{\xi}_t^+ = M_x^{\uparrow}\overline{\xi}_t - M_x^{\downarrow}\underline{\xi}_t + M_w^{\oplus}\overline{w} - M_w^{\ominus}\underline{w} + M_v^{\ominus}\overline{v} - M_v^{\oplus}\underline{v} + M_u u_t + (M_x N + L)y_t \\
\qquad +T^{\oplus}\phi_d(\overline{x}_t, \underline{x}_t) - T^{\ominus}\phi_d(\underline{x}_t, \overline{x}_t) + L^{\ominus}\psi_d(\overline{x}_t, \underline{x}_t) - L^{\oplus}\psi_d(\underline{x}_t, \overline{x}_t) \\
\qquad +N^{\ominus}\rho_d(\overline{x}_t, \overline{w}, \overline{u}, \underline{x}_t, \underline{w}, \underline{u}) - N^{\oplus}\rho_d(\underline{x}_t, \underline{w}, \underline{u}, \overline{x}_t, \overline{w}, \overline{u}), \\
\underline{x}_t = \underline{\xi}_t + (NV)^{\ominus}\underline{v} - (NV)^{\oplus}\overline{v} + Ny_t - NDu_t, \\
\overline{x}_t = \overline{\xi}_t + (NV)^{\ominus}\overline{v} - (NV)^{\oplus}\underline{v} + Ny_t - NDu_t,
\end{cases} \tag{28}
$$

where M_x, M_w, M_v and M_u are defined in (17). *Moreover, the initial conditions for $\underline{\xi}$ and $\overline{\xi}$ are $\underline{\xi}_0 = \underline{x}_0 - Ny_0 + NDu_0 + (NV)^{\oplus}\underline{v} - (NV)^{\ominus}\overline{v}$ and $\overline{\xi}_0 = \overline{x}_0 - Ny_0 + NDu_0 + (NV)^{\oplus}\overline{v} - (NV)^{\ominus}\underline{v}$, respectively.*

Proof We begin the proof by framing the auxiliary states of the equivalent system in (16), then applying Proposition 1 and Efimov et al. (2013, Lemma 1), results in (28). Hence, (28) creates a framer system for equivalent system in (16) and in turn is a framer system for the original plant $\mathcal{G}$, which results from the fact that the summation of the constituent embedding framers/systems creates an embedding framer/system for the constituent systems summation. To compute the constituent systems framers of (16), we categorize them as follows:

(i) Known terms that are independent of noise and state, which can also be viewed as having their lower and upper bounds equal to their original (known) values;

(ii) Linear terms in state and noise that can be lower and upper framed by using Efimov et al. (2013, Lemma 1), with a subtle difference in calculating the lower and upper embedding/framer systems for the DT and CT cases, which reflects in the definitions of $M^{\downarrow}$, $M^{\uparrow}$ given below (24);

(iii) Nonlinear terms with respect to the noise and state that can be lower and upper bounded/framed by applying Propositions 1 and 3 as well as Efimov et al. (2013, Lemma 1).

Then, adding up the constituent embedding framers/systems resulting from (i)–(iii) yields the embedding (observer) system (28). Finally, the embedding systems framer property Khajenejad and Yong (2023, Proposition 3) implies the correctness property.

Remark 2 (Additional Degrees of Freedom) The observer structure described above draws inspiration from Wang et al. (2012) and Degue et al. (2021), aiming to introduce additional degrees of freedom. In contrast to the most existing designs, e.g., in Khajenejad and Yong (2022), which relied on only one observer gain L, the current structure involves three to-be-designed observer gains N, T and L. While this augmentation is advantageous for enhancing performance, it was found to be not an exact replacement for coordinate transformations that could potentially render the observer gain design problem in Theorem 2 feasible (cf. Sect. 4.3). In this case, a coordinate transformation can be applied in a standard manner (cf. Sect. 4.3 below, Tahir and Açıkmeşe (2021, Sect. 5) or Mazenc and Bernard (2021) and references therein for more details).

4.2 L_1 or $\mathcal{H}_\infty$-Robust Observer Design

In this section, we address the input-to-state stability of the proposed interval framer system (28), for both DT and CT cases. To do this, we first compute the error dynamics of the observer, which is a nonlinear positive/cooperative system by construction. Then, by applying the results in Proposition 3 and (8), we compute a linear positive comparison system for the error dynamics. Finally, we provide necessary and sufficient conditions for computing the observer gains that stabilize and minimize the L_1 or $\mathcal{H}_\infty$ norm of the comparison system.

4.2.1 Error Dynamics

We start by computing the observer error system and a corresponding linear comparison system through the following lemma.

Lemma 1 *Consider system $\mathcal{G}$ in (10) and the proposed framer system (28). Then, the framer error ($\varepsilon_t \triangleq \overline{x}_t - \underline{x}_t$) dynamics, as well as a corresponding linear comparison system can be computed as follows:*

$$\begin{aligned} e_t^+ &= |M_x|^* e_t + |M_w|\delta_w + (|M_v| + \sigma|NV| - |M_x|^*|NV|)\delta_v + |T|\delta_t^\phi + |L|\delta_t^\psi + |N|\delta_t^\rho, \\ &\leq \begin{cases} |M_x|^* e_t + |M_w|\delta_w + (|NV| + |LV|)\delta_v + |T|\delta_t^\phi + |L|\delta_t^\psi + |N|\delta_t^\rho, & (DT) \\ |M_x|^* e_t + |M_w|\delta_w + ((|M_x| - M_x^m)|NV| + |LV|)\delta_v + |T|\delta_t^\phi + |L|\delta_t^\psi + |N|\delta_t^\rho, & (CT) \end{cases} \\ &\leq \tilde{A}_x e_t + \tilde{A}_w\delta_w + \tilde{A}_v\delta_v + \tilde{A}_u\delta_u = \tilde{A}_x e_t + \tilde{B}\delta_{\tilde{w}}, \end{aligned} \tag{29}$$

where $\delta_s \triangleq \overline{s} - \underline{s}$, $\forall s \in \{w, v, u\}$, $\delta_t^\mu \triangleq \mu(\overline{z}_t) - \mu(\underline{z}_t)$, $\forall \mu \in \{\phi, \psi, \rho\}$ *and* $\delta_{\tilde{w}} \triangleq \begin{bmatrix}\delta_w & \delta_v & \delta_u\end{bmatrix}^\top$, *as well as*

$$\begin{aligned} &\tilde{A}_w \triangleq |M_w| + |N|F_\rho^w, && \tilde{A}_x \triangleq |M_x|^* + |T|F_\phi + |L|F_\psi + |N|F_\rho^x, \\ &\tilde{A}_u \triangleq |N|F_\rho^u, && \tilde{A}_v \triangleq |LV| + (\sigma I + (1-\sigma)(|M_x| - M_x^{\mathrm{m}}))|NV|, \\ &\tilde{B} \triangleq \begin{bmatrix}\tilde{A}_w & \tilde{A}_v & \tilde{A}_u\end{bmatrix}, && \\ &\sigma = \begin{cases} 1 & if\ \mathcal{G}\ is\ DT, \\ 0 & if\ \mathcal{G}\ is\ CT, \end{cases} && |M_x|^* \triangleq \begin{cases} |M_x| \triangleq M_x^\oplus + M_x^\ominus & if\ \mathcal{G}\ is\ DT, \\ M_x^{\mathrm{m}} \triangleq |M_x^{\mathrm{nd}}| + M_x^{\mathrm{d}} & if\ \mathcal{G}\ is\ CT. \end{cases} \end{aligned} \tag{30}$$

Moreover, as in (17),

$$\begin{aligned} &M_x \triangleq TA - LC - NA_2,\ M_w \triangleq TW - NW_2, \\ &M_u \triangleq TB - NB_2 - (M_xN + L)D,\ M_v \triangleq (M_xN + L)V. \end{aligned}$$

Furthermore, F_ϕ, F_ψ *and* $F_\rho^s, \forall s \in \{w, w, u\}$, *for the JSS mappings* ϕ, ψ *and* ρ, *respectively, are as given below* (8) *(cf. Proposition 3).*

Proof Defining $\varepsilon_t \triangleq \overline{\xi}_t - \underline{\xi}_t$, note that the first two equations in (28) imply that:

$$\varepsilon_t^+ = |M_x|^*\varepsilon_t + |M_w|\delta_w + |M_v|\delta_v + |T|\delta_t^\phi + |L|\delta_t^\psi + |N|\delta_t^\rho, \tag{31}$$

where $|M_x|^*$, M_w and M_v are given in (30). On the other hand, from the third and fourth equations in (28) we obtain the following:

$$\begin{aligned} e_t &\triangleq \overline{x}_t - \underline{x}_t = \overline{\xi}_t - \underline{\xi}_t + |NV|\delta_v = \varepsilon_t + |NV|\delta_v \\ \Longrightarrow e_t^+ &\triangleq \begin{cases} e_{t+1} = \varepsilon_{t+1} + |NV|\delta_v = \varepsilon_t^+ + |NV|\delta_v, & \text{if } \mathcal{G} \text{ is DT}, \\ \dot{e}_t = \dot{\varepsilon}_t = \varepsilon_t^+, & \text{if } \mathcal{G} \text{ is CT}, \end{cases} \\ &= \varepsilon_t^+ + \sigma|NV|\delta_v. \end{aligned} \tag{32}$$

Combining (31) and (32) returns the error dynamics in (29). Then, the first inequality holds by the facts that in the DT case, we have $\sigma = 1$ and

$$\begin{aligned} &(|M_v| + |NV| - |M_x|^*|NV|)\delta_v \\ &= (|M_xNV + LV| - |M_x||NV| + |NV|)\delta_v \\ &\leq (|M_xNV| + |LV| - |M_x||NV| + |NV|)\delta_v = (|LV| + |NV|)\delta_v, \end{aligned} \tag{33}$$

and in the CT case, we have $\sigma = 0$ and

$$\begin{aligned}&(|M_v| - |M_x|^*|NV|)\delta_v \\ &= (|M_x NV + LV| - M_x^m|NV|)\delta_v \\ &\leq (|M_x NV| + |LV| - M_x^m|NV|)\delta_v \leq ((|M_x| - M_x^m)|NV| + |LV|)\delta_v,\end{aligned} \tag{34}$$

where the inequalities in (33) and (34) hold by the sub-multiplicative property of the $|\cdot|$ operator and the positivity of all the multiplicands. Finally, bounding the nonlinear JSS function differences $\delta_t^s \triangleq \overline{s} - \underline{s}$, $\forall s \in \{\phi, \psi, \rho\}$ based on Proposition 3 and Equation (8) yields the second inequality, i.e., the comparison system in (29).

Remark 3 The main purpose of the upper-bounding in (33) (DT case) is to remove the bilinearities at the expense of potentially adding some conservatism. Moreover, the two upper-boundings in (34) (CT case) were done to reduce the number of bilinear terms to only one single bilinear term, which we found to be unavoidable.

4.2.2 Observer Design

In this subsection, we provide mixed integer linear programs (MILPs) and mixed-integer semi-definite programs (MISDPs) for computing stabilizing observer gains, while minimizing the L_1 and $\mathcal{H}_\infty$ norms of the comparison system in (29), respectively. The provided conditions are tight, in the sense that the results are necessary and sufficient for the stability of the comparison system and its L_1 or $\mathcal{H}_\infty$-robustness, for both CT and DT cases, as summarized in the following theorem.

Theorem 2 (L_1 or $\mathcal{H}_\infty$-Robust and ISS Observer Design) *Suppose Assumptions 1 and 2 hold for the nonlinear system $\mathcal{G}$ in* (10). *Then, the following statements hold.*

(i) The correct interval framer proposed in (28) *is L_1-robust if there exists a tuple $(\gamma_*, \Delta_*, Q_*, \Omega_*, \Gamma_*, \tilde{L}_*, \tilde{N}_*, \tilde{T}_*, \tilde{M}_{x*}, \tilde{Z}_*, N_*, \Phi_*)$ that solves the following mixed-integer program (MIP):*

$$\begin{aligned}&\min_{\{\gamma,\Delta,Q,\Omega,\Gamma,\tilde{L},\tilde{N},\tilde{T},\tilde{M}_x,\tilde{Z},N,\Phi\}} \quad \gamma \\ &s.t.\ \mathbf{1}_n^\top \left[\Delta\ \Gamma\ \Phi\ \Omega\right] < \left[\ \gamma\mathbf{1}_{n_w}^\top\ \gamma\mathbf{1}_m^\top\ \gamma\mathbf{1}_{n_v}^\top\ \mathbf{1}_n^\top(\sigma Q - I)\right],\ \textit{and}\ \{\mathbf{C}\}\ \textit{holds}.\end{aligned} \tag{35}$$

(ii) The correct interval framer proposed in (28) *is $\mathcal{H}_\infty$-robust if there exists a tuple $(\gamma_*, \Delta_*, Q_*, \Omega_*, \Gamma_*, \tilde{L}_*, \tilde{N}_*, \tilde{T}_*, \tilde{M}_{x*}, \tilde{Z}_*, N_*, \Phi_*)$ that solves the following mixed-integer program (MIP):*

$$\begin{aligned}&\min_{\{\gamma,\Delta,Q,\Omega,\Gamma,\tilde{L},\tilde{N},\tilde{T},\tilde{M}_x,\tilde{Z},N,\Phi\}} \quad \gamma \\ &s.t.\ \mathbf{1}_n^\top \left[\Delta\ \Gamma\ \Phi\ \Omega\right] < \left[\ \gamma\mathbf{1}_{n_w}^\top\ \gamma\mathbf{1}_m^\top\ \gamma\mathbf{1}_{n_v}^\top\ \mathbf{1}_n^\top(\sigma Q - I)\right],\ \textit{and}\ \{\mathbf{C}\}\ \textit{holds}.\end{aligned} \tag{36}$$

$$\textit{where } \sigma = \begin{cases} 1 & \textit{if } \mathcal{G} \textit{ is DT}, \\ 0 & \textit{if } \mathcal{G} \textit{ is CT}, \end{cases}, \beta = \begin{cases} 1 & \textit{if } V \neq 0, \\ 0 & \textit{if } V = 0 \end{cases} \quad \textit{and}$$

$$\mathbf{C} = \begin{cases} \Delta = |\tilde{T}W - \tilde{N}W_2| + |\tilde{N}|F^w_\rho, \\ \Gamma = |\tilde{N}|F^u_\rho, \\ \Phi = |\tilde{L}V| + \sigma|\tilde{N}V| + (1-\sigma)Z, \\ \Omega = \sigma|\tilde{M}_x| + (1-\sigma)\tilde{M}^m_x + |\tilde{T}|F_\phi + |\tilde{L}|F_\psi + |\tilde{N}|F^x_\rho, \\ Z = (|\tilde{M}_x| - \tilde{M}^{\mathrm{m}}_x)|NV|, \\ \tilde{N} = \beta(1-\sigma)QN + \beta\sigma\tilde{N} + (1-\beta)\tilde{N}, \\ \tilde{M}_x = \tilde{T}A - \tilde{L}C - \tilde{N}A_2, \\ \tilde{T} = Q - \tilde{N}C, \\ \gamma > 0, Q \in \mathbb{D}^n_{>0}. \end{cases} \quad (37)$$

Moreover, in both DT and CT cases, the L_1 or $\mathcal{H}_\infty$-robust observer gains can be computed as $X = Q_*^{-1}\tilde{X}_*, \forall X \in \{L, T, N\}$, where the tuple (Q_*, L_*, T_*, N_*) is the optimal solution to the MIP in (35) *or* (36)*, respectively.*

Proof The comparison system in (29) can be rewritten as:

$$\tilde{\mathcal{G}} : x_t^+ = \tilde{A}_x x_t + \tilde{B}\delta_{\tilde{w}}, \quad z_t = \tilde{C}x_t + \tilde{D}\delta_{\tilde{w}}, \tilde{C} = I, \tilde{D} = 0. \quad (38)$$

Note that by construction, (38) is a positive/cooperative system since $\tilde{A}_w$, $\tilde{A}_v$ and $\tilde{A}_u$ are non-negative, whereas $\tilde{A}_x$ is non-negative in the DT case and is Metzler in the CT case. This follows from the non-negativity of the operator $|\cdot|$ and the matrices F_ϕ, F_ψ, F^x_ρ, $|\tilde{M}_x| - \tilde{M}^{\mathrm{m}}_x$, as well as the facts that in the DT case, $|\tilde{M}_x|$ is a non-negative matrix, while in the CT case, $\tilde{M}^{\mathrm{m}}_x$ is a Metzler matrix by construction, in addition to the fact that the summation of non-negative and Metzler matrices is Metzler. Now, given the positive/cooperative system (38), below we show the sufficiency and necessity of the conditions in (35) and (36) for the stability and L_1 or $\mathcal{H}_\infty$ robustness of (38).

Proof of (i): **L_1-Norm Minimization**: In the DT case, by Chen et al. (2013, Theorem 2), the positive linear system $\tilde{\mathcal{G}}$ in (38) is stable and satisfies $\|z_t\|_{\ell_1} < \gamma\|\delta_{\tilde{w}}\|_{\ell_1}$ for some $\gamma > 0$, if and only if there exist $p \in \mathbb{R}^n_{>0}$ such that

$$\begin{bmatrix} \tilde{A}_x & \tilde{B} \\ I & 0 \end{bmatrix}^\top \begin{bmatrix} p \\ \mathbf{1}_n \end{bmatrix} < \begin{bmatrix} p \\ \gamma\mathbf{1}_{\tilde{n}} \end{bmatrix} \Leftrightarrow [p^\top \ \mathbf{1}_n^\top] \begin{bmatrix} \tilde{A}_x & \tilde{B} \\ I & 0 \end{bmatrix} < \begin{bmatrix} p^\top & \gamma\mathbf{1}_{\tilde{n}}^\top \end{bmatrix} \Leftrightarrow \begin{cases} p^\top\tilde{A}_x + \mathbf{1}_n^\top < p^\top, \\ p^\top\tilde{B} < \gamma\mathbf{1}_{\tilde{n}}^\top, \end{cases} \quad (39)$$

while in the CT case, it follows from Briat (2013, Lemma 1) that the system in (38) is stable and satisfies $\|z_t\|_{\ell_1} < \gamma\|\delta_{\tilde{w}}\|_{\ell_1}$ for some $\gamma > 0$, if and only if there exist $p \in \mathbb{R}^n_{>0}$ such that

$$\begin{bmatrix} \tilde{A}_x & \tilde{B} \\ I & 0 \end{bmatrix}^\top \begin{bmatrix} p \\ \mathbf{1}_n \end{bmatrix} < \begin{bmatrix} \mathbf{0}_n \\ \gamma \mathbf{1}_{\tilde{n}} \end{bmatrix} \Leftrightarrow \begin{bmatrix} p^\top & \mathbf{1}_n^\top \end{bmatrix} \begin{bmatrix} \tilde{A}_x & \tilde{B} \\ I & 0 \end{bmatrix} < \begin{bmatrix} \mathbf{0}_n^\top & \gamma \mathbf{1}_{\tilde{n}}^\top \end{bmatrix} \Leftrightarrow \begin{cases} p^\top \tilde{A}_x + \mathbf{1}_n^\top < \mathbf{0}_n^\top, \\ p^\top \tilde{B} < \gamma \mathbf{1}_{\tilde{n}}^\top, \end{cases} \tag{40}$$

where $\tilde{n} \triangleq n_w + n_v + m$. Then, by defining the positive diagonal matrix variable $Q = Q^\top = diag(p) \Leftrightarrow p = Q\mathbf{1}_n \Leftrightarrow p^\top = \mathbf{1}_n^\top Q$, the inequalities in (39) are equivalent to:

$$\text{DT case: } \begin{cases} \mathbf{1}_n^\top Q\tilde{A}_x < \mathbf{1}_n^\top (Q - I), \\ \mathbf{1}_n^\top Q\tilde{B} < \gamma \mathbf{1}_{\tilde{n}}^\top, \end{cases} \qquad \text{CT case: } \begin{cases} \mathbf{1}_n^\top Q\tilde{A}_x < -\mathbf{1}_n^\top, \\ \mathbf{1}_n^\top Q\tilde{B} < \gamma \mathbf{1}_{\tilde{n}}^\top. \end{cases} \tag{41}$$

The conditions in (41) are equivalent to the ones in (35) in both DT ($\sigma = 1$) and CT ($\sigma = 0$) cases after plugging $\tilde{A}_x$ and $\tilde{B} \triangleq \begin{bmatrix} \tilde{A}_w & \tilde{A}_v & \tilde{A}_u \end{bmatrix}$ from (30) into (41), in addition to considering the constraint $T = I - NC \Leftrightarrow QT = Q - QNC$, defining $\tilde{X} \triangleq QX, \forall X \in \{L, T, N, M_x\}$, and given the fact that $Q|Y| = |QY|$ since Q is a positive diagonal matrix.

Proof of (ii): $\mathcal{H}_\infty$**-Norm Minimization**: In the DT case, by Najson (2013, Theorem 3) as well as applying Schur complement, we conclude that the positive comparison system (38) is stable, with its $\mathcal{H}_\infty$ norm being less than γ, if and only if there exists a positive diagonal matrix Q such that

$$\begin{bmatrix} Q & Q\tilde{A}_x & Q\tilde{B} & 0 \\ * & Q & 0 & I \\ * & * & \gamma I & 0 \\ * & * & * & \gamma I \end{bmatrix} \succ 0. \tag{42}$$

Similarly, in the CT case, Tanaka and Langbort (2011, Theorem 2) as well as applying Schur complement lead to the following necessary and sufficient conditions:

$$\begin{bmatrix} Q\tilde{A}_x + \tilde{A}_x^\top Q & Q\tilde{B} & I \\ * & -\gamma I & 0 \\ * & * & -\gamma I \end{bmatrix} \prec 0. \tag{43}$$

By similar arguments as in the proof of (i), the conditions in (42) and (43) as well as the additional constraint $T = I - NC$ can be equivalently represented in the form of (36).

Note that in the CT case and only if $V \neq 0$, the presence of the terms $(|\tilde{M}_x| - \tilde{M}_x^{\mathrm{m}})NV$ and QN in the fifth and sixth equalities in $\mathbf{C}$ (cf. Eq (37)) leads to bilinear constraints, but the MIP still remains solvable with off-the-shelf solvers, e.g., Gurobi (2015). Nonetheless, for many classes of systems, e.g., for DT systems, as well as when $V = 0$ (for both DT and CT systems), as shown in the following proposition, or by choosing $N = 0$ (at the cost of losing the extra degrees of freedom with T and

N) the MIP in (35) reduces to a mixed-integer linear program (MILP) and the one in (36) to a mixed-integer semi-definite program (MISDP).

Proposition 5 (MILP and MISDP for DT Systems or when $V = 0$) *If system* $\mathcal{G}$ *is DT, or if* $V = 0$, *then* (35) *is an MILP and* (36) *is an MISDP.*

Proof If system $\mathcal{G}$ is DT, then $\sigma = 1$. This implies that the sixth inequality in **C** becomes $\tilde{N} = \tilde{N}$ while Φ no longer contains Z, i.e., there will be no bilinearity anymore in (35) and (36). Similarly, if $V = 0$ (for both DT and CT cases) and hence $\beta = 0$, then the fifth and sixth equalities in **C** will be $Z = 0$ and $\tilde{N} = \tilde{N}$, and again no bilinearity remains.

Remark 4 (MILP/MISDP versus LP/SDP) The mixed-integer nature of the proposed programs in Theorem 2 stems from the existence of the terms involving absolute values $|M|$, as well as "Metzlerization" $M^m = M^d + |M^{nd}|$. If desired, extra positivity constraints can be imposed to remove the need for the absolute values in $|M|$ and $|M^{nd}|$, similar to the literature on SDP/LMI-based interval observer designs, which would lead to a semi-definite program (SDP) or a linear program (LP). This addition is found to sometimes not incur any conservatism, but the problem becomes infeasible in others.

Remark 5 (MILP/MISDP versus LP/SDP) Note that the difference between the two proposed observer designs in Theorem 2 is due to the system gain/norm (of the transfer function) of the observer error system that we choose to minimize. As a result of this choice (as desired/determined by the user), the L_1 design requires solving a mixed-integer linear program, while the $\mathcal{H}_\infty$ one is based on solving a mixed-integer semi-definite program, and nevertheless, both of the designs are applicable to bounded-Jacobian DT and CT nonlinear systems. Finally, in general it is hard to compare them in the sense of tightness of the interval estimates, since they are minimizing different system norms.

Remark 6 (Feasibility and Conservatism of our Designs) It is worth emphasizing that a necessary condition for finding the observer gains using Theorem 2 is that the closed-loop framer error system matrix $\tilde{A}_x$ in (29) must be stable and we do not require $\tilde{A}_x$ to be Metzler (for CT systems) or positive (for DT systems). Instead, one can observe that $\tilde{A}_x$ in (30) is already Metzler or positive by design. It is well known that simultaneously imposing Metzler/positivity and stability constraints in interval observer designs can be very conservative. To illustrate this, consider a discrete-time framer error system, $x_t^+ = Ax_t$ with $A = -0.1$. It can be observed that A is not simultaneously Schur stable and positive, but analog to our absolute value operator within $\tilde{A}_x$ in (30), $|A| = 0.1$ is Schur stable.

4.3 Coordinate Transformation

Note that in the construction of the proposed interval observer in (28) inspired by Wang et al. (2012), we introduced a linear transformation using gains T and N such that $T + NC = I$. This transformation provides additional degree of freedom, i.e., with to-be-designed gains L, N, T, in contrast to only L. While these additional observer gains resulting in a linear transformation can act as a surrogate to coordinate transformations and are helpful from the performance perspective, we could still benefit from having a coordinate transformation (which is a form of nonlinear transformation) that is commonly required for previous interval observer designs. Note that the coordinate transformation should be applied to the original system (10) and "before" defining auxiliary states, not to the equivalent system (16) that includes auxiliary states. In such cases, by introducing a coordinate transformation $z = Sx$ for invertible $S \in \mathbb{R}^{n\times n}$, the system (10) can be rewritten in the new coordinates as:

$$\mathcal{G}: \begin{cases} z_t^+ = \hat{f}(z_t) + \hat{W}w_t + \hat{B}u_t, \\ y_t = \hat{h}(z_t) + Vv_t + Du_t, \end{cases} \text{, for all } t \in \mathbb{T}, \tag{44}$$

where $\hat{f}(z_t) = Sf(S^{-1}z_t)$, $\hat{W} = SW$, $\hat{B} = SB$ and $\hat{h}(z_t) = h(S^{-1}z_t)$.

Then, for an appropriate choice of an invertible S and using the transformed system (44), the interval observer approach in Theorem 1 and Theorem 2 can be used to compute feasible observer gains T, N, L. Such an invertible S can be found for example, by selecting any L with $N = 0$ (consequently $T = I$) for A in (14), such that $A - LC$ is Schur/Hurwitz with real eigenvalues, and then choosing the invertible S that diagonalizes $A - LC$ i.e., such that $S(A - LC)S^{-1}$ is diagonal. This method for finding an invertible S is a modification of an approach discussed in Tahir and Açıkmeşe (2021).

Finally, to motivate the potential need for coordinate transformation on top of the additional degrees of freedom, consider an autonomous linear system $\dot{x} = Ax$, with observations $y = Cx$, with

$$A = \begin{bmatrix} -0.198 & 0.287 & -1.087 \\ -0.09 & 0.231 & 0.915 \\ -0.17 & 0.274 & -1.139 \end{bmatrix}, \quad C = \begin{bmatrix} 0.4 & -0.2 & -0.2 \end{bmatrix}.$$

For this system, we observe that the optimization problems in (35) and (36) are infeasible without any coordinate transformations. In contrast, feasible and optimal interval observers can be designed using the proposed approaches in previous section by transforming the system into new coordinates using the following invertible time-invariant transformation matrix

$$S = \begin{bmatrix} 3704.95 & -184.77 & -3768.21 \\ -7216.86 & 282.19 & 7252.94 \\ -3511.73 & 96.81 & 3485.65 \end{bmatrix}.$$

Hence, coordinate transformation prior to designing the interval observer can still be useful in making the design feasible for our proposed approaches, which will be further demonstrated in simulations in the next section.

5 Illustrative Examples

In this section, we consider both CT and DT examples to demonstrate the effectiveness of our approaches. The MILPs in (35) and MISDPs in (36) are solved using YALMIP (Löfberg 2004) and Gurobi (Gurobi Optimization 2015).

5.1 CT System Example 1

Consider the CT system from Dinh et al. (2014, Eq. (30)):

$$\dot{x}_1 = x_2 + w_1, \quad \dot{x}_2 = b_1 x_3 - a_1 \sin(x_1) - a_2 x_2 + w_2,$$
$$\dot{x}_3 = -\, a_3(a_2 x_1 + x_2) + \frac{a_1}{b_1}(a_4 \sin(x_1) + \cos(x_1) x_2) - a_4 x_3 + w_3,$$

with state $x = [x_1, x_2, x_3]^\top$, output $y = x_1$ and corresponding parameters: $b_1 = 15, a_1 = 35.63, a_2 = 0.25, a_3 = 36, a_4 = 200$, $\mathcal{W} = [-0.1, 0.1]^3$ and $\mathcal{X}_0 = [19.5, 9] \times [9, 11] \times [0.5, 1.5]$.

For the above CT system, any nonlinear functions of only state x_1 can be considered as measured outputs since we are measuring x_1. Consequently, the remaining nonlinear component of the dynamics is $f(x) = [0, 0, \frac{a_1}{b_1} cos(x_1)x_2]^\top$, and its corresponding Jacobian sign-stable mapping $\phi(x) = f(x) - Hx$, where $H = \begin{bmatrix} 0 & 0 & 0 \\ 0 & 0 & 0 \\ 29.692 & 2.375 & 0 \end{bmatrix}$, is computed as per Proposition 1. Therefore, $\phi_1(x) = \phi_2(x) = 0$ and for $\phi_3(x)$, as per Proposition 3, we can construct a decomposition function $\phi_{d,3}(x_1, x_2) = \phi_3(D_3 x_1 + (I - D_3)x_2)$ with $D_3 = \begin{bmatrix} 0 & 0 & 0 \\ 0 & 0 & 0 \\ 0 & 0 & 0 \end{bmatrix}$, as well as the corresponding $F_\phi = \begin{bmatrix} 0 & 0 & 0 \\ 0 & 0 & 0 \\ 59.384 & 4.751 & 0 \end{bmatrix}$ as shown below (8).

The resultant L_1-robust observer gains computed by solving the optimization in Theorem 2 are given below:

$$T = \begin{bmatrix} 0 & 0 & 0 \\ -20.506 & 1 & 0 \\ 0 & 0 & 1 \end{bmatrix}, \quad L = \begin{bmatrix} 159.384 \\ 102.531 \\ 29.692 \end{bmatrix}, \quad N = \begin{bmatrix} 1 \\ 20.506 \\ 0 \end{bmatrix}$$

and using (17), the corresponding observer matrices M_x and M_w are:

$$M_x = \begin{bmatrix} -159.384 & 0 & 0 \\ 0 & -20.756 & 15 \\ 0 & -33.624 & -200 \end{bmatrix}, \quad M_w = \begin{bmatrix} 0 & 0 & 0 \\ -20.506 & 1 & 0 \\ 0 & 0 & 1 \end{bmatrix}.$$

Similarly, we can compute the $\mathcal{H}_\infty$-robust observer gains as

$$T = \begin{bmatrix} 0 & 0 & 0 \\ -104.538 & 1 & 0 \\ 0 & 0 & 1 \end{bmatrix}, \quad L = \begin{bmatrix} 5015607.653 \\ 522.690 \\ 29.692 \end{bmatrix}, \quad N = \begin{bmatrix} 1 \\ 104.538 \\ 0 \end{bmatrix}$$

with the observer matrices

$$M_x = \begin{bmatrix} -5015607.653 & 0 & 0 \\ 0 & -104.788 & 15 \\ 0 & -33.625 & -200 \end{bmatrix}, \quad M_w = \begin{bmatrix} 0 & 0 & 0 \\ -104.538 & 1 & 0 \\ 0 & 0 & 1 \end{bmatrix}.$$

In both L_1 and $\mathcal{H}_\infty$ cases, due to the absence of a control input and measurement noise, both M_u and M_v are matrices of zeros.

From Fig. 1 (only results for state x_3 is shown for brevity; other states follow the same trends), both of the proposed L_1- and $\mathcal{H}^\infty$-robust observers have comparable performance in framing the true state and in the convergence of the observer error $\varepsilon_t = \overline{x}_t - \underline{x}_t$, and notably, both approaches provide tighter bounds than the interval observer presented in Dinh et al. (2014), $\underline{x}^{DMN}, \overline{x}^{DMN}$. Further, from Fig. 1 (right), the framer error in both the proposed cases exponentially converge to a steady-state faster than the convergence of the framer error in Dinh et al. (2014).

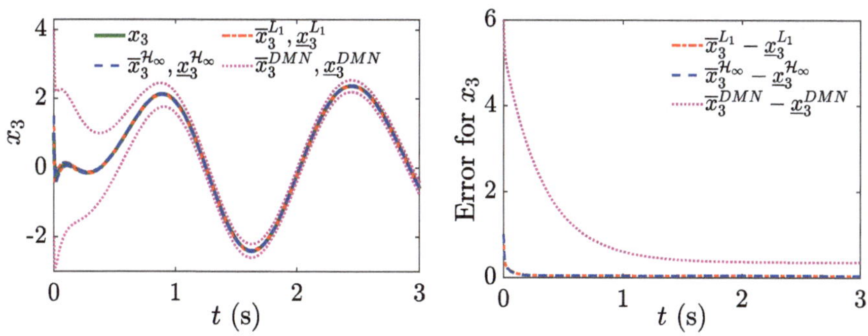

Fig. 1 CT Example 1: State, x_3, and its upper and lower framers and error of our proposed observer, $\overline{x}_3^{L_1}, \underline{x}_3^{L_1}, \varepsilon_3^{L_1}$ for the L_1-robust interval observer, $\overline{x}_3^{\mathcal{H}^\infty}, \underline{x}_3^{\mathcal{H}^\infty}, \varepsilon_3^{\mathcal{H}^\infty}$ for the $\mathcal{H}^\infty$-robust interval observer, and $\overline{x}_3^{DMN}, \underline{x}_3^{DMN}, \varepsilon_3^{DMN}$ for the observer in Dinh et al. (2014)

5.2 CT System Example 2

Consider the linear CT system from Sect. 4.3 with additive process noise, i.e., $\dot{x} = Ax + Ww$, and observations $y = Cx$, with system matrices A, C and transformation matrix S as given in Sect. 4.3 and with $W = S^{-1}$. This is an unstable linear system and as stated in Sect. 4.3, the corresponding optimization problems in (35) and (36) are infeasible without any coordinate transformations. However, with the transformation matrix S, we obtain feasible L_1-robust observer gains as given below:

$$T = \begin{bmatrix} 1 & 0 & 0 \\ -0.143 & 0.857 & 0.144 \\ 0 & 0 & 1 \end{bmatrix}, \ L = \begin{bmatrix} 86.988 \\ 64.847 \\ 147.363 \end{bmatrix}, \ N = \begin{bmatrix} 0 \\ -0.6271 \\ 0 \end{bmatrix}$$

and the corresponding observer matrices are:

$$M_x = \begin{bmatrix} -0.03 & 0 & 0 \\ 0.0029 & -0.0186 & 0 \\ 0.008 & 0.008 & -0.018 \end{bmatrix}, \ M_w = \begin{bmatrix} 1 & 0 & 0 \\ -0.143 & 0.857 & 0.144 \\ 0 & 0 & 1 \end{bmatrix}.$$

Similarly, the $\mathcal{H}_\infty$-robust observer gains and matrices are

$$T = \begin{bmatrix} 1 & 0 & 0 \\ 0 & 1 & 0 \\ 0 & 0 & 1 \end{bmatrix}, \ L = \begin{bmatrix} 86.988 \\ 65.509 \\ 147.386 \end{bmatrix}, \ N = \begin{bmatrix} 0 \\ 0 \\ 0 \end{bmatrix}$$
$$M_x = \begin{bmatrix} -0.03 & 0 & 0 \\ 0 & -0.02 & 0 \\ 0.0133 & 0.0134 & -0.0234 \end{bmatrix}, \ M_w = \begin{bmatrix} 1 & 0 & 0 \\ 0 & 1 & 0 \\ 0 & 0 & 1 \end{bmatrix}.$$

and similar to the earlier CT example, M_u and M_v are matrices of zeros.

From Fig. 2 (for brevity only x_1 results are shown, other states follow the same trend), both of the proposed observers have comparable performance in framing the true state with the L_1 robust observers error converging slightly faster.

5.3 DT System Example 1

For the DT case, consider the DT Hénon chaos system Khajenejad and Yong (2022, Sect. 5-B):

$$x_{t+1} = Ax_t + r[1 - x_{t,1}^2] + Bw_t, \quad y_t = x_{t,1} + v_t, \tag{45}$$

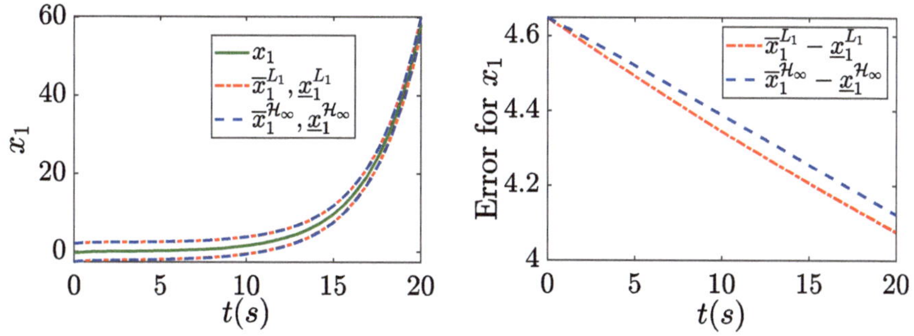

Fig. 2 CT Example 2: State, x_1, and its upper and lower framers and error of our proposed observer, $\overline{x}_1^{L_1}, \underline{x}_1^{L_1}, \varepsilon_1^{L_1}$ for the L_1-robust interval observer, $\overline{x}_1^{\mathcal{H}^\infty}, \underline{x}_1^{\mathcal{H}^\infty}, \varepsilon_1^{\mathcal{H}^\infty}$ for the $\mathcal{H}^\infty$-robust interval observer

with $A = \begin{bmatrix} 0 & 1 \\ 0.3 & 0 \end{bmatrix}$, $B = I$, $r = \begin{bmatrix} 0.05 \\ 0 \end{bmatrix}$, $\mathcal{W} = [-0.01, 0.01]^2$, $\mathcal{V} = [-0.1, 0.1]$ and $\mathcal{X}_0 = [-2, 2] \times [-1, 1]$. From (45), the nonlinear component of the dynamics is $f(x) = r[1 - x_{t,1}^2]$ and its corresponding Jacobian sign-stable mapping is $\phi(x) = f(x) - Hx$, where $H = \begin{bmatrix} 0.01 & 0 \\ 0 & 0 \end{bmatrix}$ is computed using Proposition 1. Consequently, $\phi_2(x) = 0$ and for $\phi_1(x)$, we can construct a tight decomposition function $\phi_{d,1}(x_1, x_2) = \phi_1(D_1 x_1 + (I - D_1)x_2)$ using Proposition 3 with $D_1 = \begin{bmatrix} 0 & 0 \\ 0 & 0 \end{bmatrix}$, and its corresponding $F_\phi = \begin{bmatrix} 0.11 & 0 \\ 0 & 0 \end{bmatrix}$ as shown below (8). The solution of (35) then returns the L_1-robust observer gains as:

$$T = \begin{bmatrix} 1 & 0 \\ 0 & 1 \end{bmatrix}, \quad L = \begin{bmatrix} 0 \\ 0.1 \end{bmatrix}, \quad N = \begin{bmatrix} 0 \\ 0 \end{bmatrix},$$
$$M_x = \begin{bmatrix} 0 & 1 \\ 0.2 & 0 \end{bmatrix}, \quad M_w = \begin{bmatrix} 1 & 0 \\ 0 & 1 \end{bmatrix}, \quad M_v = \begin{bmatrix} 0 \\ 0.1 \end{bmatrix}.$$

Similarly, the solution of (36) returns the $\mathcal{H}_\infty$-robust observer gains as:

$$T = \begin{bmatrix} 1 & 0 \\ 0 & 1 \end{bmatrix}, \quad L = \begin{bmatrix} 0 \\ 0.0304 \end{bmatrix}, \quad N = \begin{bmatrix} 0 \\ 0 \end{bmatrix},$$
$$M_x = \begin{bmatrix} 0 & 1 \\ 0.2696 & 0 \end{bmatrix}, \quad M_w = \begin{bmatrix} 1 & 0 \\ 0 & 1 \end{bmatrix}, \quad M_v = \begin{bmatrix} 0 \\ 0.0304 \end{bmatrix}.$$

In both cases, M_u is a zero matrix.

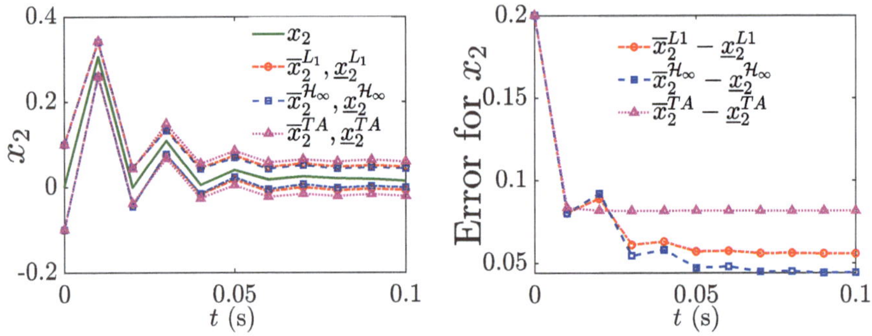

Fig. 3 DT Example 1: State, x_2, and its upper and lower framers and error of our proposed observers, $\overline{x}_2^{L_1}, \underline{x}_2^{L_1}, \varepsilon_2^{L_1}$ for the L_1-robust interval observer, $\overline{x}_2^{\mathcal{H}_\infty}, \underline{x}_2^{\mathcal{H}_\infty}, \varepsilon_2^{\mathcal{H}_\infty}$ for the $\mathcal{H}_\infty$-robust interval observer, and $\overline{x}_2^{TA}, \underline{x}_2^{TA}, \varepsilon_2^{TA}$ for the observer in Tahir and Açıkmeşe (2021)

From Fig. 3 (only x_2 is considered for brevity, x_1 shows a similar trend), the estimates from both our proposed approaches are tighter than the method in Tahir and Açıkmeşe (2021) and when compared to each other, the framers for the $\mathcal{H}_\infty$-robust observer are slightly tighter than the L_1-robust version.

5.4 DT System Example 2

For the second discrete time example, consider the slightly modified version of the discrete-time predator-prey system from Khajenejad et al. (2021, Sect. 5B):

$$\begin{bmatrix} x_{1,t+1} \\ x_{2,t+1} \\ d_{t+1} \end{bmatrix} = \begin{bmatrix} x_{1,t} \\ x_{2,t} \\ d_t \end{bmatrix} + h \begin{bmatrix} -x_{1,t}x_{2,t} - x_{2,t} + d_t + w_{1,t} \\ x_{1,t}x_{2,t} + x_{1,t} + w_{2,t} \\ 100(\cos(x_{1,t}) - \sin(x_{2,t})) + w_{3,t} \end{bmatrix}, \tag{46}$$

$$y_t = \begin{bmatrix} x_{1,t} + v_{1,t} \\ x_{2,t} + v_{2,t} \\ \sin(d_t) + v_{3,t} \end{bmatrix}, \tag{47}$$

with bounded disturbances $w_{l,t} \in [-0.1, 0.1]$, $\forall l = 1, 2, 3$, bounded noise $v_{l,t} \in [-0.01, 0.01]$, $\forall l = 1, 2, 3$, sampling time $h = 0.05$ and $\mathcal{X}_0 = [-0.35, 0] \times [-0.1, 0.6] \times [-10, 10]$.

From (46), the nonlinear component of the dynamics is

$$f(x) = \begin{bmatrix} -x_{1,t}x_{2,t} \\ -x_{1,t}x_{2,t} \\ 100(\cos(x_{1,t}) - \sin(x_{2,t})) \end{bmatrix}$$

with the corresponding Jacobian sign-stable mapping $\phi(x) = f(x) - Hx$, where $H = \begin{bmatrix} 0.001 & 0 & 0 \\ -0.001 & 0 & 0 \\ 0 & 0 & 0 \end{bmatrix}$ is computed using Proposition 1. Therefore, using Proposition 3, we can construct a decomposition function $\phi_{d,i}(x_1, x_2) = \phi_i(D_i^f x_1 + (I - D_i^f)x_2)$ for each JSS function component $\phi_i(x)$ where $D_i^f = \text{diag}(D_{x,i}^f)$ with $D_{x,i}^f$ representing the ith row of $D_x^f = \begin{bmatrix} 0 & 1 & 0 \\ 1 & 0 & 0 \\ 1 & 0 & 0 \end{bmatrix}$. Further, the corresponding $F_\phi = \begin{bmatrix} 0.0070 & 0.0035 & 0 \\ 0.0070 & 0.0035 & 0 \\ 0.3429 & 1.0000 & 0 \end{bmatrix}$ can be computed as shown below (8).

Similarly, the nonlinear component of measurement $h(x) = [0, 0, \sin(d_t)]^\top$ and its corresponding JSS function is $\psi(x) = h(x) - Cx$ with $C = \begin{bmatrix} 0 & 0 & 0 \\ 0 & 0 & 0 \\ 0 & 0 & 1 \end{bmatrix}$. Consequently, using Proposition 3, $\psi_{d,3}(x_1, x_2) = \psi_3(D_3^h x_1 + (I - D_3^h)x_2)$, where $D_3^h = \begin{bmatrix} 0 & 0 & 0 \\ 0 & 0 & 0 \\ 0 & 0 & 0 \end{bmatrix}$ and $F_\psi = \begin{bmatrix} 0 & 0 & 0 \\ 0 & 0 & 0 \\ 0 & 0 & 2 \end{bmatrix}$ can be computed as shown below (8).

Further, $\psi^+(x, w) = \sin(d_{t+1}) - d_{t+1}$ and $\rho(x, w) = \psi^+(x, w, u) - A_2 x - W_2 w$ with

$$A_2 = \begin{bmatrix} 0 & 0 & 0 \\ 0 & 0 & 0 \\ 0.3429 & -0.1747 & 0 \end{bmatrix}, \quad W_2 = \begin{bmatrix} 0 & 0 & 0 \\ 0 & 0 & 0 \\ 0 & 0 & 0 \end{bmatrix}.$$

As a consequence, the L_1-robust observer gains can be computed by solving the optimization problem in (35) to obtain

$$T = \begin{bmatrix} 0 & 0 & 0 \\ 0 & 0 & 0 \\ -20 & 17.837 & 1 \end{bmatrix}, \quad L = \begin{bmatrix} 0 & 0 & 0 \\ 0 & 0 & 0 \\ 0 & 0 & 0 \end{bmatrix}, \quad N = \begin{bmatrix} 1 & 0 & 0 \\ 0 & 1 & 0 \\ 20 & -17.837 & 0 \end{bmatrix}$$

and the corresponding observer matrices

$$M_x = \begin{bmatrix} 0 & 0 & 0 \\ 0 & 0 & 0 \\ -19.2973 & 18.837 & 0 \end{bmatrix}, \quad M_w = \begin{bmatrix} 0 & 0 & 0 \\ 0 & 0 & 0 \\ -1 & 0.8918 & 0.05 \end{bmatrix},$$

$$M_v = \begin{bmatrix} 0 & 0 & 0 \\ 0 & 0 & 0 \\ -19.2973 & 18.837 & 0 \end{bmatrix}.$$

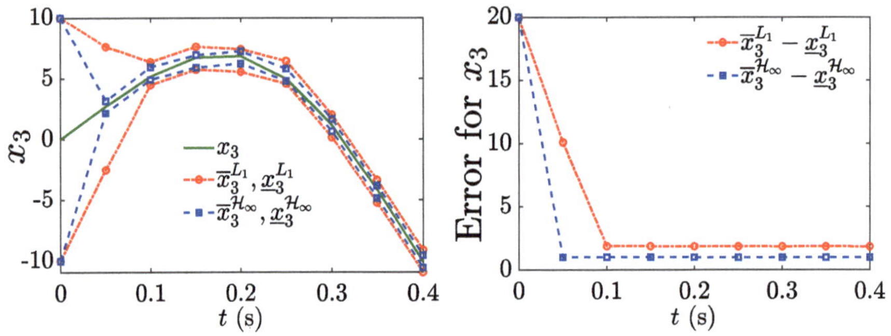

Fig. 4 DT Example 2: State, x_3, and its upper and lower framers and error of our proposed observers, $\overline{x}_3^{L_1}, \underline{x}_3^{L_1}, \varepsilon_3^{L_1}$ for the L_1-robust interval observer, $\overline{x}_3^{\mathcal{H}_\infty}, \underline{x}_3^{\mathcal{H}_\infty}, \varepsilon_3^{\mathcal{H}_\infty}$ for the $\mathcal{H}_\infty$-robust interval observer

Similarly, the $\mathcal{H}_\infty$-robust observer gains computed by solving (36) are as follows:

$$T = \begin{bmatrix} 0 & 0 & 0 \\ 0 & 0 & 0 \\ -19.9999 & 0 & 1 \end{bmatrix}, \quad L = \begin{bmatrix} 0 & 0 & 0 \\ 0 & 0 & 0 \\ -20.0999 & 1 & 0 \end{bmatrix}, \quad N = \begin{bmatrix} 1 & 0 & 0 \\ 0 & 1 & 0 \\ 19.9999 & 0 & 0 \end{bmatrix}$$

and the corresponding observer matrices are

$$M_x = \begin{bmatrix} 0 & 0 & 0 \\ 0 & 0 & 0 \\ 0 & 0 & 0.7246 \end{bmatrix} \times 10^{-5}, \quad M_w = \begin{bmatrix} 0 & 0 & 0 \\ 0 & 0 & 0 \\ -1 & 0 & 0.05 \end{bmatrix}, \quad M_v = \begin{bmatrix} 0 & 0 & 0 \\ 0 & 0 & 0 \\ -20.0997 & 1 & 0 \end{bmatrix}.$$

The corresponding framers and errors are shown in Fig. 4.

6 Conclusion and Future Work

In this chapter, we introduced a unified framework for two novel interval observer designs for uncertain discrete- and continuous-time nonlinear systems under bounded Jacobians and bounded uncertainty assumptions. Specifically, the proposed observers are correct by design without the requirement for extra positivity constraints or assumptions. This is achieved by leveraging bounding techniques from mixed-monotone decompositions and embedding systems. Moreover, we proved that the proposed observers are input-to-state stable and minimize the $\mathcal{H}_\infty$- and L_1-gains of the framer error's linear comparison systems, respectively. Our proposed approaches involve mixed-integer linear and semi-definite programs, respectively, and both methods offer additional degrees of freedom. These methods outperform state-of-the-art approaches while having comparable performance to each other, with the main advantage of the L_1-robust interval observer design involving MILPs over the $\mathcal{H}_\infty$-robust interval observer design involving MISDP being the availability of faster solvers. As for future work, we will consider extensions of our interval observer design framework to uncertain hybrid systems as well as systems with unknown inputs for obtaining resilient state observers.

Acknowledgements This work is supported in part by the National Science Foundation grant CNS-2313814.

References

Abate M, Dutreix M, Coogan S (2020) Tight decomposition functions for continuous-time mixed-monotone systems with disturbances. IEEE Control Syst Lett 5(1):139–144

Briat C (2013) Robust stability and stabilization of uncertain linear positive systems via integral linear constraints: l_1-gain and l_∞-gain characterization. Int J Robust Nonlinear Control 23(17):1932–1954

Briat C, Khammash M (2016) Interval peak-to-peak observers for continuous-and discrete-time systems with persistent inputs and delays. Automatica 74:206–213

Cacace F, Germani A, Manes C (2014) A new approach to design interval observers for linear systems. IEEE Trans Autom Control 60(6):1665–1670

Chebotarev S, Efimov D, Raïssi T, Zolghadri A (2015) Interval observers for continuous-time LPV systems with L_1/L_2 performance. Automatica 58:82–89

Chen X, Lam J, Li P, Shu Z (2013) ℓ_1-induced norm and controller synthesis of positive systems. Automatica 49(5):1377–1385

Degue KH, Efimov D, Le Ny J, Hirche S (2021) Design of interval observers for uncertain linear impulsive systems. In: IEEE conference on decision and control (CDC). IEEE, pp 3990–3995

Dinh T, Mazenc F, Niculescu S (2014) Interval observer composed of observers for nonlinear systems. In: European control conference (ECC). IEEE, pp 660–665

Efimov D, Raïssi T (2016) Design of interval observers for uncertain dynamical systems. Autom Remote Control 77(2):191–225

Efimov D, Raïssi T, Chebotarev S, Zolghadri A (2013) Interval state observer for nonlinear time varying systems. Automatica 49(1):200–205

Farina L, Rinaldi S (2000) Positive linear systems: theory and applications, vol 50. Wiley

Gurobi Optimization, Inc (2015) Gurobi optimizer reference manual

Khajenejad M, Yong SZ (2022) Simultaneous state and unknown input set-valued observers for quadratically constrained nonlinear dynamical systems. Int J Robust Nonlinear Control 32(12):6589–6622

Khajenejad M, Yong SZ (2022) $\mathcal{H}_\infty$-optimal interval observer synthesis for uncertain nonlinear dynamical systems via mixed-monotone decompositions. IEEE Control Syst Lett 6:3008–3013

Khajenejad M, Yong SZ (2023) Tight remainder-form decomposition functions with applications to constrained reachability and guaranteed state estimation. IEEE Trans Autom Control 68(12):7057–7072

Khajenejad M, Brown S, Martínez S (2023) Distributed interval observers for bounded-error LTI systems. In: 2023 American control conference (ACC). IEEE, pp 2444–2449

Khajenejad M, Brown S, Martínez S (2023) Distributed resilient interval observers for bounded-error LTI systems subject to false data injection attacks. In: 2023 American control conference (ACC). IEEE, pp 3502–3507

Khajenejad M, Brown S, Martinez S (2025) Distributed resilient interval observer synthesis for nonlinear discrete-time systems. In: IEEE transaction on automatic control, to appear

Khajenejad M, Jin Z, Dinh T, Yong SZ (2023) Resilient state estimation for nonlinear discrete-time systems via input and state interval observer synthesis. In: 2023 62nd IEEE conference on decision and control (CDC). IEEE, pp 1826–1832

Khajenejad M, Jin Z, Yong SZ (2021) Interval observers for simultaneous state and model estimation of partially known nonlinear systems. In: American control conference (ACC), pp 2848–2854

Khajenejad M, Jin Z, Yong SZ (2022) Resilient interval observer for simultaneous estimation of states, modes and attack policies. In: 2022 American control conference (ACC). IEEE, pp 322–329

Khajenejad M, Shoaib F, Yong SZ (2022) Interval observer synthesis for locally Lipschitz nonlinear dynamical systems via mixed-monotone decompositions. In: 2022 American control conference (ACC). IEEE, pp 2970–2975

Khajenejad M, Yong SZ (2020) Simultaneous input and state interval observers for nonlinear systems with full-rank direct feedthrough. In: IEEE conference on decision and control, pp 5443–5448

Khajenejad M, Yong SZ (2021) Simultaneous input and state interval observers for nonlinear systems with rank-deficient direct feedthrough. In: 2021 European control conference (ECC). IEEE, pp 2311–2316

Khajenejad M, Yong SZ (2022) Resilient state estimation and attack mitigation in cyber-physical systems. In: Security and resilience in cyber-physical systems: detection, estimation and control. Springer, pp 149–185

Kieffer M, Walter E (2006) Guaranteed nonlinear state estimation for continuous-time dynamical models from discrete-time measurements. IFAC Proc Vol 39(9):685–690

Löfberg J (2004) Yalmip: a toolbox for modeling and optimization in MATLAB, in CACSD. Taipei, Taiwan

Mazenc F, Bernard O (2011) Interval observers for linear time-invariant systems with disturbances. Automatica 47(1):140–147

Mazenc F, Bernard O (2021) When is a matrix of dimension three similar to a metzler matrix? Application to interval observer design. IEEE Trans Autom Control 67(11):6053–6059

Najson F (2013) On the Kalman-Yakubovich-Popov lemma for discrete-time positive linear systems: a novel simple proof and some related results. Int J Control 86(10):1813–1823

Pati T, Khajenejad M, Daddala S, Yong SZ (2022) L_1-robust interval observer design for uncertain nonlinear dynamical systems. IEEE Control Syst Lett 6:3475–3480

Pati T, Khajenejad M, Daddala S, Sanfelice R, Yong SZ (2023) Interval observers for hybrid dynamical systems with known jump times. In: 2023 62nd IEEE conference on decision and control (CDC). IEEE, pp 7507–7513

Tahir A, Açıkmeşe B (2021) Synthesis of interval observers for bounded Jacobian nonlinear discrete-time systems. IEEE Control Syst Lett

Tanaka T, Langbort C (2011) The bounded real lemma for internally positive systems and $\mathcal{H}_\infty$ structured static state feedback

Wang Z, Shen Y, Zhang X, Wang Q (2012) Observer design for discrete-time descriptor systems: an LMI approach. Syst Control Lett 61(6):683–687

Wang Y, Bevly D, Rajamani R (2015) Interval observer design for LPV systems with parametric uncertainty. Automatica 60:79–85

Yang L, Mickelin O, Ozay N (2019) On sufficient conditions for mixed monotonicity. IEEE Trans Autom Control 64(12):5080–5085

Sliding–Mode–Based Robust Interval Predictive Control for a Class of Uncertain and Constrained Systems

Héctor Ríos, Tarek Raïssi, and Denis Efimov

Abstract This chapter presents a review of interval observers tools and the development of a robust control strategy for a class of uncertain systems subject to both input and state constraints. The proposed approach integrates an integral sliding–mode control (ISMC) framework with an interval predictor–based state feedback controller and a Model Predictive Control (MPC) scheme. The controller is designed to handle external disturbances and parameter uncertainties while ensuring compliance with state and input constraints. The ISMC component compensates for matched perturbations from the initial time, thereby enforcing sliding–mode conditions from the onset. Once the system reaches the sliding surface, the interval predictor–based feedback controller and the MPC handle constraint satisfaction. The proposed control strategy guarantees exponential stability of the closed–loop system. Its effectiveness is demonstrated through simulation results on an academic case study.

H. Ríos
Tecnológico Nacional de México/I.T. La Laguna, Torreón, Coahuila, Mexico
e-mail: hriosb@lalaguna.tecnm.mx

SECIHTI, Investigadoras e Investigadores por México, Ciudad de México, Mexico

T. Raïssi
Conservatoire National des Arts et Métiers (CNAM), Paris, France
e-mail: tarek.raissi@cnam.fr

D. Efimov (✉)
Inria, University of Lille, CNRS, UMR 9189-CRISTAL, Lille, France
e-mail: denis.efimov@inria.fr

Department of Control Systems and Informatics, Information Technologies Mechanics and Optics University, Saint Petersburg, Russia

T. N. Dinh et al. (eds.), *Set-Valued Approaches to Control and Estimation of Uncertain Systems*, Mathematical Engineering,
https://doi.org/10.1007/978-3-031-94239-6_4

1 Introduction

Over the past few decades, the control of nonlinear systems has become a prominent area of research, motivated by its critical role in a wide spectrum of engineering applications. These applications encompass, among others, flight control systems, robotic manipulators, automated highway systems, aircraft wing structures, and high–performance fuel injection systems. A unifying characteristic of such systems is the presence of parametric uncertainties and constraints on both states and inputs. Furthermore, they are frequently subjected to external disturbances, which further complicate the control design process.

The control design of constrained dynamical systems is widely recognized as a challenging task, particularly when approached using classical feedback control strategies (Mayne et al. 2000). Given the recurrent nature of such scenarios, MPC has emerged as a powerful framework for addressing constrained optimal control problems in the presence of system uncertainties (Mayne 2014). MPC has gained significant popularity and demonstrated considerable success across a range of industrial applications (see, e.g., Borrelli et al. (2017), Falcone et al. (2007), and Funke et al. (2016)), as well as in academic research (Limon et al. 2009). The core mechanism of MPC involves solving an optimal control problem at each sampling instant, wherein a sequence of control inputs is computed and only the first input is applied to the system. An MPC scheme is deemed robust if it ensures the successful completion of the control task while guaranteeing adherence to state and input constraints across all admissible realizations of uncertainties within a specified range (Raković 2016).

Despite the advances achieved through MPC, several challenges remain in its application to uncertain systems. One of the primary difficulties lies in the reliance of MPC on predictive models that may diverge significantly from the actual system dynamics due to the presence of uncertainties, measurement noise, and external disturbances (Bemporad and Morari 2007). Additionally, traditional MPC formulations typically assume full state availability, a condition that is often impractical in real–world scenarios. This limitation necessitates the use of state estimation techniques, which further complicate the control design and introduce additional errors stemming from estimation inaccuracies (Findeisen et al. 2003).

In this context, Linear Parameter–Varying (LPV) systems are frequently employed to model systems characterized by uncertainties and/or nonlinear behaviors. The LPV framework serves as a basis for the development of MPC strategies capable of addressing such complexities. For instance, the authors in (Leurent et al. 2020), propose an MPC scheme incorporating an interval state predictor to address the stabilization problem for a class of continuous–time LPV systems, subject to state and input constraints, bounded disturbances, and parameter uncertainties. Similarly, dos Reis de Souza et al. (2022b) presents a robust output–feedback predictive control strategy for discrete–time LPV systems, accounting for system constraints as well as state and measurement disturbances. Nonetheless, both approaches in dos Reis de Souza et al. (2022b) and Leurent et al. (2020) provide only Input–to–State Stability (ISS) guarantees in the presence of external disturbances.

Among robust control methodologies, sliding–mode control (SMC) stands out as a widely adopted approach, primarily due to its inherent robustness against a class of external perturbations and its finite–time convergence properties (see, e.g., Shtessel et al. (2014) and Young et al. (1999)). In recent years, several studies have explored the integration of SMC with MPC to leverage the strengths of both techniques.

For instance, in Incremona et al. (2015), a hierarchical control framework is proposed, wherein MPC operates at a higher level and SMC at a lower level, targeting a class of nonlinear, constrained, and uncertain systems. In Incremona et al. (2017), the authors introduce a model–based event–triggered control strategy for constrained, nonlinear, continuous–time uncertain systems deployed in a networked environment. This approach combines MPC with integral sliding–mode control (ISMC), ensuring compliance with state and input constraints while addressing matched uncertainties.

Further developments include the work in Steinberger et al. (2020), where a combination of MPC and output ISMC techniques is proposed for output tracking in uncertain linear systems. The method guarantees exponential convergence of the output tracking error to zero. In Galván-Guerra et al. (2022), an optimal control scheme that integrates MPC and ISMC is presented for linear systems modeled using a multi–model representation, subject to parameter uncertainties and matched external disturbances. The authors formulate a min–max multi–model MPC problem and establish asymptotic stability of the resulting closed–loop system.

Despite these advancements, many of the aforementioned approaches rely heavily on accurate knowledge of the system model and parameters to ensure regulation performance. Moreover, most of these methods provide only ISS guarantees in the presence of disturbances.

The principal contribution of this chapter lies in the development of a robust control strategy for a class of uncertain LPV systems. The proposed approach integrates an ISMC scheme with an interval predictor–based state feedback controller and an MPC framework. This hybrid control structure is specifically designed to address external disturbances, parameter uncertainties, and state and input constraints. The ISMC component ensures robustness against matched perturbations and enforces the sliding motion from the initial time instant. Meanwhile, the interval predictor–based state feedback controller and the MPC scheme, forming a switching control architecture, are responsible for satisfying the state and input constraints. Notably, the synthesis of the interval predictor–based controller is formulated through a set of Linear Matrix Inequalities (LMIs), which are independent of the upper bounds of external disturbances due to the inherent robustness of the ISMC. It is demonstrated that the proposed control strategy ensures exponential stability of the closed–loop system. Moreover, the switching structure contributes to a reduction in computational complexity. This work extends the results presented in Leurent et al. (2020), offering complete compensation for the effects of external disturbances. A preliminary version of this approach was presented in [?]. However, the present chapter provides comprehensive theoretical developments, including detailed proofs, and introduces a distinct simulation example to further illustrate the effectiveness of the proposed method.

This chapter is organized as follows: Sect. 2 presents some preliminary concepts and results. A brief tutorial on continuous–time interval observers is given in Sect. 3. The problem statement of this chapter is introduced in Sect. 4 and the proposed controller is presented in Sect. 5. Section 6 illustrates the developed approach with a numerical example. Some concluding remarks are provided in Sect. 7. Finally, the Appendix 8 presents the proofs of the proposed results.

2 Preliminaries

First, we provide some use of notation. The symbols $|x|$ and $||x||$ represent the absolute value and the Euclidean norm of a scalar $x \in \mathbb{R}$ and a vector $x \in \mathbb{R}^n$, respectively. The matrix (induced) norm is defined as $||A|| = \sqrt{\lambda_{\max}(A^\top A)}$, for $A \in \mathbb{R}^{m\times n}$. The symbol I_n represents an identity matrix of dimension n. An integers sequence $1, \ldots, n$ is denoted as $\overline{1,n}$, for any $n \in \mathbb{N}$. The set $\mathscr{L}_\infty^p$ denotes the set of inputs $u : \mathbb{R}_{\geq 0} \to \mathbb{R}^p$ such that $||u||_\infty := ||u||_{[0,\infty]} = ess\sup_{t\geq 0}\|u(t)\| < \infty$, i.e., the L_∞ norm of the inputs u, on $[0, \infty]$, is less than infinity. The expressions $x_1 \leq x_2$ and $A_1 \leq A_2$ are understood in the component–wise sense, for vectors $x_1, x_2 \in \mathbb{R}^n$ and matrices $A_1, A_2 \in \mathbb{R}^{n\times n}$, respectively. We define $A^+ = \max\{0, A\}$, $A^- = A^+ - A$, and $|A| = A^+ + A^-$, for a matrix $A \in \mathbb{R}^{n\times n}$, in the same way for a vector. For a square matrix $A \in \mathbb{R}^{n\times n}$, the symbol $\mathrm{He}(A)$ represents the operation $A + A^\top$. For a symmetric matrix $P \in \mathbb{R}^{n\times n}$, $A \prec 0$ and $A \succeq 0$ mean P is negative and positive definite, respectively. Finally, a Metzler matrix $A \in \mathbb{R}^{n\times n}$ is a matrix such that all of its non–diagonal elements are equal to or greater than zero.

Now, we will introduce some definitions and useful results.

Consider the system

$$\dot{x} = f(t, x), \ t \in \mathbb{R}_{\geq 0}, \ x(0) = x_0, \tag{1}$$

where $x \in \mathbb{R}^n$ is the state vector; $f : \mathbb{R}_{\geq 0} \times \mathbb{R}^n \to \mathbb{R}^n$ is assumed to be locally bounded uniformly in t. For f locally measurable but discontinuous with respect to x, the solutions are understood in the sense of Filippov (Filippov 1988). Then, a solution of the system (1), for an initial condition $x_0 \in \mathbb{R}^n$ at time instant $t_0 \in \mathbb{R}$ is denoted as $x(t, t_0, x_0)$, and defined on some finite time interval $[t_0, t_0 + T)$ such that $0 \leq T \leq \infty$. Let Ω be open neighborhood of the origin in $\mathbb{R}^n$, $0 \in \Omega$.

Definition 1 (Khalil 2002; Polyakov 2011) *At the steady state $x = 0$, the system* (1) *is said to be:*

1. Uniformly Stable (US) *if for any $\varepsilon > 0$ there is $\delta(\varepsilon)$ such that for any $x_0 \in \Omega$, if $||x_0|| \leq \delta(\varepsilon)$; then, $||x(t, t_0, x_0)|| \leq \varepsilon$, for all $t \geq t_0$, and any to $t_0 \in \mathbb{R}_+$.*
2. Uniformly Exponentially Stable (UES) *if it is US and exponentially converging from Ω*, i.e., *for any $x_0 \in \Omega$, there exist $k, \gamma > 0$ such that $||x(t, t_0, x_0)|| \leq k||x_0||e^{-\gamma(t-t_0)}$, for all $t \geq t_0$ and any $t_0 \in \mathbb{R}_+$.*

3. Uniformly Finite–Time Stable (UFTS) *if it is US and finite–time converging from* Ω, i.e., *for any* $x_0 \in \Omega$, *there exists* $0 \leq T_{x_0} < +\infty$ *such that* $x(t, t_0, x_0) = 0$, *for all* $t \geq t_0 + T_{x_0}$, *and any* $t_0 \in \mathbb{R}_+$. *The function* $T_0(x_0) = \inf\{T_{x_0} \geq 0 : x(t, t_0, x_0) = 0 \; \forall t \geq t_0 + T_{x_0}\}$ *is called the settling–time of the system* (1).

The following lemma is also used.

Lemma 1 (Efimov et al. 2013) *Let* $\xi \in \mathbb{R}^n$ *be a varying vector such that* $\underline{\xi} \leq \xi \leq \overline{\xi}$, *for some* $\underline{\xi}, \overline{\xi} \in \mathbb{R}^n$. *If* $A \in \mathbb{R}^{m \times n}$ *is a constant matrix; then*

$$A^+\underline{\xi} - A^-\overline{\xi} \leq A\xi \leq A^+\overline{\xi} - A^-\underline{\xi}. \tag{2}$$

If $A \in \mathbb{R}^{m \times n}$ *is a variable matrix such that* $\underline{A} \leq A \leq \overline{A}$, *for some* $\underline{A}, \overline{A} \in \mathbb{R}^{m \times n}$; *then*

$$A_m \leq A\xi \leq A_M, \tag{3}$$

with $A_m = \underline{A}^+\underline{\xi}^+ - \overline{A}^+\underline{\xi}^- - \underline{A}^-\overline{\xi}^+ + \overline{A}^-\overline{\xi}^-$ *and* $A_M = \overline{A}^+\overline{\xi}^+ - \underline{A}^+\overline{\xi}^- - \overline{A}^-\underline{\xi}^+ + \underline{A}^-\underline{\xi}^-$.

Note that these estimates are tight (Bako and Andrieu 2019). Furthermore, if $-\overline{A} = \underline{A} \leq 0 \leq \overline{A}$; then, the inequality (3) can be simplified as $-\overline{A}(\overline{\xi}^+ + \underline{\xi}^-) \leq A\xi \leq \overline{A}(\overline{\xi}^+ + \underline{\xi}^-)$.

Any solution of the linear system

$$\dot{x}(t) = Ax(t) + Bd(t), \; d : \mathbb{R}_+ \to \mathbb{R}_+^q, \; d \in \mathcal{L}_\infty^q, \tag{4}$$

$$y(t) = Cx(t) + Dd(t), \tag{5}$$

with $x(t) \in \mathbb{R}^n$, $y(t) \in \mathbb{R}^p$ and a Metzler matrix $A \in \mathbb{R}^{n \times n}$, is element–wise nonnegative for all $t \geq 0$ provided that $x(0) \geq 0$ and $B \in \mathbb{R}_+^{n \times q}$ (see Farina and Rinaldi (2000) and Smith (1995)). The output solution $y(t)$ is nonnegative if $C \in \mathbb{R}_+^{p \times n}$ and $D \in \mathbb{R}_+^{p \times q}$. Such dynamical systems are called cooperative (monotone) or *nonnegative* if only initial conditions in $\mathbb{R}_+^n$ are considered (see Farina and Rinaldi (2000) and Smith (1995)).

For a Metzler matrix $A \in \mathbb{R}^{n \times n}$ its stability can be checked verifying a Linear Programming (LP) problem

$$A^\top \lambda < 0,$$

for some $\lambda \in \mathbb{R}_+^n \setminus \{0\}$, or a Lyapunov matrix equation

$$A^\top P + PA \prec 0,$$

for a diagonal matrix $P \in \mathbb{R}^{n \times n}$, $P > 0$ (in the general case, for non Metzler matrices, the matrix P cannot be diagonal). The L_1 and L_∞ gains for nonnegative systems (4), i.e., gains of transfer function from input to output in different norms, have been studied in Briat (2011), Ebihara et al. (2011), and they are interrelated for these kinds of systems.

Lemma 2 (Briat 2011; Ebihara et al. 2011) *Let the system (4) be nonnegative (i.e.., A is Metzler, $B \geq 0$, $C \geq 0$, and $D \geq 0$), then it is asymptotically stable if and only if there exist $\lambda \in \mathbb{R}^n_+ \setminus \{0\}$ and a scalar $\gamma > 0$ such that the following LP problem is feasible:*

$$\begin{pmatrix} A^\top \lambda + C^\top E_p \\ B^\top \lambda - \gamma E_q + D^\top E_p \end{pmatrix} < 0.$$

Moreover, in this case the L_1 gain of the operator $d \to y$ is lower than γ.

Lemma 3 (Briat 2011; Ebihara et al. 2011) *Let the system (4) be nonnegative (i.e., A is Metzler, $B \geq 0$, $C \geq 0$, and $D \geq 0$), then it is asymptotically stable if and only if there exist $\lambda \in \mathbb{R}^n_+ \setminus \{0\}$ and a scalar $\gamma > 0$ such that the following LP problem is feasible:*

$$\begin{pmatrix} A\lambda + BE_q \\ C\lambda - \gamma E_p + DE_q \end{pmatrix} < 0.$$

Moreover, in this case the L_∞ gain of the transfer $d \to y$ is lower than γ.

The conventional results and definitions on L_2/L_∞ stability for linear systems can be found in Khalil (2002).

3 Interval Observers

The problem of state vector estimation for a dynamical system can be encountered in many engineering applications and other domains of science (Besançon 2007; Fossen and Nijmeijer 1999; Meurer et al. 2005). For linear time–invariant models there are plenty of solutions providing different performances to the estimation error. The most popular are Luenberger observer and Kalman filter for deterministic and stochastic settings, respectively. Recalling other popular solutions for estimations, it is worth mentioning the high–gain observers (Esfandiari and Khalil 1992) or the high–order sliding–mode observers/differentiators (Levant 2003). In nonlinear case, observer or controller design is based on transformation of the system into a canonical form (frequently close to a linear representation (Besançon 2007; Fossen and Nijmeijer 1999; Meurer et al. 2005)). Since such a transformation may depend on uncertain parameters or may be unknown due to the model complexity, application of the transformation can be an obstruction in practice. That is why the class of LPV systems became popular in technical sciences: a wide class of nonlinear dynamical systems can be presented in the LPV form (in this case the system equations are extended, and the nonlinearity of the model is replaced by an auxiliary vector of time–varying scheduling parameters). The linear shape of LPV models allows a rich spectrum of methods developed for linear systems to be adapted and applied in the nonlinear setting (Marcos and Balas 2004; Shamma and Cloutier 1993; Shamma 2012; Tan 1997).

Complementing the model complexity, another difficulty for an estimator design consists in the uncertainty (unknown parameters or/and external disturbances, measurement noises, delays). In the presence of uncertainty in the model, the design of a conventional estimator, which has to converge to the ideal value of the state, may be rarely realized. Application of sliding–mode tools (Shtessel et al. 2014) or other disturbance cancellation approaches (robust or adaptive) may solve this issue in some cases of matched uncertainty. However, in general, the state estimation error is not approaching zero (it can be bounded or asymptotically bounded, and different versions of practical stability are used for analysis). A straightforward example deals with the presence of the measurement noise, in such a case the state estimate usually contains a noise shift providing the static errors (filtering for proper models of the noise may help at the price of the additional delay introduction). In this case, an interval or set–membership estimation is often more feasible: an observer can be constructed that, using input–output information and the bounds on the uncertain elements, evaluates a bounded set of admissible values (interval) for the state at each instant of time. It is desirable that the interval length be minimized by tuning the observer parameters, and it should be proportional to the size of the model uncertainty. It looks like such a formulation is a simplification of the initial state estimation problem (instead of the state value just a neighboring set is evaluated), but in fact it is an improvement since the interval mean value can be used as the state point–wise estimate, while the interval width gives the admissible (for the chosen model of the uncertainty) deviations from that value. Thus, an interval or set–membership estimator provides a simultaneous evaluation of accuracy for a bounded uncertainty, which may not have known statistics, similarly to the Kalman filter, but in the deterministic framework (an Interval Kalman filter has been proposed in Guanrong et al. (1997)).

There exist many methods for designing interval/set–membership estimators, see e.g., Jaulin (2002), Kieffer and Walter (2004), Milanese and Novara (2011), Moisan et al. (2009), Olivier and Gouzé (2004), Pati et al. (2022), Tahir and Açıkmeşe (2022), and the references therein, for different types of dynamical systems. This chapter is devoted to interval observers/predictors, which form a subclass of set–membership estimators and whose design is based on the monotone systems theory (Efimov et al. 2012b; Moisan et al. 2009; Olivier and Gouzé 2004; Raïssi et al. 2012, 2010). The idea of interval estimation has been proposed in Gouzé et al. (2000), and received numerous extensions. However, the main restriction for the interval observer consists in providing cooperativity (or nonnegtivity) of the interval estimation error dynamics by a proper design. Such an obstruction can be over–passed for Linear Time–Invariant or Linear Time–Varying systems, LPV and particular classes of nonlinear systems (Combastel 2013; Mazenc and Bernard 2011; Raïssi et al. 2012). The main idea of those studies is that under mild conditions, by applying a similarity transformation, a Hurwitz matrix could be transformed to a Hurwitz and Metzler one (cooperative).

In the next subsection, following Efimov and Raïssi (2016) and Raïssi and Efimov (2018), the basic theory of interval observer design is explained for continuous–time models.

3.1 Interval State Estimation Problem

Consider the following system

$$\dot{x}(t) = Ax(t) + d(t), \;\; y(t) = Cx(t) + v(t), \;\; t \in \mathbb{R}_+, \tag{6}$$

where $x(t) \in \mathbb{R}^n$ is the state, $y(t) \in \mathbb{R}^p$ is the output; $d(t) \in \mathbb{R}^n$ is the disturbance, $d \in \mathcal{L}_\infty^n$; $v(t) \in \mathbb{R}^p$ is the measurement noise, $v \in \mathcal{L}_\infty^p$; the matrices A, C have appropriate dimensions. This model has three sources of uncertainty: initial conditions for $x(0)$, instant values of d and v. The standard standing hypotheses are that all these uncertain factors belong to known intervals.

Assumption 1 Let $x(0) \in [\underline{x}_0, \overline{x}_0]$ for some known $\underline{x}_0, \overline{x}_0 \in \mathbb{R}^n$, let also two functions $\underline{d}, \overline{d} \in \mathcal{L}_\infty^n$ and a constant $V > 0$ be given such that

$$\underline{d}(t) \leq d(t) \leq \overline{d}(t), \;\; \|v(t)\| \leq V, \;\; \forall t \geq 0.$$

Thus, by this assumption, three intervals $[\underline{x}_0, \overline{x}_0]$, $[\underline{d}(t), \overline{d}(t)]$, and $[-V, V]$ determine uncertainty of $x(0)$, $d(t)$, and $v(t)$, respectively. It is required to calculate two estimates $\underline{x}(t), \overline{x}(t) \in \mathbb{R}^n$ with $\overline{x} - \underline{x} \in \mathcal{L}_\infty^n$, using the available information about these intervals and $y(t)$, such that

$$\underline{x}(t) \leq x(t) \leq \overline{x}(t), \;\; \forall t \geq 0. \tag{7}$$

3.2 Basic Structure

The following simplest interval observer is a solution to this problem:

$$\dot{\underline{x}}(t) = A\underline{x}(t) + L[y(t) - C\underline{x}(t)] - |L|E_pV + \underline{d}(t), \;\; \underline{x}(0) = \underline{x}_0, \tag{8a}$$

$$\dot{\overline{x}}(t) = A\overline{x}(t) + L[y(t) - C\overline{x}(t)] + |L|E_pV + \overline{d}(t), \;\; \overline{x}(0) = \overline{x}_0, \tag{8b}$$

where $L \in \mathbb{R}^{n \times p}$ is the observer gain to be designed. The conditions to satisfy for L are given below.

Theorem 1 (Gouzé et al. 2000) *Let Assumption 1 hold, then in the system (6) with the interval observer (8) the relations (7) are satisfied provided that the matrix* $A - LC$ *is Metzler. In addition,* $\overline{x} - \underline{x} \in \mathscr{L}_\infty^n$ *if* $A - LC$ *is Hurwitz.*

To explain the main ideas behind the conditions of this result (that is very representative for all kinds of interval observers) consider its proof.

Proof Define two estimation errors

$$\underline{e}(t) = x(t) - \underline{x}(t),\ \ \overline{e}(t) = \overline{x}(t) - x(t),$$

which yield differential equations:

$$\dot{\underline{e}}(t) = [A - LC]\underline{e}(t) + Lv(t) + |L|E_pV + d(t) - \underline{d}(t),$$
$$\dot{\overline{e}}(t) = [A - LC]\overline{e}(t) - Lv(t) + |L|E_pV + \overline{d}(t) - d(t).$$

By Assumption 1, for all $t \geq 0$

$$|L|E_pV \pm Lv(t) \geq 0,\ \ d(t) - \underline{d}(t) \geq 0,\ \ \overline{d}(t) - d(t) \geq 0.$$

If $A - LC$ is a Metzler matrix, since all inputs in the dynamics of $\underline{e}(t)$, $\overline{e}(t)$ are positive and $\underline{e}(0) \geq 0$, $\overline{e}(0) \geq 0$, then $\underline{e}(t) \geq 0$, $\overline{e}(t) \geq 0$ for all $t \geq 0$ (Farina and Rinaldi 2000; Smith 1995). The property (7) follows from these relations. If $A - LC$ is Hurwitz, since all inputs of $\underline{e}(t)$, $\overline{e}(t)$ are bounded, then $\underline{e}, \overline{e} \in \mathscr{L}_\infty^n$, and boundedness of $\overline{x} - \underline{x} = \overline{e} + \underline{e}$ is implied by construction. □

Investigating this proof we can conclude that the idea behind synthesis of an interval observer is to guarantee nonnegativity of the estimation error dynamics. The observer gain L has to be designed in a way that the matrix $A - LC$ would be Metzler and Hurwitz. In addition, in order to optimize the width of estimated interval $[\underline{x}(t), \overline{x}(t)]$ the problem of L_1 or L_∞ optimization of that gain value can be posed. Using Lemmas 2 and 3 this problem can be formulated as an LP computational procedure (see also McCarthy et al. (2014), for another LP problem formulation and solution for a more generic LPV system). For instance, selecting L_1 optimization, it is necessary to find $\lambda \in \mathbb{R}^n$, $w \in \mathbb{R}^p$, and a diagonal matrix $M \in \mathbb{R}^{n \times n}$ such that

$$\begin{pmatrix} A^\top\lambda - C^\top w + E_n \\ \lambda - \gamma E_n \end{pmatrix} < 0, \tag{9a}$$

$$\lambda > 0,\ \ M \geq 0, \tag{9b}$$

$$A^\top\lambda - C^\top w + M\lambda \geq 0, \tag{9c}$$

then, $w = L^\top\lambda$ (Chebotarev et al. 2015).

3.3 Transformations of Coordinates to Nonnegative Form

Formulation (9) provides an effective computational tool to design interval observers, but it is only a sufficient condition and in some cases this LP problem may have no solution, however, it does not imply that it is not possible to construct an interval observer. Roughly speaking in this case it is not possible to find L such that $A - LC$ is simultaneously Metzler and Hurwitz. It is well known fact that Hurwitz property of a matrix is preserved under a similarity transformation of coordinates, then to overcome the issue it is possible to design the gain L such that the matrix $A - LC$ is Hurwitz and next to find a nonsingular matrix $S \in \mathbb{R}^{n\times n}$ such that in the new coordinates $z = Sx$ the state matrix $D = S(A - LC)S^{-1}$ is Metzler (it is Hurwitz by construction). The conditions of existence of such a real transformation matrix S are given in the following lemma.

Lemma 4 (Raïssi et al. 2012) *Given the matrices $A \in \mathbb{R}^{n\times n}$, $D \in \mathbb{R}^{n\times n}$ and $C \in \mathbb{R}^{p\times n}$. If there is a matrix $L \in \mathbb{R}^{n\times p}$ such that the matrices $A - LC$ and D have the same eigenvalues, then there is a matrix $S \in \mathbb{R}^{n\times n}$ such that $D = S(A - LC)S^{-1}$ provided that the pairs $(A - LC, \chi_1)$ and (D, χ_2) are observable for some $\chi_1 \in \mathbb{R}^{1\times n}$, $\chi_2 \in \mathbb{R}^{1\times n}$.*

Note that (6) can be rewritten as follows:

$$\dot{x}(t) = (A - LC)x(t) + Ly(t) - Lv(t) + d(t).$$

Under conditions of this lemma and in the new coordinates $z = Sx$, the system (6) takes the form:

$$\dot{z}(t) = Dz(t) + SLy(t) + \delta(t),\ \delta(t) = S[d(t) - Lv(t)]. \tag{10}$$

Then, using Lemma 1 we obtain that

$$\underline{\delta}(t) \le \delta(t) \le \overline{\delta}(t),$$

where $\underline{\delta}(t) = S^+\underline{d}(t) - S^-\overline{d}(t) - |SL|E_pV$ and $\overline{\delta}(t) = S^+\overline{d}(t) - S^-\underline{d}(t) + |SL|E_pV$. Next, for the system (10) all conditions of Theorem 1 are satisfied and an interval observer similar to (8) can be designed:

$$\dot{\underline{z}}(t) = D\underline{z}(t) + SLy(t) + \underline{\delta}(t),\ \underline{z}(0) = S^+\underline{x}_0 - S^-\overline{x}_0, \tag{11a}$$

$$\dot{\overline{z}}(t) = D\overline{z}(t) + SLy(t) + \overline{\delta}(t),\ \overline{z}(0) = S^+\overline{x}_0 - S^-\underline{x}_0, \tag{11b}$$

$$\underline{x}(t) = (S^{-1})^+\underline{z}(t) - (S^{-1})^-\overline{z}(t), \tag{11c}$$

$$\overline{x}(t) = (S^{-1})^+\overline{z}(t) - (S^{-1})^-\underline{z}(t), \tag{11d}$$

where the relations (2) are used to calculate the initial conditions for $\underline{z}, \overline{z}$, and the estimates $\underline{x}, \overline{x}$. It is easy to show that in (11) the inclusion (7) is satisfied and $\overline{x} - \underline{x} = |S^{-1}|(\overline{z} - \underline{z}) \in \mathscr{L}_\infty^n$.

If conditions of Lemma 4 are not satisfied, then it is possible to consider a time–varying transformation of coordinates as in the next lemma.

Lemma 5 (Mazenc and Bernard 2011) *Let $A - LC$ be Hurwitz, then there exists an invertible matrix function $P : \mathbb{R} \to \mathbb{R}^{n \times n}$, of class C^∞ elementwise, $||P(t)|| < +\infty$ for all $t \in \mathbb{R}$, such that for all $t \in \mathbb{R}$*

$$\dot{P}(t) = DP(t) - P(t)(A - LC),$$

where $D \in \mathbb{R}^{n \times n}$ is a Hurwitz and Metzler matrix.

A variant of Lemma 5 has been proposed in Combastel (2013), where complex–valued formulation of the system is used that leads to a more simple design of the transformation matrix.

Under conditions of Lemma 5 in the new coordinates $\zeta(t) = P(t)x(t)$ we obtain

$$\dot{\zeta}(t) = D\zeta(t) + P(t)Ly(t) + \delta(t), \ \ \delta(t) = P(t)[d(t) - Lv(t)],$$

and again

$$\begin{aligned}
\underline{\delta}(t) \leq \delta(t) \leq \overline{\delta}(t), \\
\underline{\delta}(t) = P^+(t)\underline{d}(t) - P^-(t)\overline{d}(t) - |P(t)L|E_pV, \\
\overline{\delta}(t) = P^+(t)\overline{d}(t) - P^-(t)\underline{d}(t) + |P(t)L|E_pV.
\end{aligned}$$

The interval observer has a form similar to (11) (Mazenc and Bernard 2011):

$$\begin{aligned}
&\dot{\underline{z}}(t) = D\underline{z}(t) + P(t)Ly(t) + \underline{\delta}(t), \ \ \underline{z}(0) = P^+(t)\underline{x}_0 - P^-(t)\overline{x}_0, \\
&\dot{\overline{z}}(t) = D\overline{z}(t) + P(t)Ly(t) + \overline{\delta}(t), \ \ \overline{z}(0) = P^+(t)\overline{x}_0 - P^-(t)\underline{x}_0, \\
&\underline{x}(t) = (P^{-1})^+(t)\underline{z}(t) - (P^{-1})^-(t)\overline{z}(t), \\
&\overline{x}(t) = (P^{-1})^+(t)\overline{z}(t) - (P^{-1})^-(t)\underline{z}(t),
\end{aligned}$$

however, its realization needs more computations than for (11) since the obtained interval estimator is time–varying.

3.4 Intermediate Solution

There is another approach presented in Li et al. (2019) that mixes the advantages of a (partial) transformation of coordinates with simplicity of formulation in the original variables, where the following interval observer is proposed:

$$\underline{x}(t) = \underline{z}(t) + Ny(t) - |N|E_pV, \tag{12a}$$
$$\overline{x}(t) = \overline{z}(t) + Ny(t) + |N|E_pV, \tag{12b}$$
$$\dot{\underline{z}}(t) = A_{TL}\underline{x}(t) + Ly(t) - (2A_{TL}^{-}|N| + |L|)E_pV + T^{+}\underline{d}(t) - T^{-}\overline{d}(t), \tag{12c}$$
$$\dot{\overline{z}}(t) = A_{TL}\overline{x}(t) + Ly(t) + (2A_{TL}^{-}|N| + |L|)E_pV + T^{+}\overline{d}(t) - T^{-}\underline{d}(t), \tag{12d}$$
$$\underline{z}(0) = T^{+}\underline{x}_0 - T^{-}\overline{x}_0, \tag{12e}$$
$$\overline{z}(0) = T^{+}\overline{x}_0 - T^{-}\underline{x}_0, \tag{12f}$$

where $A_{TL} = TA - LC$, $L, N \in \mathbb{R}^{n\times p}$ are observer gains to select, and $T = I_n - NC$.

Theorem 2 *Let Assumption 1 hold, then in the system (6) with the interval observer (12) the relations (7) are satisfied provided that the matrix $TA - LC$ is Metzler. In addition, $\overline{x} - \underline{x} \in \mathcal{L}_\infty^n$ if $TA - LC$ is Hurwitz.*

Proof It is straightforward to check that

$$x(t) = z(t) + N(y(t) - v(t)), \;\; z(t) = Tx(t),$$
$$\dot{z}(t) = A_{TL}x(t) + L(y(t) - v(t)) + Td(t),$$

which explains the structure of the interval observer (12), and all claims of the theorem follows from the same arguments as above considering auxiliary estimation errors $\underline{\varepsilon}(t) = z(t) - \underline{z}(t)$, $\overline{\varepsilon}(t) = \overline{z}(t) - z(t)$, and observing that $\overline{x} - \underline{x} = \overline{z} - \underline{z} + 2||N||E_pV$. □

For $N = 0$ (i.e., $T = I_n$) we recover the case of Theorem 1. Thus, (12) is a natural extension of the observer (8) since the conditions of Theorem 2 impose the restrictions on the matrix $TA - LC = A - LC - NCA$, where by the choice of N it is possible to recover the Metzler property even in the case there is no matrix L providing this feature to the matrix $A - LC$.

3.5 Nonnegative Embedding

It is worth to mention that in general any system may be immersed in its internal positive representation (Cacace et al. 2015), which has dimension $2n$ (then interval observer has dimension $4n$). For example, any matrix can be decomposed as a difference of Metzler and nonnegative matrices:

$$\begin{aligned} A - LC &= (A - LC)_{\searrow} + (A - LC)_{\times} \\ &= \underbrace{(A - LC)_{\searrow} + (A - LC)_{\times}^{+}}_{\text{Metzler}} - \underbrace{(A - LC)_{\times}^{-}}_{\text{nonnegative}}, \end{aligned}$$

where $(A - LC)_{\searrow}$ is the diagonal matrix composed by the elements of $A - LC$ on the main diagonal and $(A - LC)_{\times} = A - LC - (A - LC)_{\searrow}$. Then the following interval observer can be proposed for (6):

$$\dot{\underline{x}}(t) = [A_{L\searrow} + A_{L\times}^{+}]\underline{x}(t) - A_{L\times}^{-}\overline{x}(t) + Ly(t) - |L|E_pV + \underline{d}(t), \tag{13a}$$

$$\dot{\overline{x}}(t) = [A_{L\searrow} + A_{L\times}^{+}]\overline{x}(t) - A_{L\times}^{-}\underline{x}(t) + Ly(t) + |L|E_pV + \overline{d}(t), \tag{13b}$$

$$\underline{x}(0) = \underline{x}_0,\ \overline{x}(0) = \overline{x}_0,$$

where $A_L = A - LC$.

Theorem 3 (Mazenc and Bernard 2010) *Let Assumption 1 hold, then in the system (6) with the interval observer (13) the relations (7) are satisfied. In addition,* $\overline{x} - \underline{x} \in \mathcal{L}_{\infty}^{n}$ *if*

$$R = \begin{pmatrix} A_{L\searrow} + A_{L\times}^{+} & A_{L\times}^{-} \\ A_{L\times}^{-} & A_{L\searrow} + A_{L\times}^{+} \end{pmatrix}$$

is Hurwitz.

Proof Consider again the estimation errors

$$e(t) = (\underline{e}^{\top}(t)\,\overline{e}^{\top}(t))^{\top},$$
$$\underline{e}(t) = x(t) - \underline{x}(t),\ \overline{e}(t) = \overline{x}(t) - x(t),$$

which yield for (13) differential equations:

$$\dot{e}(t) = Re(t) + \begin{pmatrix} |L|E_pV + Lv(t) + d(t) + \underline{d}(t) \\ |L|E_pV - Lv(t) + \overline{d}(t) - d(t) \end{pmatrix}.$$

The relations (7) immediately follows under Assumption 1 since R is Metzler (the rest of the proof repeats the case of Theorem 1). □

As we can conclude, the requirement on Metzler property of the matrix $A - LC$ is completely avoided, and the main difficulty in application of the last theorem consists in finding conditions under which the Metzler matrix R is Hurwitz. It is also difficult to formulate some LMIs to find L in this setting.

3.6 Comments

In this subsection, the main approaches to design interval observers for LTI systems were presented. If it is possible to find L such that $A - LC$ is Hurwitz and Metzler (looking for a solution of LP problem (9)), then the interval observer (8) is a right choice (or look for L and N in the interval observer (12)). If such a gain L does not exist, then different transformations of coordinates from Lemmas 4 and 5 can be tested. In this case L also can be found as a solution of LP problem or LMIs, and the interval observer (11) can be applied. An alternative solution (13) also can be tested, which always guarantees the interval inclusion (7), but the boundedness of the estimated interval $\overline{x} - \underline{x}$ can be an issue.

Remark 1 Note that observers (8) and (11) can be easily applied for some nonlinear systems, where all nonlinear terms are contained in $d(t)$, i.e., $d(t) = d(t, u(t), x(t))$, in a manner that Assumption 1 is satisfied for some $\underline{d}, \overline{d} \in \mathcal{L}_\infty^n$, with $\underline{d}(t) = \underline{d}(t, u(t), y(t), V)$, and $\overline{d}(t) = \overline{d}(t, u(t), y(t), V)$.

In the subsequent section, we apply the aforementioned concepts to the design of a robust control strategy for a class of LPV systems, incorporating both input and state constraints. This strategy is formulated based on an ISMC approach, utilizing an interval predictor–based state feedback controller in conjunction with an MPC scheme.

4 Interval Observer–Based Control for LPV Systems

Let us take into account the following LPV system:

$$\dot{x}(t) = A(\theta)x(t) + B(u(t) + d(t)), \tag{14}$$

where the state, input, and external disturbances are give by $x(t) \in \mathbb{R}^n$, $u(t) \in \mathbb{R}^m$, and $d : \mathbb{R}_{\geq 0} \to \mathbb{R}^m$, respectively. The vector $\theta \in \Theta \subset \mathbb{R}^p$ represents an unknown scheduling parameter vector in the matrix $A : \mathbb{R}^p \to \mathbb{R}^{n \times n}$ that belongs to a compact set $\Theta \in \mathbb{R}^p$. The known matrix $B \in \mathbb{R}^{n \times m}$ holds $\text{rank}(B) = m$.

The objective of this work is to design a robust controller that ensures the system trajectories, as described by (14)*, converge to zero. The controller must be capable of handling external disturbances and parameter uncertainties while respecting state and input constraints. Specifically, the state must remain within a bounded set,* $x(t) \in \mathbb{X} \subset \mathbb{R}^n$*, and the control input must lie within a specified set,* $u(t) \in \mathbb{U} \subset \mathbb{R}^m$*, for all* $t \geq 0$*, where both* $\mathbb{X}$ *and* $\mathbb{U}$ *are convex, bounded constraint sets. For simplicity, we assume that the input set is* $\mathbb{U} = \{u \in \mathbb{R}^m : ||u|| \leq u_{\max}\}$*, where* $u_{\max} > 0$.

To facilitate the control design, we assume that the state x is measurable, and the following assumptions are imposed on system (14).

Assumption 2 There exist matrices $A_0 \in \mathbb{R}^{n\times n}$ and $A_i \in \mathbb{R}^{n\times n}$, for $i = \overline{1, \eta}$, such that

$$A(\theta) = A_0 + \sum_{i=1}^{\eta} \lambda_i(\theta) A_i, \ \sum_{i=1}^{\eta} \lambda_i(\theta) = 1,$$

with some $\eta \in \mathbb{N}_+$, $\lambda_i(\theta) \in [0, 1]$, and A_0 being a Metzler matrix.

Assumption 3 The external disturbances are characterized as $d \in \mathbb{D} := \{d \in \mathscr{L}_\infty : ||d||_\infty \le d^+\}$, with a known $d^+ > 0$. In addition, there exists $\underline{x}_0, \overline{x}_0 \in \mathbb{X}$ such that $\underline{x}_0 \le x(0) \le \overline{x}_0$.

Note that Assumption 2 is not restrictive since we can always find a Metzler matrix A_0 by adding and subtracting some terms. Assumption 3 is common in the sliding–mode control theory and characterizes the class of admitted external disturbances.

5 Control Design

This section was partially presented in Gutiérrez-Martínez et al. (2023). However, in the present chapter, we provide the detailed proofs of the main results, and a distinct simulation example to further illustrate the effectiveness of the proposed method.

The proposed control law possesses two terms

$$u(t) = u_0(t) + u_1(t). \tag{15}$$

The component u_1 is designed using an ISMC approach to compensate for the effect of matched perturbations. In contrast, the nominal control component u_0 consists of an interval predictor–based state-feedback control law combined with an MPC scheme to handle state and input constraints. The proposed control strategy ensures exponential stability of the dynamic system.

Taking into account $||u|| \le u_{\max}$, for a given $u_{\max} > 0$, then a specific control effort can be assigned to each part of controller (15), i.e., $||u_0|| \le u_{0\max}$ and $||u_1|| \le u_{1\max}$, with $u_{0\max}, u_{1\max} > 0$, such that $u_{0\max} + u_{1\max} \le u_{\max}$.

5.1 Integral Sliding–Mode Control

The following sliding variable is considered

$$s(x) = G[x(t) - x(0)] - G\int_0^t [A_0 x(\tau) + Bu_0(\tau)]d\tau, \tag{16}$$

where $G \in \mathbb{R}^{m\times n}$ satisfies $\det(GB) \neq 0$. This matrix may be designed as $G = (B^\top B)^{-1}B^\top$ (see Castaños and Fridman (2006) and Utkin et al. (1999), for more details). Hence, the sliding variable dynamics is

$$\dot{s} = G\sum_{i=1}^{\eta} \lambda_i(\theta)A_i x + GB(u_1 + d(t)). \tag{17}$$

Thus, we propose the following sliding–mode control

$$u_1 = -\gamma(x)\frac{(GB)^\top s}{||(GB)^\top s||}, \tag{18}$$

where $\gamma(x) > 0$, for all $x \in \mathbb{X}$. The following lemma establishes the conditions under which the sliding variable converges to zero in finite–time while satisfying the input constraints.

Lemma 6 *(Gutiérrez-Martínez et al. 2023) Let the ISMC (18) be applied to the system (17), for a given* $u_{1\,\max} > 0$. *If the gain* $\gamma(x)$ *is selected as*

$$\gamma(x) = \rho + d^+ + ||G||A_{\max}||x||, \tag{19}$$

with $A_{\max} = \sum_{i=1}^{\eta} ||A_i||$, *and some* $\rho > 0$ *such that*

$$0 < \rho \leq u_{1\,\max} - d^+ - ||G||A_{\max}x_{\max}, \tag{20}$$

with some $x_{\max}$ *such that* $||x||_\infty \leq x_{\max}$, *for all* $x \in \mathbb{X}$, *is satisfied for a given* $u_{1\,\max} > 0$*; then,* $s = 0$ *is* UFTS.

The proof of the previous result, and the subsequent ones, are postponed to the Appendix.

Remark 2 The term u_1 in (18) provides a discontinuous control signal. Still, a continuous approximation is given as

$$u_1 = -\gamma(x)\frac{(GB)^\top s}{||(GB)^\top s||} \approx -\gamma(x)\frac{(GB)^\top s}{||(GB)^\top s|| + \tau},$$

where τ is a small positive constant.

Note that the robust controller u_1 is designed to compensate for the disturbance d and part of the parametric uncertainty $\sum_{i=1}^{\eta} \lambda_i(\theta) A_i x$, while satisfying the input constraint $||u_1|| \leq u_{1\,\max}$. In particular, the effect of d is completely rejected from the outset. However, the remaining parametric uncertainty term $\sum_{i=1}^{\eta} \lambda_i(\theta) Ax$ is not matched with the control input u_1, and thus only its projection onto the range space of B can be compensated by u_1.

Furthermore, it is possible to fix $u_{1\,\max}$ based on the known upper bound of the disturbances. The remaining control effort can then be allocated to the nominal controller u_0, ensuring that $u_{0\,\max} \leq u_{\max} - u_1$.

Remark 3 The sliding–mode control (18) could be also designed by means of continuous control laws, e.g., using multi–variable Super–Twisting algorithms (see, e.g., López-Caamal and Moreno (2019), Moreno et al. (2021), or Nagesh and Edwards (2014)). Nevertheless, saturation constraints must still be explicitly addressed.

Remark 4 The sliding variable (16) requires the knowledge of the initial state $x(0)$. This assumption can be relaxed by Assumption 3, i.e., only the existence of two vectors $\underline{x}_0, \overline{x}_0 \in \mathbb{X}$ such that $\underline{x}_0 \leq x(0) \leq \overline{x}_0$, which implies that $x(0) \in \mathbb{X}_0 \subset \mathbb{R}^n$ with a known set $\mathbb{X}_0$. In this case, the sliding–mode control strategy will possess a reaching phase after which the disturbances will be compensated. However, during such a reaching phase we cannot guarantee fulfilment of the system constraints.

5.2 *Interval Predictor*

Since the sliding–mode takes place, from (17), it follows that the equivalent control is

$$u_{1eq} = -(GB)^{-1} G \sum_{i=1}^{\eta} \lambda_i(\theta) A_i x - d, \tag{21}$$

and hence, the system dynamics, on the sliding surface, satisfies

$$\dot{x} = \left[A_0 + \sum_{i=1}^{\eta} \lambda_i(\theta) \tilde{A}_i \right] x + B u_0, \tag{22}$$

where $\tilde{A}_i = (I - B(GB)^{-1}G)A_i$. Then, according to Lemma 1 and dos Reis de Souza et al. (2022a), it follows that

$$-\overline{A}\underline{x}^- - \underline{A}\overline{x}^+ \leq \sum_{i=1}^{\eta} \lambda_i(\theta) \tilde{A}_i x \leq \overline{A}\overline{x}^+ + \underline{A}\underline{x}^-, \tag{23}$$

where $\overline{A} = \sum_{i=1}^{n} \tilde{A}_i^+$, $\underline{A} = \sum_{i=1}^{n} \tilde{A}_i^-$, and a couple of vectors $\underline{x}, \overline{x} \in \mathbb{R}^n$. Thus, it is possible to design the following interval predictor (Leurent et al. 2020) for the system (22)

$$\dot{\xi} = \mathscr{A}_0\xi + \mathscr{A}_1\xi^+ + \mathscr{A}_2\xi^- + \mathscr{B}u_0, \tag{24}$$

where $\xi = \left(\underline{x}^\top, \overline{x}^\top\right)^\top \in \mathbb{R}^{2n}$, and the system matrices

$$\mathscr{A}_0 = \begin{pmatrix} A_0 & 0 \\ 0 & A_0 \end{pmatrix}, \quad \mathscr{A}_1 = \begin{pmatrix} 0 & -\underline{A} \\ 0 & \overline{A} \end{pmatrix}, \quad \mathscr{A}_2 = \begin{pmatrix} -\overline{A} & 0 \\ \underline{A} & 0 \end{pmatrix}, \quad \mathscr{B} = \begin{pmatrix} B \\ B \end{pmatrix}.$$

The boundedness of ξ implies the boundedness of the state x as well. Therefore, to stabilize system (14), it is necessary to design a state–feedback controller u_0 that drives the trajectories of system (24) to zero, as discussed in Efimov et al. (2013).

5.3 State–Feedback Control Design

The control signal u_0 is designed as

$$u_0(t) = \begin{cases} \mathscr{U}_0(t), & \xi(t_i) \notin \mathbb{X}_f, \\ \bar{u}_0(t), & \xi(t_i) \in \mathbb{X}_f, \end{cases} \tag{25}$$

where $\mathscr{U}_0$ is the control signal generated by the MPC and applied over the interval $t \in [t_i, t_i + h)$, where $h \in (0, T)$ with T denoting both the control application period and the prediction horizon of the MPC. The term $\overline{u}_0$ refers to the state–feedback controller. The switching set $\mathbb{X}_f$ will be defined later.

The state–feedback control law $\bar{u}_0$ is designed as

$$\bar{u}_0 = K_0\xi + K_1\xi^+ + K_2\xi^-, \tag{26}$$

with $K_0, K_1, K_2 \in \mathbb{R}^{m\times 2n}$ being some gains to design. The following lemma provides a constructive method for designing state–feedback gains that guarantee the convergence to zero of the trajectories of system (24).

Lemma 7 *(Gutiérrez-Martínez et al. 2023) Let Assumptions 2 and 3 be satisfied and the state–feedback control law (26) be applied to the system (24), i.e., $u_0(t) = \bar{u}_0(t)$. Suppose that there exist two vectors $\underline{x}_0, \overline{x}_0 \in \mathbb{R}^n$, such that $\underline{x}_0 \le x(0) \le \overline{x}_0$, diagonal matrices $0 < \Gamma_l \in \mathbb{R}^{2n\times 2n}$, $0 \le R_1, R_2 \in \mathbb{R}^{2n\times 2n}$, some diagonal matrices $\Psi_j,\ R_0 \in \mathbb{R}^{2n\times 2n}$, and some matrices $Y_j \in \mathbb{R}^{m\times 2n}$, for $j = \overline{0,2}$, such that the following LMIs*

$$\Omega = \begin{pmatrix} \Omega_{11} & \Omega_{12} & \Omega_{13} \\ \star & \Omega_{22} & \Omega_{23} \\ \star & \star & \Omega_{33} \end{pmatrix} \preceq 0, \tag{27}$$

$$\Psi_0 + \min\{\Psi_1, \Psi_2\} + 2\min\{R_1, R_2\} > 0, \tag{28}$$

$$\begin{aligned} \Omega_{11} &= \mathrm{He}(\mathscr{A}_0\Gamma_0 + \mathscr{B}Y_0) + \Psi_0, \\ \Omega_{12} &= \mathscr{A}_1\Gamma_1 + \mathscr{B}Y_1 + \Gamma_0\mathscr{A}_0^\top + Y_0^\top\mathscr{B}^\top + R_1, \\ \Omega_{13} &= \mathscr{A}_2\Gamma_2 + \mathscr{B}Y_2 - \Gamma_0\mathscr{A}_0^\top - Y_0^\top\mathscr{B}^\top - R_2, \\ \Omega_{22} &= \mathrm{He}(\mathscr{A}_1\Gamma_1 + \mathscr{B}Y_1) + \Psi_1, \\ \Omega_{23} &= \mathscr{A}_2\Gamma_2 + \mathscr{B}Y_2 - \Gamma_1\mathscr{A}_1^\top - Y_1^\top\mathscr{B}^\top + R_0, \\ \Omega_{33} &= \Psi_2 - \mathrm{He}(\mathscr{A}_2\Gamma_2 + \mathscr{B}Y_2), \end{aligned}$$

are feasible. If the state–feedback gains are designed as $K_j = Y_j\Gamma_j^{-1}$, *for* $j = \overline{0,2}$*; then, the trajectories of the system (24) exponentially converge to zero.*

The requirement for Γ_0 to have a diagonal structure is natural, as the existence of a diagonal matrix Γ_0 satisfying the Lyapunov inequality $\mathrm{He}(\bar{\mathscr{A}_0}\Gamma_0) \preceq 0$ is equivalent to the stability of the Metzler matrix $\bar{\mathscr{A}_0} = \mathscr{A}_0 + \mathscr{B}K_0$.

Hence, the control (26) ensures the exponential convergence of the interval predictor trajectories (24) to zero. In addition, taking into account that $\mathscr{A}_0$ is Metzler and $\underline{x}_0 \leq x(0) \leq \overline{x}_0$, for two vectors $\underline{x}_0$, $\overline{x}_0 \in \mathbb{R}^n$, the inclusion property $\underline{x}(t) \leq x(t) \leq \overline{x}(t)$ holds (see, e.g., Efimov et al. (2012a) and Efimov et al. (2013)). Thus, the trajectories of system (14) will converge to zero and the considered problem will be properly solved provided that $x(t) \in \mathbb{X}$ and $u(t) \in \mathbb{U}$, for all $t \geq 0$, are satisfied.

Then, we define the set $\mathbb{X}_f$ as follows

$$\mathbb{X}_f = \left\{\xi \in \mathbb{R}^{2n} : V_\xi(\xi) \leq \kappa^{-1}\varepsilon\right\}, \tag{29}$$

where $V_\xi = \xi^\top P_0\xi + \xi^\top P_1\xi^+ - \xi^\top P_2\xi^-$, ε is a positive constant, and $\kappa = \min_{\forall i=\overline{1,2n}} \lambda_i\left[\Upsilon(P_0 + P_1^+ + P_2^+)^{-1}\right]$, with $\Upsilon = \bar{\Psi}_0 + \min\{\bar{\Psi}_1, \bar{\Psi}_2\} + 2\min\{\bar{R}_1, \bar{R}_2\} > 0$, $\Gamma_j = P_j^{-1}$, $\Psi_j = \Gamma_j\bar{\Psi}_j\Gamma_j$, $R_j = \Gamma_j\bar{R}_j\Gamma_j$, and $Y_j = K_j\Gamma_j$, for $j = \overline{0,2}$. Note that $\mathbb{X}_f$ is an invariant set for the system (24). Additionally, all the trajectories inside of $\mathbb{X}_f$, converge to zero. In this way, we can select ε to ensure that $\mathbb{X}_f \subset \mathbb{X} \times \mathbb{X}$, i.e., such that $x(t) \in \mathbb{X}$ holds, for all $t \geq t_f \geq 0$.

5.4 Model Predictive Control Design

Before proceeding with the description of the MPC scheme, it is required to introduce the following assumption.

Assumption 4 There exist $\gamma(x)$, K_0, K_1, and K_2 satisfying the conditions of Lemmas 6 and 7, such that

$$-\gamma(x(t))\frac{(GB)^\top s}{||(GB)^\top s||} + K_0\xi(t) + K_1\xi_1^+(t) + K_2\xi^-(t) \in \mathbb{U},$$

for any $\xi \in \mathbb{X}_f$ and all $t \geq t_f \geq 0$.

The above assumption ensures the existence of a controller (15), defined by (18) and (26), that stabilizes system (14). Within the set $\mathbb{X}_f$, both the state and input constraints are satisfied. It is worth noting that Assumption 4 is not restrictive, as ε can always be chosen such that the set $\mathbb{X}_f$ is sufficiently small.

We now turn to the implementation of the MPC strategy, which is designed to handle state and input constraints whenever $\xi(t_i) \notin \mathbb{X}_f$. Let T denote the prediction horizon and $h \in (0, T)$ the control application interval for the MPC. The corresponding optimal control problem is formulated as follows (see Mayne et al. (2000) and Michalska and Mayne (1993)):

Problem 1 Given matrices $0 \preceq Q_j \in \mathbb{R}^{2n\times 2n}$, $j = \overline{0, 1}$, $0 \preceq Q_2 \in \mathbb{R}^{m\times m}$, and letting $t_i = ih$ with $i \in \mathbb{N}_+$, to determine the control signals

$$\mathscr{U}_0 = \underset{u_0:[t_i,t_i+T]\to\mathbb{R}^m}{\arg\min} \; \xi^\top(t_i+T)Q_0\xi(t_i+T) + \int_{t_i}^{t_i+T}[\xi^\top(\sigma)Q_1\xi(\sigma) + u_0^\top(\sigma)Q_2u_0(\sigma)]d\sigma, \quad (30)$$

satisfying the threefold constraints:

(a) $\xi : [t_i, t_i + T] \to \mathbb{R}^{2n}$ is a solution of (24);
(b) $\xi(\sigma) \in \mathbb{X} \times \mathbb{X}$ and $u_0(\sigma) \in \mathbb{U}$, for $\sigma \in [t_i, t_i + T]$;
(c) $\xi(t_i + T) \in \mathbb{X}_f$.

Therefore, if the aforementioned optimal control problem is feasible, the trajectories of the system (26) will converge to the terminal set $\mathbb{X}_f$ under the control law $u_0(t) = \mathscr{U}_0(t)$. Once within $\mathbb{X}_f$, the control switches to $u_0(t) = \bar{u}_0(t)$, under which the trajectories converge to zero while holding both state and input constraints.

Finally, the results of Lemmas 6 and 7, together with the solution to Problem 1, lead to the main result of this section.

Theorem 4 *(Gutiérrez-Martínez et al. 2023) Suppose that Assumption 2, the conditions of Lemmas 6 and 7, and the feasibility of Problem 1 are all satisfied. If the control law (15), defined by (18) and (25), is applied to system (14) and designed in accordance with Lemmas 6 and 7, as well as the solution to Problem 1, then the origin of system (14) is* UES, *and the state and input constraints are guaranteed to be satisfied.*

6 Simulation Results

Consider the LPV system (14) with the following system matrices:

$$A(\theta) = \begin{pmatrix} -1 & 4+\theta(t) \\ \theta(t) & -1 \end{pmatrix}, \quad B = \begin{pmatrix} 0 \\ 1 \end{pmatrix},$$

where $\theta(t) = 3\sin(t)$ and $d(t) = 2\sin(3t) + 1$. The constraint sets are defined as $\mathbb{X} = [-3, 3] \times [-2, 2]$ and $\mathbb{U} = [-10, 10]$, with initial conditions $x(0) = (2, -11)^\top$. It can be verified that system satisfies Assumption 2, and the matrices A_0 and A_i, for $i = 1, 2$, can be derived using a convex polytopic approach, resulting in the following matrices:

$$A_0 = \begin{pmatrix} -1 & 4 \\ 0 & -1 \end{pmatrix}, \quad A_1 = \begin{pmatrix} 0 & 3 \\ 3 & 0 \end{pmatrix}, \quad A_2 = \begin{pmatrix} 0 & -3 \\ -3 & 0 \end{pmatrix}.$$

The disturbance term $d(t)$ satisfies Assumption 3, with $d^+ = 3$. The predictor initial conditions are $\underline{x}_0 = (2.5, -0.5)^\top$ and $\overline{x}_0 = (1.5, -1.5)^\top$.

The state–feedback controller gains are computed with the LMIs of Lemma 7 by means of SDPT3 solver among YALMIP in MATLAB obtaining the following results:

$$\begin{aligned} K_0 &= (-0.0362, -0.3493, -0.0362, -0.3493), \\ K_1 &= (-0.0908, -0.0808, -0.0908, -6.7341), \\ K_2 &= (0.0908, 6.7332, 0.0908, 0.0808). \end{aligned}$$

The prediction interval is selected as $T = 0.01$[s] and we compute the corresponding switching set $\mathbb{X}_f$ according to (29). All the simulations were conducted in MATLAB using the Euler discretization method with sampling–time of 0.001 [s]. The MPC was implemented using the `nlmpc` toolbox in MATLAB.

The simulation scenario is illustrated in Figs. 1 , 2 , and 3 . In Fig. 1, it is evident that the system trajectories do not violate the state constraints and exponentially converge to zero. Additionally, the time evolution of the interval predictor trajectories is shown, which also respect the state constraints.

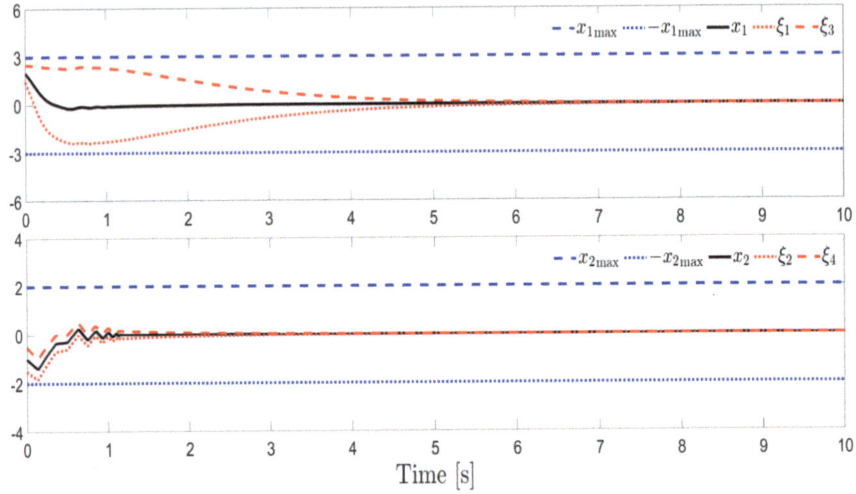

Fig. 1 Evolution of the state variables. The figure consists of two subplots illustrating the time evolution of various variables. In both plots, the blue dotted lines represent the state constraints, while the red dotted lines show the trajectories of the interval predictor states ξ_j, $j = \overline{1, 4}$. The solid black lines correspond to the actual states x_1 and x_2

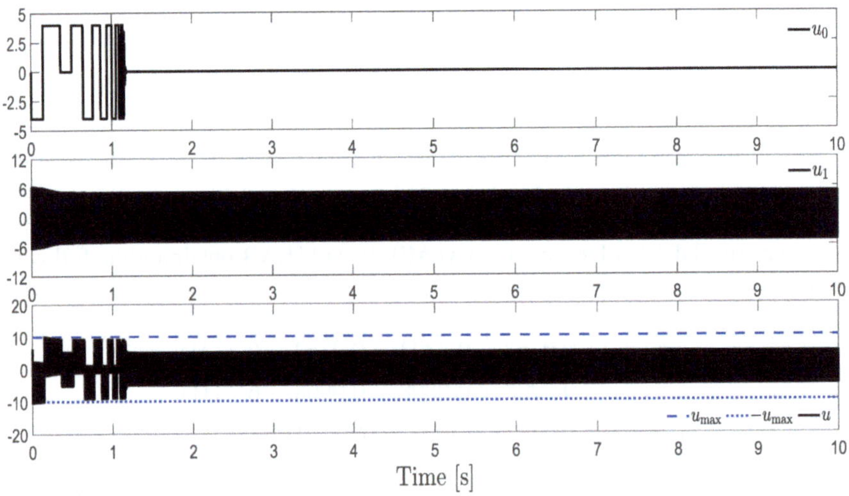

Fig. 2 Time evolution of the control signals. This figure presents three subplots showing the time evolution of the control signals. In each subplot, the blue dotted lines indicate the input constraints, while the solid black lines depict to the actual control inputs u_0, u_1, and the total input u

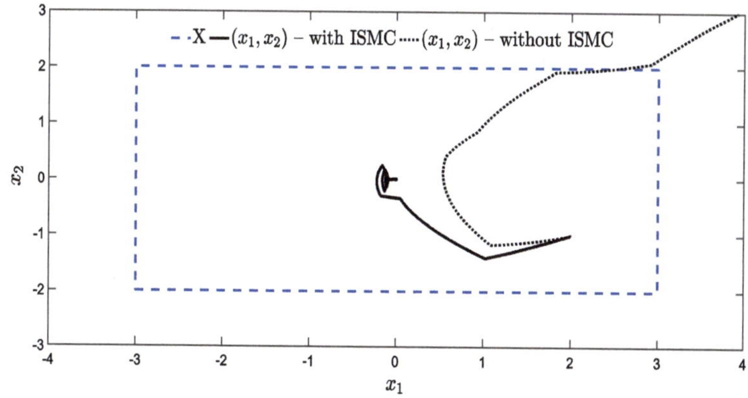

Fig. 3 Comparison of the performance with and without ISMC. The figure presents the phase portrait (x_1 vs. x_2). The blue dotted line denotes the state constraints, the black dotted line shows the state trajectories without ISMC activation, and the solid black line illustrates the state trajectories under the proposed ISMC strategy

The input u applied to the system is displayed in Fig. 2, where we observe the behaviour of each part of the control law. It is clear that the proposed approach ensures the fulfilment of the input constraints.

To illustrate the influence of the ISMC, Fig. 3 presents a comparison of the control input with and without the ISMC, i.e., with $u_1 = 0$. In the absence of ISMC, the control signal u_0 alone is insufficient to stabilize the system under external perturbations. Consequently, the system trajectories diverge and eventually violate the prescribed state constraints.

In Ríos et al. (2024), we have employed the proposed strategy to solve a trajectory tracking problem for constrained and perturbed unicycle mobile robots, which provides a more meaningful example.

7 Conclusions

In this chapter, we present first a review of interval observers tools, and then, based on these tools, we introduce the design of a robust controller for a class of uncertain and constrained LPV systems. The proposed strategy combines an ISMC approach with an interval predictor–based state feedback controller and an MPC scheme. The contributions of this work are described bellow:

1. The ISMC control strategy is designed to completely compensate the matched perturbations, ensuring the sliding–mode from the onset and dealing with the input constraints.
2. The interval predictor–based state feedback control law and the MPC handle the constraints in the state and input.

3. The proposed controller possesses robustness to parameter uncertainties.
4. The use of LMIs facilitates a constructive design of the interval predictor–based controller gains, without relying on the upper bound of the external disturbance, thanks to the robustness properties of the sliding–mode control strategy.
5. The exponential stability of the closed–loop system is guaranteed
6. The computational burden is reduced due to the control hybrid structure.

Acknowledgements This work was supported in part by the SEP–CONACYT–ANUIES–ECOS NORD Project 315597 and ECOS NORD Project M20M04. The work of Héctor Ríos was supported in part by the SECIHTI program "Investigadoras e Investigadores por México", CVU 270504 Project 922 and in part by TecNM Projects.

8 Appendix

8.1 Proof of Lemma 6

Consider the candidate Lyapunov function $V_s = s^\top s/2$. Thus, taking the time derivative of the Lyapunov function along the trajectories of the system (17), it follows that

$$\begin{aligned}\dot{V}_s &= s^\top G\left[\sum_{i=1}^{\eta}\lambda_i(\theta)A_i x + B\left(-\gamma(x)\frac{(GB)^\top s}{||(GB)^\top s||}+d\right)\right],\\ &\leq -||(GB)^\top s||\left[\gamma(x)-d^+\right]+s^\top||G||A_{\max}||x||,\end{aligned}$$

where $A_{\max} = \sum_{i=1}^{\eta}||A_i||$. Due to the fact that $G = (B^\top B)^{-1}B^\top$ and $\text{rank}(B) = m$, we have that $GB = I$, and then

$$V_s \leq -||s||\left[\gamma(x) - d^+ - ||G||A_{\max}||x||\right].$$

Hence, if the gain $\gamma(x)$ is designed according to (19), i.e., $\gamma(x) = \rho + d^+ + ||G||A_{\max}||x||$, where $\gamma(x) \leq u_{1\max}$, and given that $x \in \mathbb{X}$, this implies that $||x||_\infty \leq x_{\max}$, with some $x_{\max} > 0$, and ρ satisfies the constraint (20), which implies that $||u_1|| \leq u_{1\max}$; with $||x|| \leq x_{\max}$, for all $x \in \mathbb{X}$; thus, the time derivative of V_s is upper bounded as

$$\dot{V}_s \leq -\rho||s|| \leq -\sqrt{2}\rho V_s^{\frac{1}{2}}, \tag{31}$$

then, due to the structure of the sliding variable (16), it is proven that $s(t) = 0$, for all $t \geq 0$.

8.2 Proof of Lemma 7

The nominal closed–loop system, taking into account (26), is given as

$$\dot{\xi} = \bar{\mathscr{A}_0}\xi + \bar{\mathscr{A}_1}\xi^+ + \bar{\mathscr{A}_2}\xi^-, \tag{32}$$

where $\bar{\mathscr{A}_0} = \mathscr{A}_0 + \mathscr{B}K_0$, $\bar{\mathscr{A}_1} = \mathscr{A}_+\mathscr{B}K_1$, and $\bar{\mathscr{A}_2} = \mathscr{A}_2 + \mathscr{B}K_2$. Consider a candidate Lyapunov function as in Leurent et al. (2020), i.e.,

$$V_\xi = \xi^\top P_0\xi + \xi^\top P_1\xi^+ - \xi^\top P_2\xi^-, \tag{33}$$

with positive diagonal matrices P_0, P_1 and P_2, which implies that $P_0 + \min\{P_1, P_2\} > 0$, and $V_\xi > 0$. The candidate Lyapunov function satisfies the following bounds

$$\xi^\top(P_0 + \min\{P_1, P_2\})\xi \leq V_\xi \leq \xi^\top(P_0 + P_1^+ + P_2^+)\xi. \tag{34}$$

Thus, the time derivative of V_ξ, along the trajectories of the system (32) satisfies

$$\begin{aligned}\dot{V}_\xi = \xi^\top \mathrm{He}(P_0\bar{\mathscr{A}_0})\xi + 2\xi^\top P_0(\bar{\mathscr{A}_1}\xi^+ + \bar{\mathscr{A}_2}\xi^-)\\ + \xi^{+\top}\mathrm{He}(P_1\bar{\mathscr{A}_1})\xi^+ + 2\xi^{+\top}P_1(\bar{\mathscr{A}_0}\xi + \bar{\mathscr{A}_2}\xi^-)\\ - \xi^{+\top}\mathrm{He}(P_2\bar{\mathscr{A}_2})\xi^- - 2\xi^{-\top}P_2(\bar{\mathscr{A}_0}\xi + \bar{\mathscr{A}_1}\xi^+),\end{aligned}$$

and defining an extended vector $\varphi = (\xi^\top, \xi^{+\top}, \xi^{-\top})^\top$, it follows that $\dot{V}_\xi$ could be rewritten as

$$\begin{aligned}\dot{V}_\xi = \varphi^\top\bar{\Omega}\varphi - \xi^\top\bar{\Psi}_0\xi - \xi^{+\top}\bar{\Psi}_1\xi^+ - \xi^{-\top}\bar{\Psi}_2\xi^-\\ - 2\xi^{+\top}\bar{R}_0\xi^- - 2\xi^{+\top}\bar{R}_1\xi - 2(-\xi^-)^\top\bar{R}_2\xi,\end{aligned}$$

for some diagonal matrices $\bar{\Psi}_j$, $\bar{R}_j \in \mathbb{R}^{2n\times 2n}$, for $j = \overline{0,2}$, such that $\Upsilon = \bar{\Psi}_0 + \min\{\bar{\Psi}_1, \bar{\Psi}_2\} + 2\min\{\bar{R}_1, \bar{R}_2\} > 0$, and with

$$\bar{\Omega} = \begin{pmatrix}\bar{\Omega}_{11} & \bar{\Omega}_{12} & \bar{\Omega}_{13}\\ \star & \bar{\Omega}_{22} & \bar{\Omega}_{23}\\ \star & \star & \bar{\Omega}_{33}\end{pmatrix},$$

$$\begin{aligned}\bar{\Omega}_{11} &= \mathrm{He}(P_0\bar{\mathscr{A}_0}) + \bar{\Psi}_0,\ \bar{\Omega}_{12} = P_0\bar{\mathscr{A}_1} + \bar{\mathscr{A}_0}^\top P_1 + \bar{R}_1,\\ \bar{\Omega}_{13} &= P_0\bar{\mathscr{A}_2} - \bar{\mathscr{A}_0}^\top P_2 - \bar{R}_2,\ \bar{\Omega}_{22} = \mathrm{He}(P_1\bar{\mathscr{A}_1}) + \bar{\Psi}_1,\\ \bar{\Omega}_{23} &= P_1\bar{\mathscr{A}_2} - \bar{\mathscr{A}_2}^\top P_2 + \bar{R}_0,\ \bar{\Omega}_{33} = \bar{\Psi}_2 - \mathrm{He}(P_2\bar{\mathscr{A}_2}).\end{aligned}$$

It is worth noting that the constraints $\xi^{+\top}\bar{R}_0\xi^- = 0$, $\xi^{+\top}\bar{R}_1\xi \geq 0$, and $(-\xi^-)^\top\bar{R}_2\xi \geq 0$, hold for any diagonal matrix $\bar{R}_0$, and positive diagonal matrices $\bar{R}_1$, $\bar{R}_2 \geq 0$. Therefore, taking into account that $\bar{\Omega} \preceq 0$; then, the time derivative of V_ξ satisfies

$$\begin{aligned}\dot{V}_\xi \leq &-\xi^\top\bar{\Psi}_0\xi - \xi^{+\top}\bar{\Psi}_1\xi^+ - \xi^{-\top}\bar{\Psi}_2\xi^- - 2\xi^{+\top}\bar{R}_1\xi - 2(-\xi^-)^\top\bar{R}_2\xi \\ \leq &-\xi^\top\Upsilon\xi,\end{aligned}$$

with the diagonal matrix $\Upsilon > 0$. Therefore, considering (34), it is obtained that

$$\dot{V}_\xi \leq -\kappa V_\xi,$$

where $\kappa = \min_{\forall i=\overline{1,2n}} \lambda_i\left[\Upsilon(P_0 + P_1^+ + P_2^+)^{-1}\right]$. Thus, the trajectories of the system (32) exponentially converge to zero.

Finally, in order to obtain a constructive way to design the gains K_0, K_1, and K_2 by means of LMIs, it can pre–and post–multiply $\bar{\Omega}$ by $\mathrm{diag}(P_0^{-1}, P_1^{-1}, P_2^{-1})$, in order to obtain the LMI (27), with $\Gamma_j = P_j^{-1}$, $\Psi_j = \Gamma_j\bar{\Psi}_j\Gamma_j$, $R_j = \Gamma_j\bar{R}_j\Gamma_j$, and $Y_j = K_j\Gamma_j$, for $j = \overline{0,2}$. Note that, due to the diagonal structure, the constraint $\bar{\Psi}_0 + \min\{\bar{\Psi}_1, \bar{\Psi}_2\} + 2\min\{\bar{R}_1, \bar{R}_2\} > 0$ is equivalent to (28), i.e., $\Psi_0 + \min\{\Psi_1, \Psi_2\} + 2\min\{R_1, R_2\} > 0$. Therefore, $\Omega \preceq 0$ is equivalent to $\bar{\Omega} \preceq 0$. This concludes the proof.

8.3 Proof of Theorem 4

Recall that the controller consists of two parts: u_1, based on ISMC, that is able to perform the rejection of the uncertain matched terms affecting the system; while u_0 is a nominal stabilizing controller, which is composed of an interval predictor–based state–feedback controller and the MPC law, which will deal with the state and input constraints.

1. First, note that the interval predictor (24) generates an estimation $\xi \in \mathbb{R}^{2n}$ such that the relation $\underline{x}(t) \leq x(t) \leq \overline{x}(t)$ is satisfied. Then, consider the control u_0 and note that, for some $t_f > 0$, if the initial conditions $\xi(0) \in \mathbb{X}_f \subset \mathbb{X} \times \mathbb{X}$; then, the control (25) equals to (26). According to (29) and Assumption 4, $\xi(t) \in \mathbb{E}_f$ and $u(t) \in \mathbb{U}$, for all $t \geq t_f$, and the trajectories of system converge to zero due to the results of Lemma 7.
2. Now, let $\xi(0) \notin \mathbb{X}_f$ and assume that there is a solution of the Problem 1. Then, applying the control (30), for $t \in [0, T)$, one has that $\xi(t) \in \mathbb{X} \times \mathbb{X}$, and $u(t) \in \mathbb{U}$, on this time interval. Note that $\mathbb{X}$ is a neighborhood of the origin, and the cost function given in (30), with positive definite matrices Q_0, Q_1, and Q_2, is minimized inside $\mathbb{X} \times \mathbb{X}$. Using these arguments, since $\xi(t_i + T) \in \mathbb{X} \times \mathbb{X}$ provided that the control $\mathscr{U}$ in (30) is applied, and $[\underline{x}^\top(t_i), \overline{x}^\top(t_i)] \subset [\underline{x}^\top(t_{i-1} + T), \overline{x}^\top(t_{i-1} + T)]$, for all $i \geq 1$, there is a finite time instant $t_f \geq T$ such that

$(\underline{x}^{\top}(t_f), \overline{x}^{\top}(t_f))^{\top} \in \mathbb{X}_f$; and then, the control law (25) switches from the control signal $\mathscr{U}_0$ computed by the MPC to the state–feedback control law u_0, which provides exponential stability of $x = 0$.

3. As a result of the previous statements, the system constraints are satisfied. This concludes the proof.

References

Bako L, Andrieu V (2019) Interval-valued state estimation for linear systems: the tightest estimator and its relaxations. Automatica 106:168–177

Bemporad A, Morari M (2007) Robust model predictive control: a survey. In: Robustness in identification and control, pp 207–226. Springer

Besançon G (ed) (2007) Nonlinear observers and applications, Lecture notes in control and information sciences, vol 363. Springer

Borrelli F, Bemporad A, Morari M (2017) Predictive control for linear and hybrid systems. Cambridge University Press

Briat C (2011) Robust stability analysis of uncertain linear positive systems via integral linear constraints: l_1- and l_∞-gain characterizations. In: Proceedings of the 50th IEEE conference on decision and control and European control conference, pp 6337–6342, Orlando

Cacace F, Germani A, Manes C (2015) A new approach to design interval observers for linear systems. IEEE Trans Autom Control 99(99):1–1

Castaños F, Fridman L (2006) Analysis and design of integral sliding manifolds for systems with unmatched perturbations. IEEE Trans Autom Control 51(5):853–858

Chebotarev S, Efimov D, Raïssi T, Zolghadri A (2015) Interval observers for continuous-time LPV systems with l_1/l_2 performance. Automatica 58(8):82–89

Combastel C (2013) Stable interval observers in C for linear systems with time-varying input bounds. IEEE Trans Autom Control 58(2):481–487

dos Reis de Souza A, Efimov D, Raïssi T (2022a) Robust output feedback MPC for LPV systems using interval observers. IEEE Trans Autom Control 67(6):3188–3195

dos Reis de Souza A, Efimov D, Raïssi T, Ping X (2022b) Robust output feedback model predictive control for constrained linear systems via interval observers. Automatica 135:109951

Ebihara Y, Peaucelle D, Arzelier D (2011) L1 gain analysis of linear positive systems and its application. In: Proceedings of the 50th IEEE conference on decision and control and european control conference, pp 4029–4035, Orlando

Efimov D, Fridman L, Raïssi T, Zolghadri A, Seydou R (2012a) Interval estimation for LPV systems applying high order sliding mode techniques. Automatica 48(9):2365–2371

Efimov D, Fridman LM, Raïssi T, Zolghadri A, Seydou R (2012b) Interval estimation for LPV systems applying high order sliding mode techniques. Automatica 48:2365–2371

Efimov D, Raïssi T (2016) Design of interval observers for uncertain dynamical systems. Autom Remote Control 77(2):191–225

Efimov D, Raïssi T, Zolghadri A (2013) Control of nonlinear and LPV systems: interval observer-based framework. IEEE Trans Autom Control 58(3):773–778

Esfandiari F, Khalil HK (1992) Output feedback stabilization of fully linearizable systems. Int J Control 56:1007–1037

Falcone P, Borrelli F, Asgari J, Tseng HE, Hrovat D (2007) Predictive active steering control for autonomous vehicle systems. IEEE Trans Control Syst Technol 15(3):566–580

Farina L, Rinaldi S (2000) Positive linear systems: theory and applications. Wiley, New York

Filippov AF (1988) Differential equations with discontinuous righthand sides: control systems, vol 18. Springer

Findeisen R, Imsland L, Allgower F, Foss BA (2003) State and output feedback nonlinear model predictive control: an overview. Eur J Control 9(2–3):190–206

Fossen TI, Nijmeijer H (1999) New directions in nonlinear observer design. Springer

Funke J, Brown M, Erlien SM, Gerdes JC (2016) Collision avoidance and stabilization for autonomous vehicles in emergency scenarios. IEEE Trans Control Syst Technol 25(4):1204–1216

Galván-Guerra R, Incremona GP, Fridman L, Ferrara A (2022) Robust multi-model predictive control via integral sliding modes. IEEE Control Syst Lett 6:2623–2628

Gouzé JL, Rapaport A, Hadj-Sadok MZ (2000) Interval observers for uncertain biological systems. Ecol Model 133:46–56

Guanrong C, Jianrong W, Leang SS (1997) Interval kalman filtering. IEEE Trans Aerosp Electron Syst 33(1):250–259

Gutiérrez-Martínez SV, Ríos H, Mera M, Raïssi T, Efimov D (2023) A robust interval MPC for uncertain LPV systems via integral sliding–mode control. In: 2023 62nd IEEE conference on decision and control (CDC), pp 7899–7904, Singapore, Singapore

Incremona GP, Ferrara A, Magni L (2015) Hierarchical model predictive/sliding mode control of nonlinear constrained uncertain systems. IFAC-PapersOnLine 48(23):102–109

Incremona GP, Ferrara A, Magni L (2017) Asynchronous networked MPC with ISM for uncertain nonlinear systems. IEEE Trans Autom Control 62(9):4305–4317

Jaulin L (2002) Nonlinear bounded-error state estimation of continuous time systems. Automatica 38(2):1079–1082

Khalil H (2002) Nonlinear systems third edition. Patience Hall

Kieffer M, Walter E (2004) Guaranteed nonlinear state estimator for cooperative systems. Numer Algorithms 37:187–198

Leurent E, Efimov D, Maillard O-A (2020) Robust-adaptive interval predictive control for linear uncertain systems. In: 2020 59th IEEE conference on decision and control (CDC), pp 1429–1434

Levant A (2003) High-order sliding modes: differentiation and output feedback control. Int J Control 76(9–10):924–941

Li J, Wang Z, Zhang W, Raïssi T, Shen Y (2019) Interval observer design for continuous-time linear parameter-varying systems. Syst Control Lett 134:104541

Limon D, Ferramosca A, Alvarado I, Alamo T, Camacho EF (2009) MPC for tracking of constrained nonlinear systems. In: Magni L, Raimondo DM, Allgöwer F (eds) Nonlinear model predictive control. Lecture notes in control and information sciences, vol 384. Springer, Berlin, Heidelberg, pp 315–323

López-Caamal F, Moreno JA (2019) Generalised multivariable supertwisting algorithm. Int J Robust Nonlinear Control 29(3):634–660

Marcos A, Balas J (2004) Development of linear parameter-varying models for aircraft. J Guid, Control, Dyn 27(2):218–228

Mayne DQ (2014) Model predictive control: recent developments and future promise. Automatica 50(12):2967–2986

Mayne DQ, Rawlings JB, Rao CV, Scokaert POM (2000) Constrained model predictive control: stability and optimality. Automatica 36(6):789–814

Mazenc F, Bernard O (2010) Asymptotically stable interval observers for planar systems with complex poles. IEEE Trans Autom Control 55(2):523–527

Mazenc F, Bernard O (2011) Interval observers for linear time-invariant systems with disturbances. Automatica 47(1):140–147

McCarthy PJ, Nielsen C, Smith SL (2014) Cardinality constrained robust optimization applied to a class of interval observers. In: American Control Conference (ACC) 2014, pp 5337–5342

Meurer T, Graichen K, Gilles K (eds) (2005) Control and observer design for nonlinear finite and infinite dimensional systems, Lecture notes in control and information sciences, vol 322. Springer

Michalska H, Mayne DQ (1993) Robust receding horizon control of constrained nonlinear systems. IEEE Trans Autom Control 38(11):1623–1633

Milanese M, Novara C (2011) Unified set membership theory for identification, prediction and filtering of nonlinear systems. Automatica 47(10):2141–2151

Moisan M, Bernard O, Gouzé JL (2009) Near optimal interval observers bundle for uncertain bio-reactors. Automatica 45(1):291–295

Moreno JA, Ríos H, Ovalle L, Fridman L (2021) Multivariable super-twisting algorithm for systems with uncertain input matrix and perturbations. IEEE Trans Autom Control 67(12):6716–6722

Nagesh I, Edwards C (2014) A multivariable super-twisting sliding mode approach. Automatica 50(3):984–988

Olivier B, Gouzé JL (2004) Closed loop observers bundle for uncertain biotechnological models. J Process Control 14(7):765–774

Pati T, Khajenejad M, Daddala SP, Yong SZ (2022) L_1-robust interval observer design for uncertain nonlinear dynamical systems. IEEE Control Syst Lett 6:3475–3480

Polyakov A (2011) Nonlinear feedback design for fixed-time stabilization of linear control systems. IEEE Trans Autom Control 57(8):2106–2110

Raïssi T, Efimov D (2018) Some recent results on the design and implementation of interval observers for uncertain systems. Automatisierungstechnik 66(3):213–224

Raïssi T, Efimov D, Zolghadri A (2012) Interval state estimation for a class of nonlinear systems. IEEE Trans Autom Control 57(1):260–265

Raïssi T, Videau G, Zolghadri A (2010) Interval observers design for consistency checks of nonlinear continuous-time systems. Automatica 46(3):518–527

Raković SV (2016) Model predictive control: classical, robust, and stochastic. IEEE Control Syst Mag 36(6):102–105

Ríos H, Mera M, Raïssi T, Efimov D (2024) Robust interval predictive tracking control for constrained and perturbed unicycle mobile robots. Int J Robust Nonlinear Control 34(9):6303–6320

Shamma JS (2012) Control of linear parameter varying systems with applications, chapter An overview of LPV systems, pp 1–22. Springer

Shamma JS, Cloutier J (1993) Gain-scheduled missile autopilot design using linear parameter-varying transformations. J Guid, Control, Dyn 16(2):256–261

Shtessel Y, Edwards C, Fridman L, Levant A (2014) Sliding mode control and observation. Birkhauser, New York

Smith HL (1995) Monotone dynamical systems: an introduction to the theory of competitive and cooperative systems, vol 41. Surveys and monographs. AMS, Providence

Steinberger M, Castillo I, Horn M, Fridman L (2020) Robust output tracking of constrained perturbed linear systems via model predictive sliding mode control. Int J Robust Nonlinear Control 30(3):1258–1274

Tahir AM, Açıkmeşe B (2022) Synthesis of interval observers for bounded Jacobian nonlinear discrete-time systems. IEEE Control Syst Lett 6:764–769

Tan W (1997) Applications of linear parameter-varying control theory. PhD thesis, Dept. of Mechanical Engineering, University of California at Berkeley

Utkin V, Guldner J, Shi J (1999) Sliding modes in electromechanical systems

Young KD, Utkin VI, Ozguner U (1999) A control engineer's guide to sliding mode control. IEEE Trans Control Syst Technol 7(3):328–342

Active Model Discrimination Algorithms for Switched Piecewise Affine Inclusion Systems with Temporal Logic Constraints

Ruochen Niu, Elikplim Gah, and Sze Zheng Yong

Abstract Cyber-physical systems (CPS) play a pivotal role in modern life, particularly in industries like transportation and power/smart grids, where safety is a top priority. These systems generally exhibit uncertain nonlinear or hybrid system dynamics that can make the characterization of their nominal behavior and, by extension, detection and understanding of faulty and/or intended model behaviors very difficult. Thus, this chapter is dedicated to exploring the concept of Active Model Discrimination (AMD), which involves designing auxiliary (perturbation) input signals to improve the model discrimination and detection process amid uncertainties. Specifically, we consider the AMD problem for a set of noisy switched piecewise affine inclusion systems constrained by signal and metric temporal logic (STL and MTL) specifications. The challenge of these problems is the presence of binary/integer variables in the lower/inner level of the associated bilevel optimization problem that stem from the switching, the piecewise affine dynamics as well as the necessary STL/MTL encoding. To deal with the challenge, we propose three algorithms/solutions: (a) the parametric approach: solve the inner problem of the bilevel AMD optimization problem using mixed-integer parametric optimization, whose parametric solution is included when solving the outer/higher level problem; (b) the binary variable relocation approach: move the integer variables/constraints from the inner to the outer problem in a manner that retains feasibility, and recast the problem as a tractable mixed-integer linear programming (MILP) by leveraging Karush-Kuhn-Tucker (KKT) conditions; (c) the brute force approach: enumerate all combinations of the binary variables in the inner problem, and for each combination, solve the mixed-integer bilevel problem (without binary variables in the inner problem) by similarly recasting it as a single-level MILP, and then, selecting the optimal

R. Niu
Shanghai Jiao Tong University, Shanghai, China
e-mail: rcniu6@sjtu.edu.cn

E. Gah · S. Z. Yong (✉)
Northeastern University, Boston, USA
e-mail: s.yong@northeastern.edu

E. Gah
e-mail: gah.e@northeastern.edu

T. N. Dinh et al. (eds.), *Set-Valued Approaches to Control and Estimation of Uncertain Systems*, Mathematical Engineering,
https://doi.org/10.1007/978-3-031-94239-6_5

solution among these combinations. Further, we provide a complementary model invalidation algorithm to rule out models that are not consistent with noisy observations at run time, and by virtue of the AMD design, only one model will not be ruled out within a desired guaranteed detection time, and hence the true model can be identified. Simulation examples for parameter identification in a planetary assisted drive system and intent estimation in a highway lane changing scenario demonstrate the effectiveness of our proposed algorithms.

1 Introduction

The increased integration of engineering and cyber-physical systems with everyday life has made it all the more important to guarantee the safety of these systems. The challenge here is made even more difficult by the fact that faults or malicious intrusions often coexist with and are obscured by the presence of measurement noise, subtle input disturbances, and model inaccuracies. Consequently, there is a pressing need to devise methods that are both effective and economical for detecting and differentiating anomalous or faulty behaviors from nominal ones, especially under worst-case safety-critical scenarios. Such methods are pivotal for ensuring the safety of these systems.

Methods for model discrimination are typically categorized into passive and active strategies. Passive approaches rely only on measurements, (which are subject to noise), obtained under given/fixed inputs, as shown in Chiang et al. (2001), Harirchi et al. (2017), Jin et al. (2019), Lou and Si (2009), Smith and Doyle (1989). Nonetheless, the fault models are not guaranteed to be detected with the given input information or the detection time could be very long, prompting the introduction and investigation of model distinguishability (Grewal and Glover 1976; Harirchi et al. 2018, 2017; Niu et al. 2021; Vidal et al. 2003) to determine the amount of time that is necessary to discern faults or to distinguish among models.

To address the limitations inherent in passive model discrimination, the concept of Active Model Discrimination (AMD) has been introduced, which seeks to engineer an optimal input that aids in the prompt and precise detection of models (Cheong and Manchester 2015; Nikoukhah 1998; Scott et al. 2014; Nikoukhah and Campbell 2006; Tabatabaeipour 2015). The foundation of AMD was laid out in Nikoukhah (1998), Nikoukhah and Campbell (2006) through the application of the projection method on high-dimensional polytopes. Under the assumption of zonotopic uncertainty and initial states, more computationally tractable approaches were proposed in Raimondo et al. (2013), Scott et al. (2014), and further extended by Marseglia and Raimondo (2017) to capitalize on newly available measurements through multi-parametric programming. Techniques for managing polyhedral uncertainties were introduced in Harirchi et al. (2017), and later refined in Ding et al. (2018) by employing SOS-1 constraints (Gurobi Optimization, Inc. 2015). AMD's applicability to nonlinear systems has been expanded in Singh et al. (2018a) using a method of piecewise affine over-approximation, and has been adapted for switched nonlinear models with metric and signal temporal logic constraints in Niu et al. (2022). A tractable approach for addressing complex disjunctive and coupled constraints within this framework was also proposed in Shen et al. (2022).

On the other hand, when addressing systems that must adhere to temporal logic specifications, particularly in the context of fault and anomaly detection (Asarin et al. 2011), it has been demonstrated that such specifications can be encoded through mixed-integer constraints. This can be achieved by first translating these constraints as a nondeterministic finite automaton, referred to as a monitor, before being recast as mixed-integer constraints (Yang and Ozay 2019), or directly into mixed-integer constraints without the monitor construction (Niu et al. 2021).

In this chapter, three different methods are proposed for solving the AMD problem associated with switched piecewise affine inclusion systems that are constrained by signal and metric temporal logic (STL and MTL) specifications (Maler and Nickovic 2004; Koymans 1990): the parametric approach, the binary variable relocation approach, and the brute force approach. Specifically, after describing the modeling framework and notations in Sect. 3, the associated AMD problem, expressed as a bilevel optimization problem with binary (integral) variables in the inner problem, is stated in Sect. 4. Then, the three approaches are introduced in Sect. 5 to tackle the challenge of dealing with these binary variables/integers in the inner problem. Moreover, Sect. 5 also provides a model invalidation algorithm to eliminate models that are not compatible with noisy observations at run time such that the true model can be identified when used in tandem with the AMD solutions. Finally, examples involving parameter identification on a lunar/planetary assisted drive system and driver intent estimation in a vehicle lane changing scenario are considered to demonstrate and compare the effectiveness of our proposed methods.

2 Motivating Example

Consider the scenario as shown in Fig. 1: Two cars, A and B, approach an unsignalized 4-way intersection. To avoid a collision, the driver of car A must accurately determine the intention of the car B's driver, which in our problem setting refers to the driver of car A discriminating/distinguishing between several intent models of car B's driver. Moreover, the driver of car A may want to probe the intent of the other driver by controlling their own car in a way that reveals the intent of the other driver, which is analogous to the main objective of this paper of active model discrimination, i.e., to design a separating input to guarantee that the (intent) models are distinct. However, the driver of car A only has control of their own car (controlled inputs) and must consider the actions of the other car (uncontrolled inputs) as well as its position/speed (uncontrolled states), while ensuring that car A's states (controlled states) satisfy some constraints, e.g., speed limits and road boundaries.

We consider piecewise affine inclusion systems as our models because they have proven a very useful tool for accurately characterizing real world systems, including as a useful over-approximation of the nonlinear systems (see Azuma et al. (2010),

Fig. 1 Scenario with two cars approaching an unsignaled 4-way intersection

Dang et al. (2010), Jin et al. (2019, 2020)). Furthermore, desired behaviors such as navigation can be effectively represented using signal temporal logic (STL) formulas (Maler and Nickovic 2004) and metric temporal logic (MTL) formulas (Koymans 1990). For example, the formula $\varphi^x = \Diamond_{[0,5]}(10 \leq x \leq 20)$ constrains/requires the agent's position to reach a specific target area, $\mathcal{X}_g = \{x \mid 10 \leq x \leq 20\}$, within the given time frame of 0 to 5 s.

3 Preliminaries

3.1 Notations and Definitions

Let $x \in \mathbb{R}^n$ denote a vector and $M \in \mathbb{R}^{n\times m}$ a matrix with transpose $M^\intercal$, and $M \geq 0$ denotes element-wise non-negativity. The infinity and 1-norms of a vector x are denoted by $\|x\|_\infty \triangleq \max_i x_i$ and $\|x\|_1 \triangleq \sum_i |x_i|$, while $\mathbf{0}$, $\mathbb{1}$, and $\mathbb{I}$ represent the matrix of zeros, the vector of ones, and the identity matrix of appropriate dimension. The operators diag and vec are defined for a collection of matrices M_i, $i = 1, \ldots, n$ and matrix M as follows:

$$\operatorname{diag}_{i=1}^n\{M_i\} = \begin{bmatrix} M_1 & \mathbf{0} & \mathbf{0} \\ \mathbf{0} & \ddots & \mathbf{0} \\ \mathbf{0} & \mathbf{0} & M_n \end{bmatrix}, \quad \operatorname{vec}_{i=1}^n\{M_i\} = \begin{bmatrix} M_1 \\ \vdots \\ M_n \end{bmatrix},$$
$$\operatorname*{diag}_{k=\{i,j\}}\{M_k\} = \begin{bmatrix} M_i & \mathbf{0} \\ \mathbf{0} & M_j \end{bmatrix}, \quad \operatorname*{vec}_{k=\{i,j\}}\{M_k\} = \begin{bmatrix} M_i \\ M_j \end{bmatrix},$$
$$\operatorname{diag}_N\{M\} = \mathbb{I}_N \otimes M, \quad \operatorname{vec}_N\{M\} = \mathbb{1}_N \otimes M,$$

where $\otimes$ is the Kronecker product. The set of positive integers up to n is denoted by $\mathbb{Z}_+^n$, and the set of integers from m to n is denoted by $\mathbb{Z}_m^n$.

We will also make use of the Special Ordered Set of degree 1 (SOS-1) constraint[1] and set partitions, defined as follows:

Definition 1 (*SOS-1 Constraint* (Gurobi Optimization, Inc. 2015)) A special ordered set of degree 1 constraint (SOS-1) is a set of integer, continuous, or mixed-integer scalar variables for which at most one variable in the set may take a value other than zero, denoted as SOS-1: $\{v_1, \ldots, v_N\}$. For example, if $v_i \neq 0$, then this constraint imposes $v_j = 0$ on all $j \neq i$.

Moreover, we overload this constraint for vectors, where for a pair of vectors $\mathbf{u}, \mathbf{v} \in \mathbb{R}^n$, SOS-1:$\{\mathbf{u}, \mathbf{v}\}$ denotes element-wise constraints SOS-1:$\{u_i, v_i\}$ for all i.

Definition 2 (*Partition for Polyhedral Sets*) A partition of a polyhedral set $\mathcal{P}$ is a collection of $|S|$ disjoint subsets $\mathcal{P}_i$ such that $\bigcup_{i=1}^{|S|} \mathcal{P}_i = \mathcal{P}$, where each partition $\mathcal{P}_i$ is also a polyhedral set.

3.2 *Temporal Logic*

We consider 2 classes of temporal logic specifications or constraints:

(1) Signal Temporal Logic (STL),
(2) Metric Temporal Logic (MTL).

For improved legibility, STLs are denoted with φ, while MTLs are denoted with ψ.

3.2.1 Signal Temporal Logic (STL)

Let Σ be a finite set of predicates. The syntax of signal temporal logic (STL) formulas is given by:

$$\varphi ::= \top \mid p^{\pi} \mid \neg\varphi \mid \varphi_1 \vee \varphi_2 \mid \varphi_1 \, \mathcal{U}_{[t_1,t_2]} \, \varphi_2, \tag{1}$$

where $\top$ is the *true* predicate, $p^{\pi} \in \Sigma$ is a predicate whose truth value is determined by the sign of its underlying predicate function $\pi : \mathbb{R}^n \to \mathbb{R}$ (i.e., it is true if $\pi(x) \geq 0$

[1] Off-the-shelf solvers such as Gurobi and CPLEX (Gurobi Optimization, Inc. 2015; CPLEX II 2009) can readily handle these constraints, which can significantly reduce the search space for integer variables in branch and bound algorithms.

and false otherwise), while $\neg$, $\vee$, and $\mathcal{U}_{[t_1,t_2]}$ are the negation, disjunction, and time-constrained until operators, respectively, and $[t_1, t_2] \subset [0, \infty)$ is an interval of reals. Applying the grammar given in (1), we can also define next ($\bigcirc$; for discrete-time systems), conjunction ($\wedge$), implication ($\Rightarrow$), eventually in $[t_1, t_2]$ ($\Diamond_{[t_1,t_2]}$), and always in $[t_1, t_2]$ ($\Box_{[t_1,t_2]}$) as $\bigcirc\varphi = \top\mathcal{U}_{[0,1]}\varphi, \varphi_1 \wedge \varphi_2 = \neg(\neg\varphi_1 \vee \neg\varphi_2), \varphi_1 \Rightarrow \varphi_2 = \neg\varphi_1 \vee \varphi_2$, $\Diamond_{[t_1,t_2]}\varphi = \top\mathcal{U}_{[t_1,t_2]}\varphi = \bigvee_{\tau=t_1}^{t_2} \bigcirc^\tau\varphi$, and $\Box_{[t_1,t_2]}\varphi = \neg\Diamond_{[t_1,t_2]}\neg\varphi = \bigwedge_{\tau=t_1}^{t_2} \bigcirc^\tau\varphi$, respectively. Further, we abbreviate $\mathcal{U}_{[0,\infty)}$, $\Diamond_{[0,\infty)}$, $\Box_{[0,\infty)}$ as $\mathcal{U}$, $\Diamond$, $\Box$.

Definition 3 (*STL Semantics*) Let $\boldsymbol{x}$ be an ω-word of a real-valued vector signal over $\mathcal{X} \subseteq \mathbb{R}^n$, i.e., $\boldsymbol{x} \in \mathcal{X}^\omega$ and let $\boldsymbol{\sigma}$ be the ω-word over Σ corresponding to the predicate of a discrete-time STL formula φ, i.e., $\boldsymbol{\sigma} \in \Sigma^\omega$. Further, let $x(t)$ and $\sigma(t)$ be the t-th element of $\boldsymbol{x}$ and $\boldsymbol{\sigma}$, respectively. The STL semantics is defined as follows:

1. $(\boldsymbol{x}, t) \models p^\pi \Leftrightarrow \pi(x(t)) \geq 0$ (i.e., $\sigma(t) = p^\pi = \top$),
2. $(\boldsymbol{x}, t) \models \neg\varphi \Leftrightarrow (\boldsymbol{x}, t) \not\models \varphi$,
3. $(\boldsymbol{x}, t) \models \varphi_1 \vee \varphi_2 \Leftrightarrow (\boldsymbol{x}, t) \models \varphi_1$ or $(\boldsymbol{x}, t) \models \varphi_2$,
4. $(\boldsymbol{x}, t) \models \varphi_1 \wedge \varphi_2 \Leftrightarrow (\boldsymbol{x}, t) \models \varphi_1$ and $(\boldsymbol{x}, t) \models \varphi_2$,
5. $(\boldsymbol{x}, t) \models \varphi_1\mathcal{U}_{[t_1,t_2]}\varphi_2 \Leftrightarrow \exists t' \in [t + t_1, t + t_2] : (\boldsymbol{x}, t') \models \varphi_2$ and $\forall t'' \in [t, t'] : (\boldsymbol{x}, t'') \models \varphi_1$,
6. $(\boldsymbol{x}, t) \models \Diamond_{[t_1,t_2]}\varphi \Leftrightarrow \exists t' \in [t + t_1, t + t_2], (\boldsymbol{x}, t') \models \varphi$,
7. $(\boldsymbol{x}, t) \models \Box_{[t_1,t_2]}\varphi \Leftrightarrow \forall t' \in [t + t_1, t + t_2], (\boldsymbol{x}, t') \models \varphi$.

We write $\boldsymbol{x} \models \varphi$ if $(\boldsymbol{x}, 0) \models \varphi$.

Definition 4 (*Valid Signal and Predicate Subtraces of an STL*) A length-T signal $\boldsymbol{x}_T \in \mathcal{X}^T$ (and its corresponding length-T predicate $\boldsymbol{\sigma}_T \in \Sigma^T$) is called a *valid subtrace* from k_0 of an STL formula φ, if there exist a k_0-length prefix $\boldsymbol{v} \in \mathcal{X}^{k_0}$ and a suffix $\boldsymbol{w} \in \mathcal{X}^\omega$ such that their ω-word concatenation $\boldsymbol{v}\boldsymbol{x}_T\boldsymbol{w}$ satisfies φ, i.e., $\boldsymbol{v}\boldsymbol{x}_T\boldsymbol{w} \models \varphi$. Further, the sets of length-T valid signal and predicate subtraces from k_0 are denoted as $\mathcal{X}_{k_0}^T(\varphi)$ and $\Sigma_{k_0}^T(\varphi)$, respectively.

Definition 5 (*Temporal Bound of an STL formula* (Belta and Sadraddini 2019; Dokhanchi et al. 2014)) The (temporal) bound of an STL formula φ, denoted by b^φ, is the time length required to evaluate the satisfaction of φ and is recursively computed as follows: (1) $b^{p^\pi} = 0$; (2) $b^{\neg\varphi} = b^\varphi$; (3) $b^{\varphi_1\wedge\varphi_2} = \max(b^{\varphi_1}, b^{\varphi_2})$; (4) $b^{\Diamond_{[t_1,t_2]}\varphi} = b^{\Box_{[t_1,t_2]}\varphi} = t_2 + b^\varphi$; and (5) $b^{\varphi_1\mathcal{U}_{[t_1,t_2]}\varphi_2} = t_2 + \max(b^{\varphi_1}, b^{\varphi_2})$.

Mixed-Integer Encoding of STL Formulas. Next, we present the mixed-integer encoding of STL formulas for the satisfaction of STL semantics, i.e., $(\boldsymbol{x}, t) \models \varphi$. First, we equivalently express $(\boldsymbol{x}, t) \models p^\pi$, i.e.,

$$P_{p^\pi} = 1 \Leftrightarrow \pi(x) \geq 0$$

using the following mixed-integer encoding:

$$\begin{aligned} \pi(x) + s^+ \geq 0, \quad \pi(x) < s^-, \\ \text{SOS-1:}\{P_{p^\pi}\mathbb{1}, s^+\}, \quad \text{SOS-1:}\{(1 - P_{p^\pi})\mathbb{1}, s^-\}, \end{aligned} \tag{2}$$

where s^+ and s^- are unconstrained real-valued slack variables. Then, we present constraints for the predicate truth values (cf. Niu et al. (2021)) to satisfy the following operators of the STL semantics, where p^π, p^ϕ and p^{π_i} are predicates, and P^t_φ is the truth value of formula φ at time t.

Negation: The formula $\varphi = \neg p^\pi$ can be modeled as:

$$P^t_\varphi = (1 - P^t_{p^\pi}). \tag{3}$$

Disjunction: The formula $\varphi = \bigvee_{i=1}^k p^{\pi_i}$ can be modeled as:

$$P^t_\varphi \leq \Sigma_{i=1}^k P^t_{p^{\pi_i}}; \quad P^t_\varphi \geq P^t_{p^{\pi_i}}, i \in \mathbb{Z}_+^k. \tag{4}$$

Conjunction: The formula $\varphi = \bigwedge_{i=1}^k p^{\pi_i}$ can be modeled as:

$$P^t_\varphi \geq \Sigma_{i=1}^k P^t_{p^{\pi_i}} - (k-1); \quad P^t_\varphi \leq P^t_{p^{\pi_i}}, i \in \mathbb{Z}_+^k. \tag{5}$$

Next: The formula $\varphi = \bigcirc p^\pi$ can be modeled as:

$$P^t_\varphi = P^{t+1}_{p^\pi}. \tag{6}$$

Until: The formula $\varphi = p^\pi \,\mathcal{U}_{[t_1,t_2]}\, p^\phi$ can be modeled as:

$$\begin{aligned} \alpha_{tj} \geq P^j_{p^\phi} + \Sigma_{\tau=t}^{j-1} P^\tau_{p^\pi} - (j-t), j \in \mathbb{Z}_{t+t_1}^{t+t_2}; \\ \alpha_{tj} \leq P^j_{p^\phi}, \alpha_{tj} \leq P^\tau_{p^\pi}, j \in \mathbb{Z}_{t+t_1}^{t+t_2}, \tau \in \mathbb{Z}_t^{j-1}; \\ P^t_\varphi \leq \Sigma_{j=t+t_1}^{t+t_2} \alpha_{tj}, P^t_\varphi \geq \alpha_{tj}, j \in \mathbb{Z}_{t+t_1}^{t+t_2}. \end{aligned} \tag{7}$$

Eventually: The formula $\varphi = \Diamond_{[t_1,t_2]} p^\pi$ can be modeled as:

$$P^t_\varphi \leq \Sigma_{\tau=t+t_1}^{t+t_2} P^\tau_{p^\pi}; \quad P^t_\varphi \geq P^\tau_{p^\pi}, \tau \in \mathbb{Z}_{t+t_1}^{t+t_2}. \tag{8}$$

Always: The formula $\varphi = \Box_{[t_1,t_2]} p^\pi$ can be modeled as:

$$P^t_\varphi \geq \Sigma_{\tau=t+t_1}^{t+t_2} P^\tau_{p^\pi} - (t_2 - t_1); \; P^t_\varphi \leq P^\tau_{p^\pi}, \tau \in \mathbb{Z}_{t+t_1}^{t+t_2}. \tag{9}$$

3.2.2 Metric Temporal Logic (MTL)

Definition 6 (*Atomic Proposition*) An atomic proposition is an assertion whether a system variable or signal is either *True* (1 *or* $\top$) or *False* (0 *or* $\bot$).

Metric temporal logic (MTL) is a special case of STL, where the predicates (with their corresponding predicate functions on real-valued signals) are simply replaced by atomic propositions (over binary-valued truth values for the switching signals). Hence, with slight abuse of notation, we overload Σ to be also a finite set of atomic propositions such that the syntax of MTL formulas over Σ is

$$\psi ::= \top \mid p \mid \neg\psi \mid \psi_1 \vee \psi_2 \mid \psi_1 \mathcal{U}_{[t_1,t_2]} \psi_2, \tag{10}$$

where $p \in \Sigma$.

Moreover, since MTL is a special case of STL, its semantics, definitions and optimization-based encoding are similar and briefly summarized below.

MTL Semantics. Let $\boldsymbol{\sigma}$ represent an ω-length word over Σ, i.e., $\boldsymbol{\sigma} \in \Sigma^{\omega}$, and let $\sigma(t)$ be the t-th element of $\boldsymbol{\sigma}$. The MTL semantics follows the same semantics as in Definition 3 with the first statement replaced by:

1b. $(\boldsymbol{\sigma}, t) \models p \Leftrightarrow \sigma(t) = p$,

and the rest (i.e., statements 2. through 7. in Definition 3) remaining the same, except that we use $(\boldsymbol{\sigma}, t)$, ψ, ψ_1 and ψ_2 instead of $(\boldsymbol{x}, t)$, φ, φ_1 and φ_2. Similarly, we write $\boldsymbol{\sigma} \models \psi$ if $(\boldsymbol{\sigma}, 0) \models \psi$.

Valid Subtraces of an MTL. A length-T signal $\boldsymbol{\sigma} \in \Sigma^T$ is a *valid subtrace* of an MTL ψ from k_0 when $\boldsymbol{\sigma}_T \in \Sigma_{k_0}^T(\psi)$.

Temporal Bound of an MTL. The (temporal) bound of an MTL formula ψ, denoted by b^{ψ}, is as defined in Definition 5.

Mixed-Integer Encoding of MTL Formulas. We similarly present an integer encoding of MTL formulas that was first introduced in Niu et al. (2021) for the satisfaction of MTL semantics, i.e., $(\boldsymbol{\sigma}, t) \models \psi$. Specifically, since we do not have real-valued predicate functions as with STL that was encoded using (2), we only have integer constraints for the satisfaction of the MTL semantics, i.e., $(\boldsymbol{\sigma}, t) \models \psi$ for the *negation, disjunction, conjunction, next, until, eventually* and *always* operators, which are identical to integer constraints in (3)–(9) with p, q and p_i as atomic propositions and P_{ψ}^t as the truth value of formula ψ at time t instead.

3.3 Modeling Framework

Consider N discrete-time switched piecewise affine inclusion system models $\{\mathcal{H}_\ell\}_{\ell=1}^N$, each with states $\boldsymbol{x}_\ell \in \mathbb{R}^n$, measurements/outputs $z_\ell \in \mathbb{R}^p$, controlled and uncontrolled inputs $u \in \mathbb{R}^m$, $d_\ell \in \mathbb{R}^{m_d}$, process noise $w_\ell \in \mathbb{R}^{m_w}$, measurement noise $v_\ell \in \mathbb{R}^{m_v}$, and a switching signal $\sigma_\ell \in \mathbb{Z}^+_{|\Sigma|}$. The system dynamics for each piece/subregion $i \in \mathbb{Z}_+^{q_\ell}$ of model $\mathcal{H}_\ell$ (with its domain partitioned into q_ℓ subregions) is given by:

$$\begin{pmatrix} \underline{A}_{\ell,i}^{\sigma_\ell(k)}\boldsymbol{x}_\ell(k)+\underline{B}_{\ell,i}^{u,\sigma_\ell(k)}u(k)+ \\ \underline{B}_{\ell,i}^{d,\sigma_\ell(k)}d_\ell(k)+\underline{B}_{\ell,i}^{w,\sigma_\ell(k)}w_\ell(k) \\ +\underline{f}_{\ell,i}^{\sigma_\ell(k)} \end{pmatrix} \le \boldsymbol{x}_\ell(k+1) \le \begin{pmatrix} \overline{A}_{\ell,i}^{\sigma_\ell(k)}\boldsymbol{x}_\ell(k)+\overline{B}_{\ell,i}^{u,\sigma_\ell(k)}u(k)+ \\ \overline{B}_{\ell,i}^{d,\sigma_\ell(k)}d_\ell(k)+\overline{B}_{\ell,i}^{w,\sigma_\ell(k)}w_\ell(k) \\ +\overline{f}_{\ell,i}^{\sigma_\ell(k)} \end{pmatrix}, \tag{11}$$

$$z_\ell(k) = C_\ell \boldsymbol{x}_\ell(k) + D_\ell^u u(k) + D_\ell^d d_\ell(k) + V_\ell v_\ell(k) + g_\ell, \tag{12}$$

with known constant matrices and vectors of appropriate dimensions. $\boldsymbol{x}_\ell(k+1)$ denotes the state at the next time instant, and their corresponding polytopic subregions $\mathcal{P}_{\ell,i}^{\sigma_\ell(k)}$ are given by the following linear constraints:

$$S_{\ell,i}^{x,\sigma_\ell(k)}\boldsymbol{x}_\ell(k) + S_{\ell,i}^{u,\sigma_\ell(k)}u(k) + S_{\ell,i}^{d,\sigma_\ell(k)}d_\ell(k) + S_{\ell,i}^{w,\sigma_\ell(k)}w_\ell(k) + \beta_{\ell,i}^{\sigma_\ell(k)} \le 0, \tag{13}$$

with $S_{\ell,i}^{x,\sigma_\ell(k)}$, $S_{\ell,i}^{u,\sigma_\ell(k)}$, $S_{\ell,i}^{d,\sigma_\ell(k)}$, $S_{\ell,i}^{w,\sigma_\ell(k)}$, and $\beta_{\ell,i}^{\sigma_\ell(k)}$ of appropriate dimensions. Note that the above switched piecewise affine inclusion models could also represent an over-approximation/abstraction of switched nonlinear models that can be obtained via affine abstraction methods such as presented in Jin et al. (2020), Singh et al. (2018b); hence, the results in this chapter can also be applied when we have switched nonlinear models.

The states $\boldsymbol{x}_\ell$ can be divided into controlled states $x_\ell \in \mathbb{R}^{n_x}$ and uncontrolled states $y_\ell \in \mathbb{R}^{n_y}$ with $n_y = n - n_x$. In the motivating example, the ego car's displacement and velocity are controlled states, while the states of the other car, namely its displacement and velocity, are uncontrolled states. As a consequence, we have

$$\boldsymbol{x}_\ell(k) \triangleq \left[x_\ell(k)^\top \; y_\ell(k)^\top\right]^\top.$$

The initial condition for model $\mathcal{H}_\ell$, denoted by $\boldsymbol{x}_\ell^0 = \boldsymbol{x}_\ell(k_0)$, is constrained to a polyhedral set with c_0 inequalities:

$$\boldsymbol{x}_\ell^0 \in \mathcal{X}_0 := \{\boldsymbol{x} \in \mathbb{R}^n : P_0\boldsymbol{x} \le p_0\}, \; \forall \ell \in \mathbb{Z}_+^N. \tag{14}$$

Moreover, the states x_ℓ and y_ℓ satisfy polyhedral state constraints with c_x and c_y inequalities:

$$x_\ell(k) \in \mathcal{X}_{x,\ell} := \{x \in \mathbb{R}^{n_x} : P_{x,\ell}x \le p_{x,\ell}\}, \tag{15}$$

$$y_\ell(k) \in \mathcal{X}_{y,\ell} := \{y \in \mathbb{R}^{n_y} : P_{y,\ell}y \le p_{y,\ell}\}. \tag{16}$$

The constraint in (15), coupled with the STL specification φ_ℓ^x for the controlled states, makes up the responsibility of the controlled input u. On the other hand, the constraint in (16), the STL specification φ_ℓ^y for the uncontrolled states and the MTL specification ψ_ℓ for the switching signal σ_ℓ are constraints on the models that do not fall under the purview/responsibility of the controlled input u. In the motivating example, the ego car must satisfy certain constraints and specifications, e.g., for the ego car to stay within the lanes and speed limits, but is not responsible for satisfying the set of constraints and specifications that must be satisfied by the other car, e.g., for the other car to stay within the lanes.

In addition, we assume that each given temporal logic specification $\xi_\ell \in \{\varphi_\ell^x, \varphi_\ell^y, \psi_\ell\}$ is of the (unbounded global/safety) form

$$\xi_\ell = \mu^{\xi_\ell} \wedge \Box \phi^{\xi_\ell}, \tag{17}$$

where μ^{ξ_ℓ} and ϕ^{ξ_ℓ} are bounded negation-free MTL/STL formulas with (temporal) bounds $b^{\mu^{\xi_\ell}}$ and $b^{\phi^{\xi_\ell}}$ (cf. Definition 5). This assumption allows us to formulate the problem and the proposed algorithms using a finite number of variables.

Further, the controlled inputs u and uncontrolled inputs d_ℓ, e.g., the gas pedal positions of the ego and other cars, respectively, in the motivating example, must also satisfy the following polyhedral constraints with c_u and c_d inequalities, respectively:

$$u(k) \in \mathcal{U} := \{u \in \mathbb{R}^{m_u} : Q_u u \le q_u\}. \tag{18}$$

$$d_\ell(k) \in \mathcal{D}_\ell := \{d \in \mathbb{R}^{m_d} : Q_{d,\ell} d \le q_{d,\ell}\}. \tag{19}$$

The process and measurement noises, w_ℓ and v_ℓ, are also polyhedrally constrained with c_w and c_v inequalities, respectively:

$$w_\ell(k) \in \mathcal{W}_\ell := \{w \in \mathbb{R}^{m_w} : Q_{w,\ell} w \le q_{w,\ell}\}, \tag{20}$$

$$v_\ell(k) \in \mathcal{V}_\ell := \{v \in \mathbb{R}^{m_v} : Q_{v,\ell} v \le q_{v,\ell}\}. \tag{21}$$

Remark 1 It is important to ensure that the models are meaningful in the sense that, over the considered time horizon τ and for each model $\ell \in \mathbb{Z}_+^N$, (16) holds for any given $\boldsymbol{x}_\ell^0 \in \mathcal{X}_0$ that satisfies $y_\ell(0) \in \mathcal{X}_{y,\ell}$, and for any given $u(k) \in \mathcal{U}$ for all $k \in \mathbb{Z}_{k_0}^{k_0+\tau-1}$. In the motivating example, this assumption implies that the other car through their own uncontrolled inputs must be able to satisfy their own constraints and specifications regardless of the controlled inputs of the ego car. If the considered switched model satisfies this assumption, we refer to it as *well-posed* (Ding et al. 2018). We shall assume throughout the paper that the given switched models are always well-posed.

3.4 Matrix Concatenation

In the interest of keeping our equations and expressions concise, we establish the notation for concatenating matrices and vectors in the following.

3.4.1 Time Concatenation

Given a time interval $\mathbb{Z}_{\underline{\tau}}^{\overline{\tau}}$ where $\underline{\tau}, \overline{\tau}$ are integers, we first introduce some useful time-concatenated notations for the system variables/signals for each model ℓ. The time-concatenated states, outputs and switching signals are as follows:

$$\boldsymbol{x}_{\underline{\tau},\overline{\tau}}^{\ell} = \mathrm{vec}_{k=\underline{\tau}}^{\overline{\tau}}\{\boldsymbol{x}_{\ell}(k)\},\ z_{\underline{\tau},\overline{\tau}}^{\ell} = \mathrm{vec}_{k=\underline{\tau}}^{\overline{\tau}-1}\{z_{\ell}(k)\},$$
$$\boldsymbol{\sigma}_{\underline{\tau},\overline{\tau}}^{\ell} = \{\sigma_{\ell}(k)\}_{k=\underline{\tau}}^{\overline{\tau}-1}.$$

Moreover, the time-concatenated inputs and noises are defined as

$$u_{\underline{\tau},\overline{\tau}} = \mathrm{vec}_{k=\underline{\tau}}^{\overline{\tau}-1}\{u(k)\},\quad d_{\underline{\tau},\overline{\tau}}^{\ell} = \mathrm{vec}_{k=\underline{\tau}}^{\overline{\tau}-1}\{d_{\ell}(k)\},$$
$$w_{\underline{\tau},\overline{\tau}}^{\ell} = \mathrm{vec}_{k=\underline{\tau}}^{\overline{\tau}-1}\{w_{\ell}(k)\},\quad v_{\underline{\tau},\overline{\tau}}^{\ell} = \mathrm{vec}_{k=\underline{\tau}}^{\overline{\tau}-1}\{v_{\ell}(k)\}.$$

In general, planning problems are evaluated over some horizon of T time steps starting at k_0. However, with the introduction of MTL/STL specifications, there is a need to extend this horizon (forwards and backwards) to ensure that the MTL/STL constraint is satisfied. To this end, we define $k^{\mu} = \max_{\ell,\xi_{\ell}\in\{\varphi_{\ell}^{x},\varphi_{\ell}^{y},\psi_{\ell}\}}\{b^{\mu^{\xi_{\ell}}}\}$ and $\underline{k}, \overline{k}$ such that:

$$\begin{aligned}
\underline{k} &= \begin{cases} k_0 - \max_{\ell,\xi_{\ell}\in\{\varphi_{\ell}^{x},\varphi_{\ell}^{y},\psi_{\ell}\}}\{b^{\phi^{\xi_{\ell}}}\}, & \text{if } k_0 > k^{\mu}, \\ k_0, & \text{otherwise}, \end{cases} \\
\overline{k} &= \begin{cases} k_0 + T - 1 + \max_{\ell,\xi_{\ell}\in\{\varphi_{\ell}^{x},\varphi_{\ell}^{y},\psi_{\ell}\}}\{b^{\phi^{\xi_{\ell}}}\}, & \text{if } k_0 > k^{\mu}, \\ \max\{k_0 + T - 1 + \max_{\ell,\xi_{\ell}\in\{\varphi_{\ell}^{x},\varphi_{\ell}^{y},\psi_{\ell}\}}\{b^{\phi^{\xi_{\ell}}}\}, k^{\mu}\}, & \text{otherwise}. \end{cases}
\end{aligned} \tag{22}$$

Then, using this extended time horizon, the switching signals must satisfy $\boldsymbol{\sigma}_{\underline{k},\overline{k}}^{\ell} \in \Sigma_{\underline{k}}^{\overline{k}-\underline{k}+1}(\psi_{\ell})$, where $\Sigma_{\underline{k}}^{\overline{k}-\underline{k}+1}(\psi_{\ell})$ is the valid subtrace set of the MTL constraint on the switching signal. Similarly, the valid signal subtrace sets of the STL constraint on the controlled and uncontrolled states are $\mathcal{X}_{\underline{k}}^{\overline{k}-\underline{k}+1}(\varphi_{\ell}^{x})$ and $\mathcal{Y}_{\underline{k}}^{\overline{k}-\underline{k}+1}(\varphi_{\ell}^{y})$, respectively.

Note that since the MTL/STL constraints only relate to the states and switching signals (and not the outputs), this time extension does not apply to z^{ℓ} (and by extension v^{ℓ}). Thus, using the corresponding time-concatenated inputs, noise, states and outputs, for all $\ell \in \mathbb{Z}_{+}^{N}$, $i \in \mathbb{Z}_{+}^{q_{\ell}}$, the corresponding time-concatenated state and subregion inequalities in (11) and (13) as well as the output equation in (12) can be written as:

$$M_{\ell,i}\boldsymbol{x}^{\ell}_{\underline{k},\overline{k}} + \Gamma^{u}_{\ell,i}u_{\underline{k},\overline{k}} + \Gamma^{d}_{\ell,i}d^{\ell}_{\underline{k},\overline{k}} + \Gamma^{w}_{\ell,i}w^{\ell}_{\underline{k},\overline{k}} + F^{\ell,i}_{\underline{\tau},\overline{\tau}} + \alpha_{\ell,i} \leq 0, \tag{23}$$

$$\Omega^{x}_{\ell,i}\boldsymbol{x}^{\ell}_{\underline{k},\overline{k}} + \Omega^{u}_{\ell,i}u_{\underline{k},\overline{k}} + \Omega^{d}_{\ell,i}d^{\ell}_{\underline{k},\overline{k}} + \Omega^{w}_{\ell,i}w^{\ell}_{\underline{k},\overline{k}} + \beta^{\ell,i}_{\underline{\tau},\overline{\tau}} + \zeta_{\ell,i} \leq 0, \tag{24}$$

$$z^{\ell}_{k_0,k_0+T} = E_{\ell}\boldsymbol{x}^{\ell}_{k_0,k_0+T} + F^{u}_{\ell}u_{k_0,k_0+T} + F^{d}_{\ell}d^{\ell}_{k_0,k_0+T} + F^{v}_{\ell}v^{\ell}_{k_0,k_0+T} + G_{\ell}, \tag{25}$$

with $\alpha_{\ell,i} = \mathrm{vec}_{k=\underline{k}}^{\overline{k}}\{(s_{\ell,i}(k) + \tilde{s}^{\sigma}_{\ell,\star}(k))\mathbb{1}\}$ and $\zeta_{\ell,i} = \mathrm{vec}_{k=\underline{k}}^{\overline{k}}\{(s_{\ell,i}(k) + \tilde{s}^{\sigma}_{\ell,\star}(k))\mathbb{1}\}$, where $s_{\ell,i}(k)$, $\tilde{s}^{\sigma}_{\ell,\star}(k)$ are unconstrained continuous slack variables. All the matrices and vectors in (23)–(25) are defined in the Appendix with suitable dimensions.

Specifically, when $s_{\ell,i}(k) = 0$, the subregion i is valid at k time step, while when $s_{\ell,i}(k)$ is free/unconstrained, the partition i and corresponding inequalities in (23) and (24) hold trivially. The same holds true for $\tilde{s}^{\sigma}_{\ell,\star}(k)$ to indicate the active mode at a given time step k. Thus, for each $k \in \mathbb{Z}^{\overline{k}}_{\underline{k}}$, we introduce (binary) indicator variables $a_{\ell,i}(k), \tilde{a}^{\sigma}_{\ell,\star}(k) \in \{0, 1\}$, where $a_{\ell,i(k)} = 1$ indicates that subregion i is active and $\tilde{a}^{\sigma}_{\ell,\star}(k) = 1$ indicates that mode $\star$ is active, with the constraints that exactly one subregion and one mode can be active at each time step k:

$$\sum_{i} a_{\ell,i}(k) = 1, \ \sum_{\star} \tilde{a}^{\sigma}_{\ell,\star}(k) = 1.$$

Further, for the active subregion i and active mode $\star$, the states of the switched piecewise affine inclusion model must satisfy:

$$\begin{aligned} a_{\ell,i(k)} = 1 &\implies s_{\ell,i}(k) = 0, \\ \tilde{a}^{\sigma}_{\ell,\star}(k) = 1 &\implies \tilde{s}^{\sigma}_{\ell,\star}(k) = 0, \end{aligned}$$

which can be encoded using SOS-1 constraints as follows:

$$\text{SOS-1:}\{a_{\ell,i}(k), s_{\ell,i}(k)\}, \ \text{SOS-1:}\{\tilde{a}^{\sigma}_{\ell,\star}(k), \tilde{s}^{\sigma}_{\ell,\star}(k)\}.$$

Putting the above together, we obtain the following constraints $\forall \ell \in \mathbb{Z}^{N}_{+}$:

$$\begin{aligned} \forall i \in \mathbb{Z}^{q_\ell}_{+}, \forall \star \in \mathbb{Z}^{|\Sigma|}_{+} &: a_{\ell,i} \in \{0, 1\}^{\overline{k}-\underline{k}+1}, \\ \textstyle\sum_i a_{\ell,i} &= \mathbb{1}, \ \text{SOS-1:}\{a_{\ell,i}, s_{\ell,i}\}, \end{aligned} \tag{26a}$$

$$\begin{aligned} \forall i \in \mathbb{Z}^{q_\ell}_{+}, \forall \star \in \mathbb{Z}^{|\Sigma|}_{+} &: \tilde{a}^{\sigma}_{\ell,\star} \in \{0, 1\}^{\overline{k}-\underline{k}+1}, \\ \textstyle\sum_\star \tilde{a}^{\sigma}_{\ell,\star} &= \mathbb{1}, \ \text{SOS-1:}\{\tilde{a}^{\sigma}_{\ell,\star}, \tilde{s}^{\sigma}_{\ell,\star}\}, \end{aligned} \tag{26b}$$

with the time-concatenated $a_{\ell,i} = \mathrm{vec}_{k=\underline{k}}^{\overline{k}}\, a_{\ell,i}(k)$, $\tilde{a}^{\sigma}_{\ell,\star} = \mathrm{vec}_{k=\underline{k}}^{\overline{k}}\, \tilde{a}^{\sigma}_{\ell,\star}(k)$, $s_{\ell,i} = \mathrm{vec}_{k=\underline{k}}^{\overline{k}}\, s_{\ell,i}(k)$, and $\tilde{s}^{\sigma}_{\ell,\star} = \mathrm{vec}_{k=\underline{k}}^{\overline{k}}\, \tilde{s}^{\sigma}_{\ell,\star}(k)$.

Finally, we define the length-T behavior of each switched piecewise affine inclusion model $\mathcal{H}_{\ell}$.

Definition 7 (*Length-T Behavior of $\mathcal{H}_\ell$*) The length-T behavior of the switched piecewise affine inclusion model $\mathcal{H}_\ell$ at time t_0 is the set of all length-T input-mode-output trajectories compatible with $\mathcal{H}_\ell$, given by:

$$\mathscr{B}^T(\mathcal{H}_\ell) := \{u_{t_0,t_0+T}, \boldsymbol{\sigma}_{t_0,t_0+T}, z_{t_0,t_0+T} \mid u_{\underline{k},\overline{k}} \in \mathcal{U}^{\overline{k}-\underline{k}+1},$$
$$\boldsymbol{\sigma}_{\underline{k},\overline{k}} \in \Sigma_{\underline{k}}^{\overline{k}-\underline{k}+1}(\psi_\ell), x_{\underline{k},\overline{k}} \in \mathcal{X}_x^{\overline{k}-\underline{k}+1} \cap \mathcal{X}_{\underline{k}}^{\overline{k}-\underline{k}+1}(\varphi_\ell^x),$$
$$y_{\underline{k},\overline{k}} \in \mathcal{X}_y^{\overline{k}-\underline{k}+1} \cap \mathcal{Y}_{\underline{k}}^{\overline{k}-\underline{k}+1}(\varphi_\ell^y), d_{\underline{k},\overline{k}} \in \mathcal{D}^{\overline{k}-\underline{k}+1}, w_{\underline{k},\overline{k}} \in \mathcal{W}^{\overline{k}-\underline{k}+1},$$
$$v_{t_0,t_0+T} \in \mathcal{V}^T, \text{ s.t. (11)–(13) hold}\},$$

where $\underline{k}, \overline{k}$ are as given in (22) with $k_0 = t_0$.

3.4.2 Pair-Wise Concatenation

Further, given N discrete-time switched piecewise affine inclusion models, there are $I = \binom{N}{2}$ model pairs and we let $\iota \in \{1, \ldots, I\}$ denote the pair of models $(\dagger, \ddagger)$. Moreover, we define the pair-concatenated state, output and noises:

$$\boldsymbol{x}^\iota_{\underline{\tau},\overline{\tau}} = \underset{\ell=\{\dagger,\ddagger\}}{\text{vec}} \{\boldsymbol{x}^\ell_{\underline{\tau},\overline{\tau}}\}, \quad z^\iota_{\underline{\tau},\overline{\tau}} = \underset{\ell=\{\dagger,\ddagger\}}{\text{vec}} \{z^\ell_{\underline{\tau},\overline{\tau}}\}, \quad d^\iota_{\underline{\tau},\overline{\tau}} = \underset{\ell=\{\dagger,\ddagger\}}{\text{vec}} \{d^\ell_{\underline{\tau},\overline{\tau}}\},$$
$$w^\iota_{\underline{\tau},\overline{\tau}} = \underset{\ell=\{\dagger,\ddagger\}}{\text{vec}} \{w^\ell_{\underline{\tau},\overline{\tau}}\}, \quad v^\iota_{\underline{\tau},\overline{\tau}} = \underset{\ell=\{\dagger,\ddagger\}}{\text{vec}} \{v^\ell_{\underline{\tau},\overline{\tau}}\}.$$

The states, corresponding subregions and outputs over the entire time horizon in (23), (24) and (25) for each pair ι can then be written as simple functions of the state $\boldsymbol{x}^\iota_{\underline{k},\overline{k}}$, input vectors $u_{\underline{k},\overline{k}}$, $d^\iota_{\underline{k},\overline{k}}$, and noise $w^\iota_{\underline{k},\overline{k}}$, $v^\iota_{k_0,k_0+T}$:

$$M^\iota \boldsymbol{x}^\iota_{\underline{k},\overline{k}} + \Gamma^\iota_u u_{\underline{k},\overline{k}} + \Gamma^\iota_d d^\iota_{\underline{k},\overline{k}} + \Gamma^\iota_w w^\iota_{\underline{k},\overline{k}} + F^\iota_{\underline{k},\overline{k}} \le 0, \tag{27}$$
$$z^\iota_{k_0,k_0+T} = E^\iota \boldsymbol{x}^\iota_{k_0,k_0+T} + F^\iota_u u_{k_0,k_0+T} + F^\iota_d d^\iota_{k_0,k_0+T} + F^\iota_v v^\iota_{k_0,k_0+T} + G^\iota, \tag{28}$$

where the matrices and vectors $M^\iota, \Gamma^\iota_u, \Gamma^\iota_d, \Gamma^\iota_w, F^\iota_{\underline{k},\overline{k}}, E^\iota, F^\iota_u, F^\iota_d, F^\iota_v$ and G^ι are defined in the Appendix. Moreover, the uncertain variables for each pair ι are concatenated as $\bar{x}^\iota = \text{vec}\{\boldsymbol{x}^\iota_{\underline{k},\overline{k}}, d^\iota_{\underline{k},\overline{k}}, w^\iota_{\underline{k},\overline{k}}, v^\iota_{k_0,k_0+T}\}$.

We then rewrite the polyhedral state constraints in (15) and (16) in terms of time and pair. For $* \in \{x, y\}$, define

$$\bar{P}_{*,\ell} = \text{diag}_{(\overline{k}-\underline{k})}\{P_{*,\ell}\}, \quad \bar{p}_{*,\ell} = \text{vec}_{(\overline{k}-\underline{k})}\{p_{*,\ell}\},$$
$$\bar{P}^\iota_* = [\text{diag}_{\ell=\{\dagger,\ddagger\}}\{\bar{P}_{*,\ell}\}\ \mathbf{0}], \quad \bar{p}^\iota_* = \text{vec}_{\ell=\{\dagger,\ddagger\}}\{\bar{p}_{*,\ell}\}.$$

Using this, we can rewrite the polyhedral constraints as:

$$\bar{P}^\iota_* \bar{x}^\iota \le \bar{p}^\iota_*.$$

Further, the initial state constraint in (14) can be written as $\bar{P}_0^\iota \bar{x}^\iota \leq \bar{p}_0^\iota$ with

$$\bar{P}_0^\iota = \mathrm{diag}_2\{[P_0\ \mathbf{0}]\}, \quad \bar{p}_0^\iota = \mathrm{vec}_2\{p_0\}.$$

Similarly, we define

$$\begin{aligned}
\bar{Q}_u &= \underset{(\bar{k}-\underline{k})}{\mathrm{diag}}\{Q_u\}, \ \bar{Q}_d^\iota = \underset{\ell=\{\dagger,\ddagger\}}{\mathrm{diag}}\ \underset{(\bar{k}-\underline{k})}{\mathrm{diag}}\{Q_{d,\ell}\}, \\
\bar{Q}_w^\iota &= \underset{\ell=\{\dagger,\ddagger\}}{\mathrm{diag}}\ \underset{(\bar{k}-\underline{k})}{\mathrm{diag}}\{Q_{w,\ell}\}, \ \bar{Q}_v^\iota = \underset{\ell=\{\dagger,\ddagger\}}{\mathrm{diag}}\ \underset{T}{\mathrm{diag}}\{Q_{v,\ell}\}, \\
\bar{q}_u &= \underset{(\bar{k}-\underline{k})}{\mathrm{vec}}\{q_u\}, \ \bar{q}_d^\iota = \underset{\ell=\{\dagger,\ddagger\}}{\mathrm{vec}}\ \underset{(\bar{k}-\underline{k})}{\mathrm{vec}}\{q_{d,\ell}\}, \\
\bar{q}_w^\iota &= \underset{\ell=\{\dagger,\ddagger\}}{\mathrm{vec}}\ \underset{(\bar{k}-\underline{k})}{\mathrm{vec}}\{q_{w,\ell}\}, \ \bar{q}_v^\iota = \underset{\ell=\{\dagger,\ddagger\}}{\mathrm{vec}}\ \underset{T}{\mathrm{vec}}\{q_{v,\ell}\}.
\end{aligned}$$

Then, the input and uncertainty constraints in (18)–(21) for all k are concatenated as: $\bar{Q}_u u_{\underline{k},\bar{k}} \leq \bar{q}_u$, $\bar{Q}_d^\iota d_{\underline{k},\bar{k}}^\iota \leq \bar{q}_d^\iota$, $\bar{Q}_w^\iota w_{\underline{k},\bar{k}}^\iota \leq \bar{q}_w^\iota$, and $\bar{Q}_v^\iota v_{k_0,k_0+T}^\iota \leq \bar{q}_v^\iota$. Hence, in terms of $\bar{x}^\iota$, we have a polyhedral constraint of the form $H_{\bar{x}}^\iota \bar{x}^\iota \leq h_{\bar{x}}^\iota$, with

$$H_{\bar{x}}^\iota = \begin{bmatrix} \bar{P}_0^\iota & \mathbf{0} & \mathbf{0} \\ \mathbf{0} & \bar{P}_y^\iota & \mathbf{0} \\ \mathbf{0} & \mathbf{0} & \mathrm{diag}\{\bar{Q}_d^\iota, \bar{Q}_w^\iota, \bar{Q}_v^\iota\} \end{bmatrix}, \quad h_{\bar{x}}^\iota = [\bar{p}_0^{\iota\top}\ \bar{p}_y^{\iota\top}\ \bar{q}_d^{\iota\top}\ \bar{q}_w^{\iota\top}\ \bar{q}_v^{\iota\top}]^\top.$$

Note that the above definitions of matrices and vectors for the *switched piecewise affine inclusion model* in (11) are different from the ones in Ding et al. (2018), Singh et al. (2018a), which were defined for *affine (inclusion) models* (i.e., a special case of our considered model in (11) with only one piece/subregion and one mode).

4 Problem Formulation

The goal of this section is to present a formulation for the active model discrimination (AMD) problem for switched piecewise affine inclusions with temporal logic constraints, as well as the complementary model identification (MI) problem. The AMD problem is intended to be solved offline/at design time to find the optimal separating input for guaranteeing that the models are distinct. The model identification problem, in turn, allows an observer to accurately determine at run time which model from a given set of models generated the observed noisy data.

$$\{y(k)\}_{k=\underline{k}}^{\overline{k}} \in \mathcal{Y}_{\underline{k}}^{\overline{k}-\underline{k}+1}(\varphi^y), \{P_{p^{\pi^x}}(k)\}_{k=\underline{k}}^{\overline{k}} \in \Sigma_{\underline{k}}^{\overline{k}-\underline{k}+1}(\varphi^x), \tag{32m}$$

$$\{\sigma(k)\}_{k=\underline{k}}^{\overline{k}} \in \Sigma_{\underline{k}}^{\overline{k}-\underline{k}+1}(\psi), \tag{32n}$$

The signals $s_*(k)$ and $\tilde{s}_\star^\sigma(k)$, as well as $s_j^{x+}(k)$ and $s_j^{x-}(k)$ are (unconstrained real-valued) slack variables, whereas $\tilde{a}_\star^\sigma(k) = 1$ corresponds to $\sigma(k, \star) = \sigma$. The valid subtrace sets $\mathcal{Y}_{\underline{k}}^{\overline{k}-\underline{k}+1}(\varphi^y)$ and $\Sigma_{\underline{k}}^{\overline{k}-\underline{k}+1}(\psi)$ corresponding to the uncontrolled states and switching sequences, respectively, are constructed recursively using MILP formulations of MTL/STL specifications in (2)–(9), whereas the valid subtrace set $\mathcal{X}_{\underline{k}}^{\overline{k}-\underline{k}+1}(\varphi^x)$ corresponding to the STL specification φ^x is subdivided into the continuous predicate functions $\pi^x(x(k))$ according to (2) and binary predicates $P_{p^{\pi^x}}(k)$ satisfying $\Sigma_{\underline{k}}^{\overline{k}-\underline{k}+1}(\varphi^x)$ according to (3)–(9). Further, note that the functions $\pi_j(x(k))$ are linear functions of $\boldsymbol{x}(k_0)$, $d(k)$, $w(k)$ and $u(k)$ for all k that are also decision variables in this optimization problem.

Proof $a_*(k) = 1$ and $\tilde{a}_\star^\sigma(k) = 1$ in (32a)–(32c) imply that the former inequalities in (23)–(25) hold, since the SOS-1 constraints in (32j) ensure that $s_*(k) = 0$ and $\tilde{s}_\star^\sigma(k) = 0$. On the contrary, if $a_*(k) = 0$ and/or $\tilde{a}_\star^\sigma(k) = 0$, it means that $s_*(k)$ and $\tilde{s}_\star^\sigma(k)$ are free and then (32a)–(32c) hold trivially. In addition, (32k) ensures that, at each time step k, only one partition is valid for each of the state and output equations, and only one switching signal/mode is valid (cf. (26)). Moreover, to completely express the constraints induced by φ^y on $\sigma(k)$ and $\tilde{a}_\star^\sigma(k)$ using an MILP reformulation, the time horizon of $\sigma(k)$ and $\tilde{a}_\star^\sigma(k)$ is extended both backwards and forwards by $\max(b^{\phi_i^y}, b^{\phi_j^y})$ time steps via the definitions of $\underline{k}$ and $\overline{k}$. Finally, (31), in combination with (32e)–(32h), guarantees that the state constraint (15) and the continuous component of the STL φ^x hold despite all feasible/realizable uncertainties. □

Proposition 2 (Separability Condition Reformulation) *The separability condition in (29c) with (23)–(25) for each model pair $i, j \in \mathbb{Z}_+^{\mathrm{N}}, i < j$ of Problem 1 is equivalent to*

$$\delta^{(i,j)} \geq \epsilon, \tag{33}$$

where $\delta^{(i,j)}$ (implicitly/parametrically dependent on $u_{\underline{k},\overline{k}}$) is the solution to:

$$\delta^{(i,j)} = \underset{\substack{\boldsymbol{x}^\dagger_{\underline{k}.\overline{k}}, d^\dagger_{\underline{k}.\overline{k}}, w^\dagger_{\underline{k}.\overline{k}}, v^\dagger_{k_0,k_0+T}, z^\dagger_{k_0,k_0+T}, s_{\dagger,*}(k), \\ \tilde{s}_{\dagger,\star}(k), a_{\dagger,*}(k), \tilde{a}_\star(k),}}{\operatorname{argmin}} \quad \eta \tag{34a}$$

$$s.t.\ \dagger \in \{i, j\}, \forall i, j \in \mathbb{Z}_+^N, i \neq j, \forall * \in \mathbb{Z}_+^{q_\dagger}, k \in \mathbb{Z}_{\underline{k}-1}^{\overline{k}}, k' \in \mathbb{Z}_{k_0}^{k_0+T-1}: \tag{34b}$$

$$M_{\dagger,*}\boldsymbol{x}^\dagger_{\underline{k}.\overline{k}} + \Gamma^u_{\dagger,*}u_{\underline{k},\overline{k}} + \Gamma^d_{\dagger,*}d^\dagger_{\underline{k}.\overline{k}} + \Gamma^w_{\dagger,*}w^\dagger_{\underline{k}.\overline{k}} + F^{\dagger,*}_{\underline{k}.\overline{k}} + \alpha_{\dagger,*} \leq 0, \tag{34c}$$

$$\Omega^x_{\dagger,*}\boldsymbol{x}^\dagger_{\underline{k}.\overline{k}} + \Omega^u_{\dagger,*}u_{\underline{k},\overline{k}} + \Omega^d_{\dagger,*}d^\dagger_{\underline{k}.\overline{k}} + \Omega^w_{\dagger,*}w^\dagger_{\underline{k}.\overline{k}} + \beta^{\dagger,*}_{\underline{k}.\overline{k}} + \zeta_{\dagger,*} \leq 0, \tag{34d}$$

$$z^\dagger_{k_0,k_0+T} = E_\ell\boldsymbol{x}^\dagger_{k_0,k_0+T} + F^u u_{k_0,k_0+T} + F^d_\dagger d^\dagger_{k_0,k_0+T} + F^v_\dagger v^\dagger_{k_0,k_0+T} + G_\dagger, \tag{34e}$$

$$\|z_i(k') - z_j(k')\|_\infty \leq \eta, \tag{34f}$$

$$P_{y,\star} y_\star(k) \leq p_{y,\star}, \tag{34g}$$

$$a_{\dagger,*}(k) \in \{0,1\}, \tilde{a}_\dagger(k) \in \{0,1\}, \tag{34h}$$

$$\textit{SOS-1}{:}(a_{\dagger,*}(k), s_{\dagger,*}(k)), \textit{SOS-1}{:}(\tilde{a}^\sigma_{\dagger,\star}(k), \tilde{s}^\sigma_{\dagger,\star}(k)), \tag{34i}$$

$$\textstyle\sum_{*=1}^{q} a_{\dagger,*}(k) = 1, \sum_{\star=1}^{|\Sigma|} \tilde{a}^\sigma_{\dagger,\star}(k) = 1, \tag{34j}$$

$$d_\dagger(k) \in \mathcal{D}, w_\dagger(k) \in \mathcal{W}, v_\dagger(k) \in \mathcal{V}, u(k) \in \mathcal{U}, \tag{34k}$$

$$\{y_\dagger(k)\}_{k=\underline{k}}^{\overline{k}} \in \mathcal{Y}_{\underline{k}}^{\overline{k}-\underline{k}+1}(\varphi_\dagger^y) \tag{34l}$$

$$\{\sigma_\dagger(k)\}_{k=\underline{k}}^{\overline{k}} \in \Sigma_{\underline{k}}^{\overline{k}-\underline{k}+1}(\psi_\dagger), \tag{34m}$$

where $s_{\dagger,*}(k)$ *and* $\tilde{s}_{\dagger,\star}(k)$ *are slack variables,* $\tilde{a}^\sigma_{\dagger,\star}(k) = 1$ *corresponds to* $\sigma_\dagger(k) = \sigma_\dagger$, *and the valid subtrace sets* $\mathcal{Y}_{\underline{k}}^{\overline{k}-\underline{k}+1}(\varphi_\dagger^y)$ *and* $\Sigma_{\underline{k}}^{\overline{k}-\underline{k}+1}(\psi)$ *corresponding to the uncontrolled states and switching sequences, respectively, are constructed recursively using MILP formulations of MTL/STL specifications in (2)–(9).*

Proof The construction follows similar steps to Proposition 1, but with two different models i, j and with the separation condition in (34f). Note that the constraints on the controlled states in the form of (15) and the STL specification φ^x are not needed because they have been robustly enforced by the state constraint reformulation in Proposition 1. □

Using Propositions 1 and 2, it is straightforward to show that the active model discrimination problem in Problem 1 can be recast as the following bilevel optimization problem:

Theorem 1 (Parametric AMD) *Given a separability index* ϵ, *the AMD problem in Problem 1 is equivalent to*

$$u^*_{\underline{k},\overline{k}} = \arg\min_{u_{\underline{k},\overline{k}}} J(u_{\underline{k},\overline{k}}) \tag{35a}$$

$$\textit{s.t.}\ (31)\ \textit{and}\ (33)\ \textit{hold}, \tag{35b}$$

where ρ *and* $\delta^{(i,j)}$ *are solutions to the inner problems given by (32) and (34), respectively.*

Since the inner problems (32) and (34) contain integer variables, standard reformulation methods using KKT or robust optimization, which were employed in Ding et al. (2018), Singh et al. (2018a), do not apply. Thus, we propose to parametrically solve the inner (optimization) problems given in Propositions 1 and 2 using a multiparametric programming solver, e.g., MPT 3.0 (Herceg et al. 2013). Their solutions are then incorporated into the outer problem in Theorem 1 to find the minimal separating input.

Strategies for Tractable Parametric AMD

However, since even small parametric optimization problems can often be computationally demanding, we propose the following two strategies to address this issue.

Horizon Truncation. One way to make the multi-parametric optimization problems in Propositions 1 and 2 more tractable is by reducing the number of binary variables involved in some way, as they are the major reasons leading to intractable problems. The following horizon truncation strategy is one such solution that can be applied to obtain a potentially sub-optimal but tractable problem.

Proposition 3 (Tractable Parametric AMD with Truncated Horizon) *In Propositions 1 and 2, a sufficient tractable problem can be obtained by reducing the extended time horizon, by setting the upper limit to* $\overline{k} = k_0 + T - 1$ *when* $k_0 > k^{\mu}$*, and* $\overline{k} = \max\{k_0 + T - 1, k^{\mu}\}$ *otherwise, with* $k^{\mu} = \max_{\ell, \xi_\ell \in \{\varphi_\ell^x, \varphi_\ell^y, \psi_\ell\}}\{b^{\mu^{\xi_\ell}}\}$.

Proof By eliminating all variables for all $k > T - 1$, the resulting optimization problem will have relaxed constraints (and also fewer decision variables). Thus, when the relaxed problem is solved, we obtain a smaller ρ, denoted as $\underline{\rho}$. If the solution to the relaxed problem exists, it provides a lower bound to the cost of the original inner problem in Proposition 1 as $\rho \geq \underline{\rho}$, hence the separating input that satisfies $\underline{\rho} \geq 0$ also satisfies $\rho \geq 0$. A similar argument applies to its sufficiency for Proposition 2. □

Although the strategy proposed in Proposition 3 leads to a sub-optimal solution in general, there are specific cases where applying the strategy will not result in the loss of optimality. One such case arises when we have only MTL constraints on the switching modes and the future truth conditions for $\sigma_\ell(k)$, $\forall k > T - 1$ are independent of the choice of the separating input sequence.

Input Domain Partitioning. When the input domain is large, existing multi-parametric tools can have difficulties dealing with a large number of critical regions (e.g., in the order of thousands). One way to tackle this issue is to shrink/reduce the input region by partitioning the input set (cf. Definition 2), and solve each individual problem with the corresponding input subset separately (or in parallel, if desired). In other words, we still consider the same problem but we instead solve a finite number of "smaller" tractable problems. Finally, within the results from these tractable problems, we choose the best feasible solution.

Remark 4 (*Tractable Parametric AMD with Input Domain Partitioning*) Using Definition 2, the input domain $\mathcal{U}$ in (32l) and (34k) in Propositions 1 and 2 can be partitioned into $\mathcal{U}_i$, where $i \in \mathbb{Z}_+^{|U|}$. When using the partitioned input region $\mathcal{U}_i$ in lieu of $\mathcal{U}$, the multi-parametric optimization problems in Propositions 1 and 2 typically result in much fewer critical regions, thus making the problems tractable.

5.1.2 Binary Variable Relocation

As mentioned previously, the main difficulty with solving Problem 1 is the presence of binary variables in the inner problems of (29b) and (29c). So in this second approach, we move these binary variables into the outer problem by indirectly choosing them through robust satisfaction of the domains of the subregions or the predicate functions of the STL specifications. This reduces the difficulty, as we could in principle, now employ tools such as KKT. However, this comes at a slight cost of sub-optimality due to the search space restriction of the optimization problem.

It is also important to note that since MTL specifications do not have associated predicate functions, the proposed method in this section does not work for such problems. Nonetheless, this approach does still apply to switched systems with (worst-case) arbitrary switching.

Problem 3 (*Relocation-Based Active Model Discrimination for* $\{\mathcal{H}_\ell\}_{\ell=1}^N$) Given N well-posed switched piecewise affine inclusion models $\{\mathcal{H}_\ell\}_{\ell=1}^N$, and state, input and noise constraints, i.e., (14)–(21), such that for all possible initial states $\boldsymbol{x}_\ell^0$, uncontrolled input $d_{\underline{k},\overline{k}}^\ell$, process noise $w_{\underline{k},\overline{k}}^\ell$ and measurement noise v_{k_0,k_0+T}^ℓ, the output trajectories of each model pair have to differ by a threshold ϵ in at least one time instance. The optimization problem can be formally written as:

$$u_{\underline{k},\overline{k}}^* = \arg\min_{u_{\underline{k},\overline{k}}} J(u_{\underline{k},\overline{k}})$$

$$\text{s.t.} \qquad \forall k \in \mathbb{Z}_{\underline{k}}^{\overline{k}} : (18) \text{ holds}, \tag{36a}$$

$$\forall \ell \in \mathbb{Z}_+^N, \forall i \in \mathbb{Z}_+^{q_\ell} : \begin{array}{l} a_{\ell,i} \in \{0,1\}^{\overline{k}-\underline{k}+1}, \sum_i a_{\ell,i} = \mathbb{1}, \\ \text{SOS-1:}\{a_{\ell,i}, s_{\ell,i}\}, \end{array} \tag{36b}$$

$$\forall \ell \in \mathbb{Z}_+^N, \forall \star \in \mathbb{Z}_+^{|\Sigma|} : \tilde{a}_{\ell,\star}^\sigma = \mathbb{1},\ \text{SOS-1:}\{\tilde{a}_{\ell,\star}^\sigma, \tilde{s}_{\ell,\star}^\sigma\}, \tag{36c}$$

$$\left.\begin{array}{r} \forall\, \boldsymbol{x}_\ell^0, d_{\underline{k},\overline{k}}^\iota, w_{\underline{k},\overline{k}}^\iota, \alpha_\ell, \zeta_\ell : \\ (14), (16), (20), (23)\ \textit{hold} \end{array}\right\} : \begin{array}{l} (15), (24) \text{ holds}, \\ x_\ell(k) \models \varphi^x \text{ and } y_\ell(k) \models \varphi^y, \end{array} \tag{36d}$$

$$\left.\begin{array}{r} \forall \boldsymbol{x}_\ell^0, d_{\underline{k},\overline{k}}^\iota, w_{\underline{k},\overline{k}}^\iota, v_{k_0,k_0+T}^\iota, \alpha_\ell, \zeta_\ell : \\ (14), (16), (20), (21), (27), (28)\ \textit{hold} \end{array}\right\} : \begin{array}{l} \{\exists k' \in \mathbb{Z}_{k_0}^{k_0+T-1} : \\ \|z_\dagger(k') - z_\ddagger(k')\|_\infty \geq \epsilon\}, \end{array} \tag{36e}$$

with the ι-th model pair being $(\dagger, \ddagger)$, where $\dagger, \ddagger \in \mathbb{Z}_+^N, \ddagger \neq \dagger$.

Note that in comparison with the original Problem 1, the STL satisfaction of $y_\ell(k) \models \varphi^y$ has been now relocated/moved to the right-hand side of (36d), while the subregions of the switched piecewise affine dynamics in (26a) are relocated to the right-hand side of (36b) and their domains (24) to the right-hand side of (36d). This relocation allows us to robustly enforce the domains of the subregions (36d) and the predicate functions of φ^y. On the other hand, to allow (worst-case) arbitrary switching of modes, i.e., all switching signals are possible, instead of the constraints in (26b), we impose the constraint in (36c). Note that this constraint in (36c) (and also later in (37c) and (39j)) essentially means that $\tilde{s}^{\sigma}_{\ell,\star} = 0$ for all $\ell \in \mathbb{Z}_+^N$ and $\star \in \mathbb{Z}_+^{|\Sigma|}$.

Lemma 1 (Bilevel Optimization Formulation) *Given a separability index ϵ, the active model discrimination problem in Problem 3 is equivalent to a bilevel optimization problem with the following outer problem:*

$$u^*_{\underline{k},\overline{k}} = \arg\min_{u_{\underline{k},\overline{k}}} J(u_{\underline{k},\overline{k}}) \quad (P_{Outer})$$

$$
\begin{aligned}
\text{s.t.} \quad & \forall k \in \mathbb{Z}_{\underline{k}}^{\overline{k}} : (18) \textit{ holds}, && (37a)\\
& \forall \ell \in \mathbb{Z}_+^N, \forall i \in \mathbb{Z}_+^{q_\ell}, : \begin{array}{l} a_{\ell,i} \in \{0,1\}^{\overline{k}-\underline{k}+1}, \sum_i a_{\ell,i} = \mathbb{1}, \\ \textit{SOS-1}:\{a_{\ell,i}, s_{\ell,i}\}, \end{array} && (37b)\\
& \forall \ell \in \mathbb{Z}_+^N, \forall \star \in \mathbb{Z}_+^{|\Sigma|} : \tilde{a}^{\sigma}_{\ell,\star} = \mathbb{1},\ \textit{SOS-1}:\{\tilde{a}^{\sigma}_{\ell,\star}, \tilde{s}^{\sigma}_{\ell,\star}\}, && (37c)\\
& \left.\begin{array}{r} \forall k \in \mathbb{Z}_{\underline{k}}^{\overline{k}}, \forall\, \boldsymbol{x}^0_\ell, d^\iota_{\underline{k},\overline{k}}, w^\iota_{\underline{k},\overline{k}}, \alpha_\ell : \\ (14), (16), (20), (23) \textit{ hold} \end{array}\right\} : \begin{array}{l} (15), (24) \textit{ holds}, \\ x_\ell(k) \vDash \varphi^x \textit{ and } y_\ell(k) \vDash \varphi^y, \end{array} && (37d)\\
& \forall \iota \in \mathbb{Z}_+^I : \delta^{\iota *}(u_{\underline{k},\overline{k}}) \geq \epsilon, && (37e)
\end{aligned}
$$

where $\delta^{\iota *}$ is the solution of the inner problem:

$$\delta^{\iota *}(u_{\underline{k},\overline{k}}) = \min_{\delta^\iota, \boldsymbol{x}^\iota_{\underline{k},\overline{k}}, d^\iota_{\underline{k},\overline{k}}, w^\iota_{\underline{k},\overline{k}}, v^\iota_{k_0,k_0+T}} \delta^\iota \quad (P_{Inner})$$

$$
\begin{aligned}
\text{s.t.} \quad & \forall \iota \in \mathbb{Z}_+^I,\ k \in \mathbb{Z}_{\underline{k}}^{\overline{k}} : (27), (28) \textit{ hold}, && (38a)\\
& \forall \iota \in \mathbb{Z}_+^I,\ k \in \mathbb{Z}_{k_0}^{k_0+T-1} : \|z_\dagger(k) - z_\ddagger(k)\|_\infty \leq \delta^\iota, && (38b)\\
& \forall\, \boldsymbol{x}^0_\ell, d^\iota_{\underline{k},\overline{k}}, w^\iota_{\underline{k},\overline{k}}, v^\iota_{k_0,k_0+T} : (14), (16), (20), (21) \textit{ hold}, && (38c)
\end{aligned}
$$

with the ι-th model pair being $(\dagger, \ddagger)$, where $\dagger, \ddagger \in \mathbb{Z}_+^N$, $\ddagger \neq \dagger$.

Proof The proof follows a similar procedure as in Ding et al. (2018). □

This bilevel problem can then be recast into a single-level MILP that can be readily solved using off-the-shelf solvers, given as follows:

Theorem 2 (Relocation-Based Active Model Discrimination for $\{\mathcal{H}_\ell\}_{\ell=1}^N$) *Given a separability index ϵ, the bilevel optimization problem in Lemma 1 is equivalent to the following mixed-integer optimization problem:*

$$u^*_{\underline{k},\overline{k}} = \underset{\substack{u_{\underline{k},\overline{k}},\tilde{x}^\ell,\nu_1^\ell,\nu_2^\ell,a_{\ell,i}(k),s_{\ell,i}(k),\\ \bar{x}^\iota,\mu_1^\iota,\mu_2^\iota,\mu_3^\iota,b^\iota,\bar{s}^\iota,\delta^\iota}}{\arg\min} J(u_{\underline{k},\overline{k}})$$

$$\text{s.t.} \quad \bar{Q}_u u_{\underline{k},\overline{k}} \leq \bar{q}_u, \tag{39a}$$

$$\forall \ell \in \mathbb{Z}_+^N, \forall a \in \mathbb{Z}_+^{c_x} : \text{(40a)–(40e) } hold, \tag{39b}$$

$$\forall \iota \in \mathbb{Z}_+^{\mathcal{I}} : \begin{cases} \delta^\iota \geq \epsilon, \\ 0 = 1 - \mu_3^{\iota\top}\mathbb{1}, \\ 0 = \sum_{i=1}^{i=\kappa} \mu_{1,i}^\iota H_{\bar{x}}^\iota(i,m) + \sum_{j=1}^{j=\xi} \mu_{2,j}^\iota R_1^\iota(j,m) \\ \quad + \sum_{k=1}^{k=\rho} \mu_{3,k}^\iota R_2^\iota(k,m), \forall m = 1, \cdots, \eta, \\ \widetilde{H}_{\bar{x},i}^\iota \bar{x}^\iota - h_{\bar{x},i}^\iota \leq 0, \mu_{1,i}^\iota \geq 0, \forall i = 1, \ldots \kappa, \\ \widetilde{R}_{1,j}^\iota \bar{x}^\iota - r_{1,j}^\iota - S_{1,j}^\iota u_{k_0,k_0+T} \leq 0, \\ \mu_{2,j}^\iota \geq 0, \forall j = 1, \ldots \xi, \\ \widetilde{R}_{2,k}^\iota \bar{x}^\iota - \delta^\iota - r_{2,k}^\iota - S_{2,k}^\iota u_{k_0,k_0+T} \leq 0, \mu_{3,k}^\iota \geq 0, \\ \forall k = 1, \ldots \rho, \end{cases} \tag{39c}$$

$$\forall \ell \in \mathbb{Z}_+^N, \forall m \in \mathbb{Z}_+^\alpha : \textit{SOS-1} : \{\nu_{1,a,m}^\ell, \widetilde{P}_{x,m}^\ell \tilde{x}_a^\ell - \widetilde{P}_{u,m}^\ell u_{\underline{k},\overline{k}} - \bar{p}_m^\ell\}, \tag{39d}$$

$$\forall \ell \in \mathbb{Z}_+^N, \forall n \in \mathbb{Z}_+^\gamma : \textit{SOS-1} : \{\nu_{2,a,n}^\ell, \widetilde{H}_n^\ell \tilde{x}_a^\ell - \tilde{h}_n^\ell\}, \tag{39e}$$

$$\forall \iota \in \mathbb{Z}_+^{\mathcal{I}}, \forall i \in \mathbb{Z}_+^\kappa : \textit{SOS-1} : \{\mu_{1,i}^\iota, \widetilde{H}_{\bar{x},i}^\iota \bar{x}^\iota - h_{\bar{x},i}^\iota\}, \tag{39f}$$

$$\forall \iota \in \mathbb{Z}_+^{\mathcal{I}}, \forall j \in \mathbb{Z}_+^\xi : \textit{SOS-1} : \{\mu_{2,j}^\iota, \widetilde{R}_{1,j}^\iota \bar{x}^\iota - r_{1,j}^\iota - \widetilde{S}_{1,j}^\iota u_{k_0,k_0+T}\}, \tag{39g}$$

$$\forall \iota \in \mathbb{Z}_+^{\mathcal{I}}, \forall k \in \mathbb{Z}_+^\rho : \textit{SOS-1} : \{\mu_{3,k}^\iota, \widetilde{R}_{2,k}^\iota \bar{x}^\iota - \delta^\iota - r_{2,k}^\iota - \widetilde{S}_{2,k}^\iota u_{k_0,k_0+T}\}, \tag{39h}$$

$$\forall \ell \in \mathbb{Z}_+^N, \forall i \in \mathbb{Z}_+^{q_\ell}, : \begin{cases} \textit{SOS-1} : \{a_{\ell,i}, s_{\ell,i}\} \\ a_{\ell,i} \in \{0,1\}^{\overline{k}-\underline{k}+1}, \sum_{i=1}^{q_\ell} a_{\ell,i} = \mathbb{1}, \end{cases} \tag{39i}$$

$$\forall \ell \in \mathbb{Z}_+^N, \forall \star \in \mathbb{Z}_+^{|\Sigma|} : \tilde{a}_{\ell,\star}^\sigma = \mathbb{1}, \ \textit{SOS-1}:\{\tilde{a}_{\ell,\star}^\sigma, \tilde{s}_{\ell,\star}^\sigma\}, \tag{39j}$$

$$\forall \ell \in \mathbb{Z}_N^+ : \begin{array}{l} \forall \boldsymbol{x}_\ell(k_0) \in \mathcal{X}_0, \forall d_\ell(k) \in \mathcal{D}, \forall w_\ell(k) \in \mathcal{W}, \forall k \in \mathbb{Z}_{\underline{k}}^{\overline{k}} : \\ \pi_\ell^x(x(k)) + s_\ell^{x+}(k) \geq 0, \ \pi_\ell^x(x(k)) < s_\ell^{x-}(k), \\ \pi_\ell^y(y(k)) + s_\ell^{y+}(k) \geq 0, \ \pi_\ell^y(y(k)) < s_\ell^{y-}(k), \\ \{P_{p^{\pi_\ell^x}}(k)\}_{k=\underline{k}}^{\overline{k}} \in \Sigma_{\underline{k}}^{\overline{k}-\underline{k}+1}(\varphi_\ell^x), \\ \{P_{p^{\pi_\ell^y}}(k)\}_{k=\underline{k}}^{\overline{k}} \in \Sigma_{\underline{k}}^{\overline{k}-\underline{k}+1}(\varphi_\ell^y), \\ \textit{SOS-1}:\{P_{p^{\pi_\ell^x}}(k)\mathbb{1}, s_\ell^{x+}(k)\}, \\ \textit{SOS-1}:\{(1 - P_{p^{\pi_\ell^x}}(k))\mathbb{1}, s_\ell^{x-}(k)\}, \\ \textit{SOS-1}:\{P_{p^{\pi_\ell^y}}(k)\mathbb{1}, s_\ell^{y+}(k)\}, \\ \textit{SOS-1}:\{(1 - P_{p^{\pi_\ell^y}}(k))\mathbb{1}, s_\ell^{y-}(k)\}, \end{array} \tag{39k}$$

where $\nu_{1,a,m}^\ell$, $\nu_{2,a,n}^\ell$, $\mu_{1,i}^\iota$, $\mu_{2,j}^\iota$ *and* $\mu_{3,k}^\iota$ *are dual variables from KKT conditions, while* $\widetilde{P}_{*,m}^\ell$ *is the m-th row of* $\overline{P}_*^\ell$, $* \in \{x, u\}$, $\widetilde{H}_n^\ell$ *is the n-th row of* $\overline{\overline{H}}_\ell$, $\widetilde{H}_{\bar{x},i}^\iota$ *is the i-th row of* $H_{\bar{x}}^\iota$, $\widetilde{R}_{1,j}^\iota$ *and* $\widetilde{S}_{1,j}^\iota$ *are the j-th rows of* R_1^ι *and* S_1^ι*, respectively,* $\widetilde{R}_{2,k}^\iota$ *and* $\widetilde{S}_{2,k}^\iota$ *are the k-th rows of* R_2^ι *and* S_2^ι*, respectively,* $\eta = \mathcal{I}|\Sigma|T(n + m_d + m_w + m_v)$ *is the number of columns of* $H_{\bar{x}}^\iota$, $\kappa = 2\mathcal{I}|\Sigma|T(c_0 + c_d + c_w + c_v)$ *is the number of rows* *of $H^\iota_{\bar{x}}$, $\xi = 2\mathcal{I}|\Sigma|T(c_x + c_y)$ is the number of rows of R^ι_1, and $\rho = 2\mathcal{I}|\Sigma|Tp$ is the number of rows of R^ι_2. Moreover, in the above, the valid subtrace sets $\mathcal{X}^{\bar{k}-\underline{k}+1}_{\underline{k}}(\varphi^x_\ell)$ and $\mathcal{Y}^{\bar{k}-\underline{k}+1}_{\underline{k}}(\varphi^y_\ell)$ corresponding to the STL specification φ^x_ℓ and φ^y_ℓ are subdivided into the continuous predicate functions $\pi^x_\ell(x(k))$ and $\pi^y_\ell(y(k))$ according to (2) and binary predicates $P_{p^{\pi^x_\ell}}(k)$ and $P_{p^{\pi^y_\ell}}(k)$ satisfying $\Sigma^{\bar{k}-\underline{k}+1}_{\underline{k}}(\varphi^x_\ell)$ and $\Sigma^{\bar{k}-\underline{k}+1}_{\underline{k}}(\varphi^y_\ell)$ according to (3)–(9). Further, note that the functions $\pi^x_\ell(x(k))$ and $\pi^y_\ell(y(k))$ are linear functions of $\boldsymbol{x}_\ell(k_0)$, $d_\ell(k)$, $w_\ell(k)$ and $u_\ell(k)$ for all k. The satisfaction of the first two lines of (39k) for all realizations of these uncertainties can be recast as (mixed-integer) linear constraints using robust optimization or a KKT reformulation in Shen et al. (2022, Lemma 1).*

Proof We formulate the responsibility with regard to the polyhedral constraint (15) of the controlled input and controlled states in (36d) as a semi-infinite constraint that can be written as:

$$\overline{\overline{P}}^\ell_x \bar{x}^\ell \le \overline{\overline{P}}^\ell_u u_{\underline{k},\bar{k}} + \overline{\overline{p}}^\ell, \forall \bar{x}^\ell \in \left\{\bar{x}^\ell : \overline{P}^\ell_x \bar{x}^\ell \le \overline{P}^\ell_u u_{\underline{k},\bar{k}} + \bar{p}^\ell, \overline{\overline{H}}^\ell \bar{x}^\ell \le \overline{\overline{h}}^\ell\right\}.$$

From Shen et al. (2022, Lemma 1, Sect. 4), we can derive the equivalent mixed-integer constraints as:

$$0 = \sum_{m=1}^{\alpha} \nu^\ell_{1,a,m} \overline{P}^\ell_x(m,\dagger) + \sum_{n=1}^{\gamma} \nu^\ell_{2,a,n} \overline{\overline{H}}^\ell(n,\dagger) - \overline{\overline{P}}^\ell_x(a,\dagger),\ \forall \dagger = 1, \dots \psi, \tag{40a}$$

$$\pi_a = \overline{\overline{P}}^\ell_{x,a} \tilde{x}^\ell_a - \overline{\overline{P}}^\ell_{u,a} u_{\underline{k},\bar{k}} - \overline{\overline{p}}^\ell_a, \pi_a \le 0, \tag{40b}$$

$$\widetilde{P}^\ell_{x,m} \tilde{x}^\ell_a - \widetilde{P}^\ell_{u,m} u_{\underline{k},\bar{k}} - \bar{p}^\ell_m \le 0, \tag{40c}$$

$$\nu^\ell_{1,a,m} \ge 0, \forall m = 1, \dots \alpha, \tag{40d}$$

$$\widetilde{H}^\ell_n \tilde{x}^\ell_a - \tilde{h}^\ell_n \le 0, \nu^\ell_{2,a,n} \ge 0, \forall n = 1, \dots \gamma, \tag{40e}$$

$$\nu^\ell_{1,a,m}(\widetilde{P}^\ell_{x,m} \tilde{x}^\ell_a - \widetilde{P}^\ell_{u,m} u_{\underline{k},\bar{k}} - \bar{p}^\ell_m) = 0, \tag{40f}$$

$$\nu^\ell_{2,a,n}(\widetilde{H}^\ell_n \tilde{x}^\ell_a - \tilde{h}^\ell_n) = 0, \tag{40g}$$

where $\nu^\ell_{1,a,m}$, $\nu^\ell_{2,a,n}$ are dual variables from the KKT conditions, while $\widetilde{P}^\ell_{*,m}$ is the m-th row of $\overline{P}^\ell_*$, $* \in \{x, u\}$, and $\widetilde{H}^\ell_n$ is the n-th row of $\overline{\overline{H}}^\ell$. The complementary slackness constraints given in (40f) and (40g) are imposed using SOS-1 conditions in (39e) and (39f), respectively.

Furthermore, consider $a_{\ell,i^*}(k) = 1$ for some $k \in \mathbb{Z}^{\bar{k}}_{\underline{k}}$, where the additional constraint $\sum_{i=1}^{q_l} a_{\ell,i}(k) = 1$ ensures that $a_{\ell,i}(k) = 0$ for all $i \neq i^*$. Then, it follows that $s_{\ell,i}(k)$ is unconstrained for all $i \neq i^*$ due to the SOS-1 constraints. Since this holds for any $k \in \mathbb{Z}^{\bar{k}}_{\underline{k}}$, then for each model pair, the set of feasible states for both models lies within the same subregion.

In addition, the constraints in (39k) can be obtained from the valid subtrace sets $\mathcal{X}_{\underline{k}}^{\overline{k}-\underline{k}+1}(\varphi_\ell^x)$ and $\mathcal{Y}_{\underline{k}}^{\overline{k}-\underline{k}+1}(\varphi_\ell^y)$ corresponding to the STL specification φ_ℓ^x and φ_ℓ^y according to (2)–(9). Specifically, by moving the constraints related to $\pi_\ell^y(y(k))$ to the outer problem, we also need to enforce the satisfaction of this constraint for all realizations of uncertainties, similar to the robust satisfaction of the constraints corresponding to $\pi_\ell^y(y(k))$, which can be reformulated as (mixed-integer) linear constraints using robust optimization or a KKT reformulation in Shen et al. (2022, Lemma 1).

To obtain the remaining parts of the MILP formulation, we refer the reader to similar steps in Ding et al. (2018, Sect. IV). □

5.1.3 Brute Force Approach

In this subsection, we introduce a naive but optimal solution strategy to address Problem 1, which is challenging to solve due to the presence of binary variables in the lower level of the bilevel optimization framework. The core difficulty in tackling Problem 1 stems from the binary (integral) variables, which hinder the formulation of a dual problem using KKT conditions, compounded by the fact that the original problem is NP-hard. To overcome this, we propose a brute force approach (cf. Algorithm 1) that involves enumerating all possible combinations of the binary variables within the lower-level problems. This method essentially exhaustively explores all combinations of binary variables and subsequently, exploits the KKT conditions for each binary combination, transforming the problem into a set of (parallel) Linear Programming (LP) problems. Then, using the solutions from all these combinations and some additional rules for eliminating certain combinations that do not correspond to the worst-case/robust solution for active model discrimination, Algorithm 1 offers a pathway that converts the original NP-hard problem into a tractable set of LPs to find the global optimal solution. Note however that the exponential number of subproblems retains the underlying complexity/NP-hardness of the problem; thus, this approach may be limited to smaller problems only.

Specifically, by considering each possible subregion selection denoted as $a_{\ell,i}^\star(k)$, the component of (26) corresponding to the subregion selection can be modified as

$$\forall \ell \in \mathbb{Z}_+^N, \forall k \in \mathbb{Z}_{\underline{k}}^{\overline{k}} \forall i \in \mathbb{Z}_+^{q_\ell} : \; a_{\ell,i}(k) = a_{\ell,i}^\star(k), \tag{41a}$$

$$\textstyle\sum_{i=1}^{q_\ell} a_{\ell,i}(k) = 1, \;\; \text{SOS-1:}\{a_{\ell,i}(k), s_{\ell,i}(k)\}, \tag{41b}$$

where $\star \in [1, 2, \ldots, {q_\ell}^{N\times(\overline{k}-\underline{k})}]$, and ${q_\ell}^{N(\overline{k}-\underline{k})}$ is the total number of possible combination of subregions for each time step and model. Moreover, we define the subregion enumeration set as $\Xi = \{\Xi^1, \Xi^2, \ldots, \Xi^{{q_\ell}^{N(\overline{k}-\underline{k})}}\}$, where $\Xi^\star = \{\forall \ell \in \mathbb{Z}_+^N, \forall k \in \mathbb{Z}_0^{\overline{T}}, \forall i \in \mathbb{Z}_+^{q_\ell}, : a_{\ell,i}^\star(k)\}$.

Similarly, by considering each possible combination of truth values of the predicate of the STL specification on the controlled and uncontrolled states denoted as $\{P^{\oplus}_{p^{\pi^x_\ell}}(k)\}_{k=\underline{k}}^{\overline{k}} \in \Sigma_{\underline{k}}^{\overline{k}-\underline{k}+1}(\varphi^x_\ell)$ and $\{P^{\ominus}_{p^{\pi^y_\ell}}(k)\}_{k=\underline{k}}^{\overline{k}} \in \Sigma_{\underline{k}}^{\overline{k}-\underline{k}+1}(\varphi^y_\ell)$, respectively, by setting

$$\{P_{p^{\pi^x_\ell}}(k)\}_{k=\underline{k}}^{\overline{k}} = \{P^{\oplus}_{p^{\pi^x_\ell}}(k)\}_{k=\underline{k}}^{\overline{k}}, \tag{42a}$$

$$\{P_{p^{\pi^y_\ell}}(k)\}_{k=\underline{k}}^{\overline{k}} = \{P^{\ominus}_{p^{\pi^y_\ell}}(k)\}_{k=\underline{k}}^{\overline{k}}, \tag{42b}$$

where $\oplus \in [1, 2, \ldots, N_x]$, $\ominus \in [1, 2, \ldots, N_y]$ and N_x, N_y are the total numbers of possible combinations of $\Sigma_{\underline{k}}^{\overline{k}-\underline{k}+1}(\varphi^x_\ell)$, $\Sigma_{\underline{k}}^{\overline{k}-\underline{k}+1}(\varphi^y_\ell)$, respectively. We also define the corresponding enumeration set as $\{P^{\oplus}_{p^{\pi^x_\ell}}(k)\}_{k=\underline{k}}^{\overline{k}}$, $\{P^{\ominus}_{p^{\pi^y_\ell}}(k)\}_{k=\underline{k}}^{\overline{k}}$.

Further, we also consider each combination of truth values of the MTL specification on the switching mode $\{\sigma^{\triangleright}(k)\}_{k=\underline{k}}^{\overline{k}} \in \Sigma_{\underline{k}}^{\overline{k}-\underline{k}+1}(\psi)$ by letting

$$\{\sigma(k)\}_{k=\underline{k}}^{\overline{k}} = \{\sigma^{\triangleright}(k)\}_{k=\underline{k}}^{\overline{k}}, \tag{43}$$

where $\triangleright \in [1, 2, \ldots, N_\sigma]$, and N_σ is the total number of possible combinations of $\Sigma_{\underline{k}}^{\overline{k}-\underline{k}+1}(\psi)$. We again define the corresponding enumeration set as $\{\sigma^{\triangleright}(k)\}_{k=\underline{k}}^{\overline{k}}$.

Using the above definitions, the first step of the brute force approach starts from the problem reformulation for each combination as follows:

Problem 4 (*Active Model Discrimination for $\{\mathcal{H}_\ell\}_{\ell=1}^N$ with Brute Forced Combination*) Given the subregion enumeration set Ξ, STL valid subtrace set $\Sigma_{\underline{k}}^{\overline{k}-\underline{k}+1}(\varphi^x_\ell)$ and $\Sigma_{\underline{k}}^{\overline{k}-\underline{k}+1}(\varphi^y_\ell)$, MTL valid subtrace set $\Sigma_{\underline{k}}^{\overline{k}-\underline{k}+1}(\psi)$, N switched piecewise affine inclusion models $\{\mathcal{H}_\ell\}_{\ell=1}^N$, and state, input and noise constraints, i.e., (11)–(21), find an optimal input sequence $u^{*,\diamond}_{\underline{k},\overline{k}}$ to minimize a given cost function $J(u_{\underline{k},\overline{k}})$ such that only one model is valid, i.e., the output trajectories of any pair of models have to differ by a threshold ϵ in at least one time step. For each combination from the enumeration sets $\Xi^\star$, $\{P^{\oplus}_{p^{\pi^x_\ell}}(k)\}_{k=\underline{k}}^{\overline{k}}$, $\{P^{\ominus}_{p^{\pi^y_\ell}}(k)\}_{k=\underline{k}}^{\overline{k}}$ and $\{\sigma^{\triangleright}(k)\}_{k=\underline{k}}^{\overline{k}}$, the corresponding AMD subproblem can be stated as:

$$u^{*,\diamond}_{\underline{k},\overline{k}} = \arg\min_{u_{\underline{k},\overline{k}}} J(u_{\underline{k},\overline{k}})$$

$$\text{s.t.} \quad \forall k \in \mathbb{Z}_{\underline{k}}^{\overline{k}} : (18)\ \textit{holds}, \tag{44a}$$

$$\left.\begin{array}{r} \forall \ell \in \mathbb{Z}_+^N, \forall k \in \mathbb{Z}_{\underline{k}}^{\overline{k}}, \forall \boldsymbol{x}_\ell(0), d^\ell_{\underline{k},\overline{k}}, w^\ell_{\underline{k},\overline{k}} : \\ (11), (13), (14), (16), (20), (41), (42b), (43)\ \textit{hold} \end{array}\right\} : \begin{array}{l} (15)\ \textit{and} \\ (42a)\ \textit{hold}, \end{array} \tag{44b}$$

$$\left.\begin{array}{r}\forall \ell \in \{\dagger, \ddagger\}, \forall \dagger, \ddagger \in \mathbb{Z}_{+}^{\mathrm{N}}, \dagger \neq \ddagger, \\ \forall k \in \mathbb{Z}_{k_0}^{k_0+T-1}, \forall \boldsymbol{x}_{\ell}(0), d_{k_0,k_0+T}^{\ell}, w_{k_0,k_0+T}^{\ell}, v_{k_0,k_0+T}^{\ell}: \\ (11)\text{–}(14), (16), (20), (21), (41), (42b), (43), \ hold\end{array}\right\} : \begin{array}{l}\{\exists k' \in \mathbb{Z}_{k_0}^{k_0+T-1}: \\ \|z_{\dagger}(k') - z_{\ddagger}(k')\|_{\infty} \geq \epsilon\},\end{array} \tag{44c}$$

where $J(\cdot)$ is a convex objective function of the inputs $u_{\underline{k},\overline{k}}$.

Note the new reformulated problem is similar to Problem 1 with some modifications. In particular, (26) in Problem 1 is substituted by (41) and STL constraints and MTL constraints are substituted by (42) and (43), respectively. The values of the binary variables in (41a) are set to correspond to the set $\Xi^{\star}$, which is one possible subregion configuration for the entire time horizon. The STL and MTL are also set to correspond to $\{P^{\oplus}_{p^{\pi_{\ell}^{x}}}(k)\}_{k=\underline{k}}^{\overline{k}}$, $\{P^{\ominus}_{p^{\pi_{\ell}^{y}}}(k)\}_{k=\underline{k}}^{\overline{k}}$, and $\{\sigma^{\rhd}(k)\}_{k=\underline{k}}^{\overline{k}}$. Then, the KKT condition can be used in (44b), (44c) following the same procedure as in Theorem 2.

Next, using the subproblem in Problem 4 as line 3 within Algorithm 1, the algorithm selects the global optimal solution among the solutions of all these combinations. Specifically, within the for loop in lines 2–14, Problem 4 is solved ${q_{\ell}}^{N(\overline{k}-\underline{k})} N_x N_y N_{\sigma}$ times for each fixed combination of binary variables. If the problem is infeasible (line 4), line 5 checks whether the reason for the infeasibility is because of the outer problems (the right-hand sides of (44b) and (44c)). If the check is infeasible (line 6), then this combination of binary variables can be chosen in the inner problem such that the outer problem can never be satisfied, and thus, in the worst-case setting, this AMD problem has no solution (line 7); otherwise, we do not save this combination but continue with the next combination, indicating that the AMD problem may still have a solution but that this specific combination of binary variables is infeasible. On the other hand, if Problem 4 returns a feasible solution, we save the optimal cost $J(u)$ (line 10) within a list and its corresponding binary variable combination and input. After obtaining the results of all combinations, the costs are sorted from low to high as a list $B = \{J(u_1), \ldots, J(u_{|B|})\}$ (line 13), where $|B|$ is the cardinality/size of B. Lines 14–16 indicate that if none of the combinations return a feasible solution, then once again, we know that the AMD problem has no feasible solution.

Next, lines 17–26 provide the strategy of choosing the global optimal input from among all the feasible solutions. Specifically, starting from the lowest cost $J(u_1)$ among all costs of the combinations, we solve and compare the costs ρ_{α} and $\delta_{\alpha}^{(i,j)}$ of the inner problems (32) and (34) for the fixed combination and all other combinations. If the costs of the inner problems for the specific combination under consideration are always better than costs of all other combinations, then, it means that the inner problems do not have any "incentive" to switch to a different combination for the worst-case setting; thus, the input for this combination is globally optimal (lines 21–23); otherwise, we move on to the next combination and the associated input in the list (lines 19–24). Finally, once the for loop is broken by line 22 or if we reached the end

Algorithm 1: Brute Force Active Model Discrimination

Data: Models $\{\mathcal{H}_\ell\}_{\ell=1}^N$, Enumeration Set Ξ, $\Sigma_{\underline{k}}^{\overline{k}-\underline{k}+1}(\varphi_\ell^x)$, $\Sigma_{\underline{k}}^{\overline{k}-\underline{k}+1}(\varphi_\ell^y)$, and $\Sigma_{\underline{k}}^{\overline{k}-\underline{k}+1}(\psi)$

1 **Function** BruteForce($\{\mathcal{H}_\ell\}_{\ell=1}^N, \Xi, \Sigma_{\underline{k}}^{\overline{k}-\underline{k}+1}(\varphi_\ell^x), \Sigma_{\underline{k}}^{\overline{k}-\underline{k}+1}(\varphi_\ell^y), \Sigma_{\underline{k}}^{\overline{k}-\underline{k}+1}(\psi)$)
  /* For each (fixed) combination of binary variables */
2   **for** $i = 1 : {q_\ell}^{N(\overline{k}-\underline{k})} N_x N_y N_\sigma$ **do**
3   Solve Problem 4;
4   **if** Problem 4 is infeasible **then**
5   Check feasibility of the left-hand side constraints in (29) subject to robust ("for all") constraints of the right-hand side constraints in (29);
6   **if** Line 5 is infeasible **then**
7   **return** *Infeasible*; // This AMD problem has no solution
8   **end**
9   **else**
10   Save cost $J(u)$ from the solution of (29);
11   **end**
12 **end**
13 Sort cost $J(u)$ from low to high in a list $B = \{J(u_1), \ldots, J(u_{|B|})\}$; // $|B|$ is the cardinality/size of B
14 **if** $|B| = 0$ **then**
15   **return** *Infeasible*; // This AMD problem has no solution
16 **end**
  /* For each combination of binary variables and the associated inputs in the list */
17 **for** $\alpha = 1 : |B|$ **do**
18   Solve the inner problems in (32) and (34) with solutions ρ_α and $\delta_\alpha^{(i,j)}$;
    /* For each other combination of binary variables and the associated inputs in the list */
19   **for** $\beta = 1 : |B|$ **do**
20   Solve the inner problems in (32) and (34) with solutions ρ_β and $\delta_\beta^{(i,j)}$;
21   **if** *($\rho_\beta \le \rho_\alpha$ or $\delta_\beta^{(i,j)} \le \delta_\alpha^{(i,j)}$ for any (i, j)) and $J(u_\beta) > J(u_\alpha)$* **then**
22   Break;
23   **end**
24   **end**
25   **return** u_α;
26 **end**

of the list, we return the corresponding input as the optimal separating input in line 25. This algorithm provides an optimal solution similar to the parametric approach, rather than a sub-optimal solution obtained by the binary variable relocation approach from Theorem 2.

5.2 Model Invalidation

Using the computed optimal input and the observed measurements (corresponding to the unknown true model and valid sequence of subregions), the goal of model invalidation (MI) is to eliminate all models that are incompatible with the observed input-mode-output trajectory at run time. We propose to achieve this goal (i.e., to solve Problem 2) by using an MI approach to *invalidate/eliminate* each model that is inconsistent with the observed input-mode-output trajectory, as described next.

Theorem 3 *Given a switched piecewise affine inclusion model $\mathcal{H}_\ell$ and a length-T input-mode-output sequence $\{u(k), \sigma(k), z(k)\}_{k=k_0-T+1}^{k_0}$ at time k_0, the model is invalidated if the following problem is infeasible for any $k_0 \in \mathbb{Z}_0^\infty$:*

$$
\begin{aligned}
&\textit{Find}\ \ \boldsymbol{x}_\ell(k), d_\ell(k), w_\ell(k), v_\ell(k), a_{\ell,i}(k), s_{\ell,i}(k)\\
&s.t.\ \ \forall k \in \mathbb{Z}_{k_0}^{k_0+T-1}, i \in \mathbb{Z}_+^{q_l}:\\
&x_\ell(k+1) \le \overline{A}_{\ell,i}^{\sigma(k)} \boldsymbol{x}_\ell(k) + \overline{B}_{\ell,i}^{u,\sigma(k)} u(k) + \overline{B}_{\ell,i}^{d,\sigma(k)} d_\ell(k)\\
&\qquad\qquad + \overline{B}_{\ell,i}^{w,\sigma(k)} w_\ell(k) + \overline{f}_{\ell,i}^{\sigma(k)} + s_{\ell,i}(k)\mathbb{1}, &&\text{(45a)}\\
&x_\ell(k+1) \ge \underline{A}_{\ell,i}^{\sigma(k)} \boldsymbol{x}_\ell(k) + \underline{B}_{\ell,i}^{u,\sigma(k)} u(k) + \underline{B}_{\ell,i}^{d,\sigma(k)} d_\ell(k)\\
&\qquad\qquad + \underline{B}_{\ell,i}^{w,\sigma(k)} w_\ell(k) + \underline{f}_{\ell,i}^{\sigma(k)} - s_{\ell,i}(k)\mathbb{1}, &&\text{(45b)}\\
&S_{\ell,i}^{x,\sigma(k)} \boldsymbol{x}_\ell(k) + S_{\ell,i}^{u,\sigma(k)} u(k) + S_{\ell,i}^{d,\sigma(k)} d_\ell(k)\\
&\qquad + S_{\ell,i}^{w,\sigma(k)} w_\ell(k) + \beta_{\ell,i}^{\sigma(k)} \le s_{\ell,i}(k)\mathbb{1}, &&\text{(45c)}\\
&z(k) = C_\ell \boldsymbol{x}_\ell(k) + D_\ell^u u(k) + D_\ell^d d_\ell(k) + V_\ell v_\ell(k) + g_\ell, &&\text{(45d)}\\
&a_{\ell,i}(k) = a_{\ell,i}^\star(k),\ \textit{SOS-1:}(a_{\ell,i}(k), s_{\ell,i}(k)), &&\text{(45e)}\\
&x_\ell(k) \models \varphi^x\ \textit{and}\ y_\ell(k) \models \varphi^y, &&\text{(45f)}\\
&d_\ell(k) \in \mathcal{D}, w_\ell(k) \in \mathcal{W},\ v_\ell(k) \in \mathcal{V},\ \boldsymbol{x}_\ell(k) \in \mathcal{X}. &&\text{(45g)}
\end{aligned}
$$

Proof If the above optimization problem is infeasible, it means that the input-mode-output sequence $\{u(k), \sigma(k), z(k)\}_{k=k_0-T+1}^{k_0}$ cannot be consistent with the model $\mathcal{G}_\ell$, hence the model is invalidated. □

The above MILP formulation is applicable for the model discrimination problem when using the optimal separating input from either of the three AMD approaches in the previous section. However, if the separating input was obtained using the binary variable relocation approach, we can further simplify the MI problem into a linear program (LP), which is much easier to solve, especially at run time. This is possible because in the binary variable relocation approach, we robustly enforce the satisfaction of the subregions and the STL specifications and thus, the solution of the relocation-based AMD problem from Theorem 2 also provides knowledge of the subregions and truth values of the STL specifications. In this case, the (simpler) LP formulation for the MI problem is then given by:

Theorem 4 *Given a switched piecewise affine inclusion model $\mathcal{H}_\ell$ and a length-T input-mode-output sequence $\{u(k), \sigma(k), z(k)\}_{k=k_0-T+1}^{k_0}$ as well as the information of the subregions and STL truth values $\{a(k), P_{p^{\pi^x}}(k), P_{p^{\pi^y}}(k)\}_{k=k_0-T+1}^{k_0}$ at time k_0, define $p_\ell(k)$ such that $\forall k \in \mathbb{Z}_{k_0}^{k_0+T-1}, \forall i \in \mathbb{Z}_+^{q_\ell} : a_i(k) = 1 \Rightarrow p_\ell(k) = i$. Then, the model is invalidated if the following problem is infeasible for any $k_0 \in \mathbb{Z}_0^\infty$:*

$$\begin{aligned}
&\textit{Find}\ \ \boldsymbol{x}_\ell(k), d_\ell(k), w_\ell(k), v_\ell(k)\\
&s.t.\ \ \forall k \in \mathbb{Z}_{k_0}^{k_0+T-1}, i \in \mathbb{Z}_+^{q_\ell}:\\
&x_\ell(k+1) \leq \overline{A}_{\ell,p_\ell(k)}^{\sigma(k)} \boldsymbol{x}_\ell(k) + \overline{B}_{\ell,p_\ell(k)}^{u,\sigma(k)} u(k) + \overline{B}_{\ell,p_\ell(k)}^{d,\sigma(k)} d_\ell(k)\\
&\qquad + \overline{B}_{\ell,p_\ell(k)}^{w,\sigma(k)} w_\ell(k) + \overline{f}_{\ell,p_\ell(k)}^{\sigma(k)}, && \text{(46a)}\\
&x_\ell(k+1) \geq \underline{A}_{\ell,p_\ell(k)}^{\sigma(k)} \boldsymbol{x}_\ell(k) + \underline{B}_{\ell,p_\ell(k)}^{u,\sigma(k)} u(k) + \underline{B}_{\ell,p_\ell(k)}^{d,\sigma(k)} d_\ell(k)\\
&\qquad + \underline{B}_{\ell,p_\ell(k)}^{w,\sigma(k)} w_\ell(k) + \underline{f}_{\ell,p_\ell(k)}^{\sigma(k)}, && \text{(46b)}\\
&S_{\ell,p_\ell(k)}^{x,\sigma(k)} \boldsymbol{x}_\ell(k) + S_{\ell,p_\ell(k)}^{u,\sigma(k)} u(k) + S_{\ell,p_\ell(k)}^{d,\sigma(k)} d_\ell(k)\\
&\qquad + S_{\ell,p_\ell(k)}^{w,\sigma(k)} w_\ell(k) + \beta_{\ell,p_\ell(k)} \leq 0, && \text{(46c)}\\
&z(k) = C_\ell \boldsymbol{x}_\ell(k) + D_\ell^u u(k) + D_\ell^d d_\ell(k) + V_\ell v_\ell(k) + g_\ell, && \text{(46d)}\\
&\pi(x(k)) + P_{p^{\pi^x}}(k) M \geq 0, \quad \pi(x(k)) + (P_{p^{\pi^x}}(k) - 1) M < 0, && \text{(46e)}\\
&\pi(y(k)) + P_{p^{\pi^y}}(k) M \geq 0, \quad \pi(y(k)) + (P_{p^{\pi^y}}(k) - 1) M < 0, && \text{(46f)}\\
&d_\ell(k) \in \mathcal{D}, w_\ell(k) \in \mathcal{W},\ v_\ell(k) \in \mathcal{V},\ \boldsymbol{x}_\ell(k) \in \mathcal{X}, && \text{(46g)}
\end{aligned}$$

where M is any large enough constant (big M).

Proof This simplification from Theorem 3 can be immediately obtained by substituting the known binary variables into (45) and simplifying. □

Equipped with the above theorems, we propose Algorithm 2 below to eliminate or rule out all invalid models, i.e., to solve Problem 2. Since the output trajectories are guaranteed to be distinct by the application of the separating input from either of the three AMD approaches, this MI algorithm is guaranteed to return only the true model within T time steps, i.e., all other models will be ruled out/eliminated before the end of T time steps.

Algorithm 2: Length-T Model Invalidation

Data: Models $\{\mathcal{H}_\ell\}_{\ell=1}^{N}$, Input-Output Trajectory $\{u(k), \sigma(k), z(k)\}_{k=k_0-T+1}^{k_0}$,
or $\{u(k), \sigma(k), z(k), a(k), P_{p^{\pi^x}}(k), P_{p^{\pi^y}}(k)\}_{k=k_0-T+1}^{k_0}$

function findModel($\{\mathcal{H}_\ell\}_{\ell=1}^{N}$, $\{u(k), a(k), \sigma(k), z(k)\}_{k=k_0-T+1}^{k_0}$)
  valid $\leftarrow \{\mathcal{H}_\ell\}_{\ell=1}^{N}$;
  for $\ell = 1 : N$ **do**
    Check Feasibility of Theorem 3 or 4;
    if *infeasible* **then**
      Remove ℓ from valid;
    end
  end
  return valid

6 Simulation Examples

In this section, we apply the proposed approaches to a lunar/planetary assisted drive system and a vehicle lane changing scenario. Both examples are implemented in MATLAB 2022b with Gurobi v9.5 on an Intel Core i7-7Y75 CPU @ 1.3 GHz dual-core machine with 16 GB RAM.

6.1 *Lunar/Planetary Assisted Drive*

First, we consider a parameter identification problem for a lunar/planetary rover navigating an environment with uncertain terrain parameters.

6.1.1 Lunar Assisted Drive Model

The rover dynamics evolves according to a simple viscoelastic model with the damping coefficient being the uncertain parameter, as follows:

$$
\begin{aligned}
x(k+1) &= x(k) + \mathrm{v}_x(k)\delta t, \\
\mathrm{v}_x(k+1) &= -\frac{k}{M}x(k) + \left(1 - \frac{c_\ell}{M}\delta t\right)\mathrm{v}_x(k) + \frac{\delta t}{M}u(k) + w(k)\delta t,
\end{aligned}
$$

where x and v_x are the position and velocity of the rover in m and $\frac{m}{s}$ respectively, u is the acceleration input in $\frac{m}{s^2}$, w is the process noise signal in $\frac{m}{s^2}$, and M, k_ℓ, and c_ℓ are the mass, spring constant, and damping coefficient in kg, $\frac{N}{m}$, and $\frac{Ns}{m}$ respectively. δt is the sampling time in s. For this example, we have:

$$u(k) \in [-20, 20]\tfrac{m}{s^2},\ \delta t = 0.5s,\ k_\ell = 0.002\frac{N}{m},$$
$$w(k) \in [-0.001, 0.001]\tfrac{m}{s^2},\ M = 5\,kg.$$

We consider 2 uncertain models defined such that each model describes the robot dynamics given $c_\ell \in C_\ell$, where $C_\ell = [c_{l,\ell}, c_{u,\ell}]$. For each model $\ell \in \{1, 2\}$, we compute an abstraction, evaluated over two subregions as in Jin et al. (2020), that describes the rover's dynamics with $C_1 = [0.05, 0.2]$, $C_2 = [0.6, 0.8]$ as well as

$$\underline{A}_{\ell,2} = \overline{A}_{\ell,1} = \begin{bmatrix} 1 & \delta t \\ -\frac{k_\ell}{m} & (1 - \frac{c_{u,\ell}}{m}\delta t) \end{bmatrix},\ \underline{A}_{\ell,1} = \overline{A}_{\ell,2} = \begin{bmatrix} 1 & \delta t \\ -\frac{k_\ell}{m} & (1 - \frac{c_{l,\ell}}{m}\delta t) \end{bmatrix},$$
$$B_\ell^u = B_\ell^w = \begin{bmatrix} 0 \\ \frac{\delta t}{m} \end{bmatrix},\ S_{\ell,1}^x = S_{\ell,2}^x = \begin{bmatrix} 1 & 0 \\ -1 & 0 \\ 0 & 1 \\ 0 & -1 \end{bmatrix},\ \beta_{\ell,1} = \begin{bmatrix} 180 \\ 90 \\ 180 \\ 0 \end{bmatrix},\ \beta_{\ell,2} = \begin{bmatrix} 180 \\ 0 \\ 180 \\ 90 \end{bmatrix},$$

The system output is given by: $z_\ell(k) = \begin{bmatrix} x(k) \\ v_x(k) \end{bmatrix} + \begin{bmatrix} v_1(k) \\ v_2(k) \end{bmatrix}$ with measurement noise bounds $v_i(k) \in [-0.01, 0.01]$, for all $i \in \{1, 2\}$. The system is initialized with $x(0) = 0$ and $v_x(0) \in [0.5,\ 0.6]$. Additionally, an STL specification φ^x is imposed such that:

$$\varphi^x = \Diamond_{[0,5]}(10 \le x \le 20).$$

6.1.2 Results and Discussion

We first solve the AMD problem with $J(\cdot) = \|\cdot\|_\infty$, $\epsilon = 0.2$, for the binary variable relocation method considering the responsibility and separation condition, which yields the result $\{u(k)\}_{k=0}^4 = \{20,\ 20,\ 20,\ 14.0864,\ 0\}$. The result from executing this input is shown in Fig. 2, which shows the output reachable sets (shown with the larger pale outlines) of each model. Here, we see that the sets (yellow for model 1 and red for model 2) achieve the desired separation. We also observe that the STL specification (of eventually a goal region of $10 \le x \le 20$) is achieved at $k = 5$.

Moreover, a sample run-time trajectory obtained by random realizations of noise and other uncertainties (depicted in Fig. 3a) is checked against the two models using the model invalidation algorithm. Figure 3b (right) shows that the MI algorithm is able to determine the true model at $k = 1$, which is earlier than the guaranteed detection horizon of 5 time steps.

Furthermore, because of the extended duration required by the parametric approach for resolution, this example confines its comparison to the binary variable relocation method and the brute force method, taking into account the separation condition. While the binary variable relocation method does not assure optimality, it yields an identical outcome as the brute force method under the same parameters,

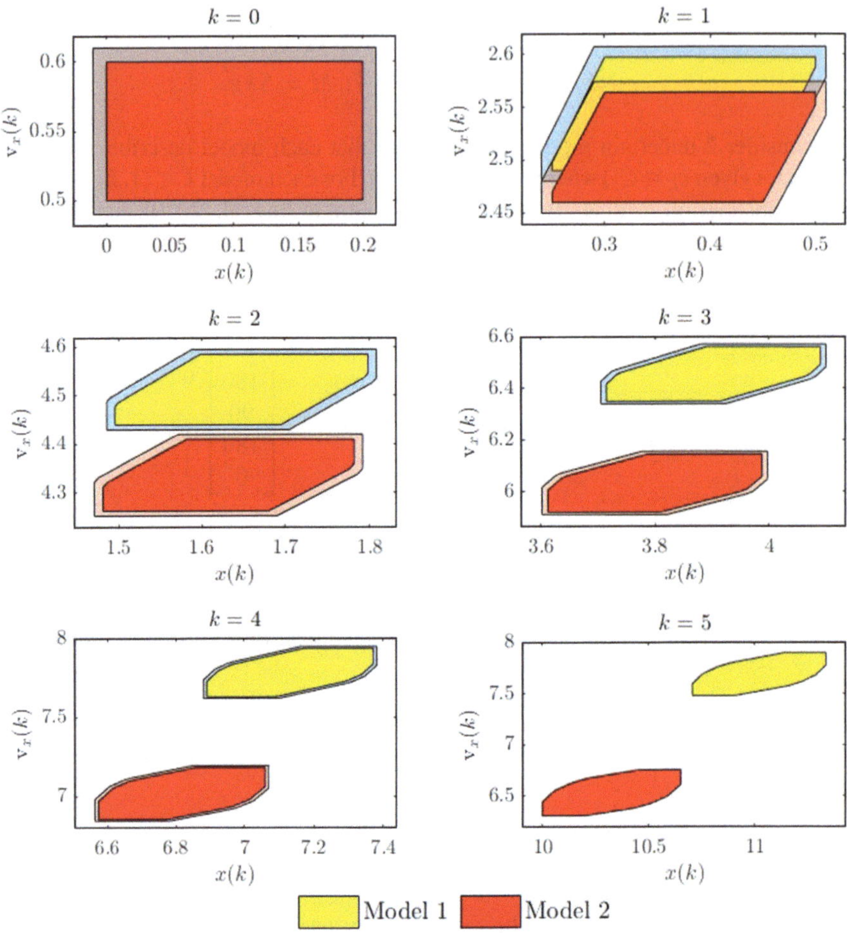

Fig. 2 Output reachable sets given the designed inputs and STL satisfaction for each model. For $k = 0$ to $k = 4$, the state sets are the solid small sets (yellow or red) drawn over their respective output reachable sets (shown in pale blue or red). For $k = 5$, only the state sets are plotted

specifically $u(k)_{k=0}^{4} = \{-4.899,\ -0.0005,\ 0,\ 4.8990,\ 4.8990\}$, in the presented example. The deployment of the parametric approach is elucidated in the subsequent simulation example.

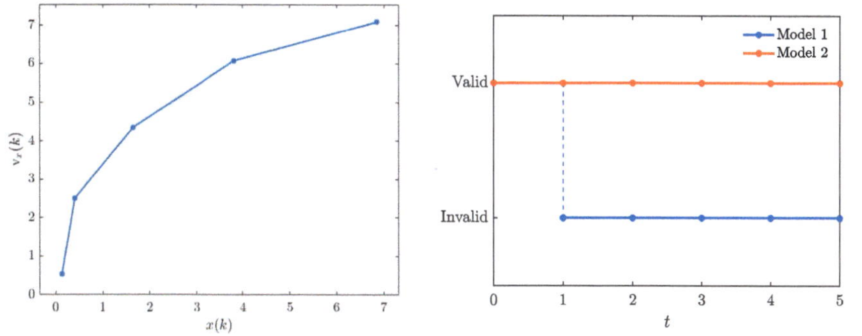

(a) Sample trajectory generated from random realizations of noise for testing the model invalidation algorithm.

(b) Result of the model invalidation algorithm with accurate determination made after $t = 1$.

Fig. 3 Sample implementation of AMD algorithm and results of the model invalidation algorithm

6.2 *Vehicle Lane Changing Example*

Next, we consider an intent estimation problem in a vehicle lane changing scenario with two possible intent models for the other car. We apply the active model discrimination approaches to actively identify the intentions of other vehicles.

6.2.1 Vehicle and Intent Model

The dynamics of the ego vehicle is of the Dubins car (Dubins 1957) with acceleration input as follows:

$$\dot{x}_e = v_e \cos\phi_e, \quad \dot{y}_e = v_e \sin\phi_e, \quad \dot{v}_e = u_1 + w_1, \quad \dot{\phi}_e = u_2 + w_2,$$

while the other car dynamics is described by:

$$\dot{x}_o = v_o, \quad \dot{v}_o = d_\ell + w_3,$$

where x_e and y_e are the longitudinal and lateral coordinates of the ego vehicle, v_e is its speed and ϕ_e is its heading angle, while x_o is the longitudinal coordinate of the other car with its speed given by v_o (no lateral movement), with process noise signals $w_j \in [-0.01, 0.01]$, $j \in \{1, 2, 3\}$. The control inputs are $u_1 \in \mathcal{U}_a \triangleq [-7.848, 3.968]$ and $u_2 \in \mathcal{U}_s \triangleq [-0.44, 0.44]$, representing the acceleration and steering inputs of the ego car, while d_ℓ is the (uncontrolled) acceleration input of the other car for each intention $\ell \in \{C, M\}$ (the *Cautious* driver tends to yield the lane to the ego car, while the *Malicious* driver does not want to yield the lane to the ego car). The output measurement is considered as $z_m = v_o + v$, with measurement noise $v \in [-0.01, 0.01]$.

To handle the nonlinear system dynamics, we utilize the affine abstraction approach in Jin et al. (2020) to first over-approximate the models as approximated by piecewise affine inclusion models with two subregions given by:

$$\begin{aligned}
&A_1 = [(1,3) : 0.9048; (2,4) : 21.781]_{6\times 6},\\
&A_2 = [(1,3) : 0.9048; (2,4) : 26.6212]_{6\times 6},\\
&B_1^u = B_2^u = B_1^w = B_2^w = B = [(3,1) : 1; (4,2) : 1; (6,3)1]_{6\times 3},\\
&F_1 = [\underline{f}_1, \overline{f}_1] = \begin{bmatrix}(1,1) : [-3.4680, 5.8193];\\ (1,2) : [-9.2068, 9.2068]\end{bmatrix}_{6\times 1},\\
&F_2 = [\underline{f}_2, \overline{f}_2] = \begin{bmatrix}(1,1) : [-4.1616, 6.9832];\\ (1,2) : [-10.8352, 10.8352]\end{bmatrix}_{6\times 1},
\end{aligned}$$

where we used a sparse matrix notation to conserve space, with the size indicated in the subscript. Note that the subscripts $i \in \{1, 2\}$ in the above matrices and vectors denote the indices for the subregions and when using the affine abstraction approach in Jin et al. (2020), the obtained results are such that $\overline{A}_i = \underline{A}_i = A_i$, $\overline{B}_i^u = \underline{B}_i^u = B_i^u$ and $\overline{B}_i^w = \underline{B}_i^w = B_i^w$ for all $i \in \{1, 2\}$.

Combining the abstraction with the intention models and using Euler method for time discretization with sampling time $\delta t = 0.4s$, we have the following models with subregion indices $i \in \{1, 2\}$:

Cautious Driver $(\ell = C)$:

$$\begin{aligned}
&\tilde{A}_C = [(6,3) : -K_{d,C}; (6,4) : L_{p,C}; (6,6) : K_{d,C}]_{6\times 6},\\
&\tilde{B}_C = [(6,2) : L_{d,C}]_{6\times 3}, \underline{A}_{C,i} = \overline{A}_{C,i} = \mathbb{I} + \delta t(A_i + \tilde{A}_C),\\
&\underline{B}^u_{C,i} = \overline{B}^u_{C,i} = \underline{B}^w_{C,i} = \overline{B}^w_{C,i} = \mathbb{I} + \delta t(B + \tilde{B}_C),\\
&\underline{C}_C = \overline{C}_C = [(1,6) : 1]_{1\times 6}, \underline{D}_C = \overline{D}_C = 0, \underline{V}_C = \overline{V}_C = 1,\\
&\underline{f}_{C,i} = \delta t \underline{f}_i, \overline{f}_{C,i} = \delta t \overline{f}_i, S^x_{C,i} = [(1,3) : 1; (2,3) : -1]_{2\times 6},\\
&\beta_{C,1} = [25; -20], \beta_{C,2} = [30; -25];
\end{aligned}$$

Malicious Driver $(\ell = M)$:

$$\begin{aligned}
&\tilde{A}_M = [(6,3) : -K_{d,M}; (6,4) : L_{p,M}; (6,6) : K_{d,M}]_{6\times 6},\\
&\tilde{B}_M = [(6,2) : L_{d,M}]_{6\times 3}, \underline{A}_{M,i} = \overline{A}_{M,i} = \mathbb{I} + \delta t(A_i - \tilde{A}_M),\\
&\underline{B}^u_{M,i} = \overline{B}^u_{M,i} = \underline{B}^w_{M,i} = \overline{B}^w_{M,i} = \mathbb{I} + \delta t(B - \tilde{B}_M),\\
&\underline{C}_M = \overline{C}_M = [(1,6) : 1]_{1\times 6}, \underline{D}_M = \overline{D}_M = 0, \underline{V}_M = \overline{V}_M = 1,\\
&\underline{f}_{M,i} = \delta t \underline{f}_i, \overline{f}_{M,i} = \delta t \overline{f}_i, S^x_{M,i} = [(1,3) : 1; (2,3) : -1]_{2\times 6},\\
&\beta_{M,1} = [25; -20], \beta_{M,2} = [30; -25],
\end{aligned}$$

Table 1 Comparison of computation times in the driver intent estimation example

Methods	Total time (s)	$\|u\|_\infty$	$\|u\|_1$
Parametric approach	Over a week	0.1541	0.2360
Binary variable relocation	7.2	0.1541	0.2360
Brute force approach	1982.6	0.1541	0.2360

where $K_{d,C} = 1$, $K_{d,M} = 0.9$, $L_{p,C} = L_{p,M} = 12$, $L_{d,C} = L_{d,M} = 14$. Further, the initial conditions and the state constraints are constrained, as follows:

$$v_e(0) \in [26, 28]\,\frac{m}{s},\ v_o(0) \in [26, 28]\,\frac{m}{s},\ y_e(0) \in [12, 14]\,m,\ x_o(0) \in [10, 12]\,m,$$
$$v_e(k) \in [20, 30]\,\frac{m}{s},\ v_o(k) \in [20, 30]\,\frac{m}{s},\ y_e(k) \in [0, 26]\,m,\ \phi_e(k) \in [-0.44, 0.44]rad,$$

while the other two initial states are $x_e(0) = 0$, $\phi_e(0) = 0$, and the bounds on the longitudinal position of ego car and other car are $x_e(k) \in [0, 1000]m$ and $x_o(k) \in [0, 1000]m$. The separation threshold is set as $\epsilon = 0.5$.

6.2.2 Results and Discussion

Due to the extensive computational time required for Theorem 1 to converge, this example is restricted to the 'separability condition' of the AMD problem. By relieving the ego vehicle of its responsibility, we effectively decrease the count of binary variables, enhancing the tractability of the problem. Employing the cost $J(\cdot) = \|\cdot\|_\infty$, we tackle the driver intent estimation problem using three distinct methodologies. The calculated control inputs with $J(\cdot) = \|\cdot\|_\infty$ are obtained as follows: $\{u_1(k)\}_{k=0}^{2} = \{0.1541, 0, 0\}$, $\{u_2(k)\}_{k=0}^{2} = \{0.1541, -0.1541, 0\}$ (cf. Fig. 4a). We also perform a comparison of the total computational times, as presented in Table 1. Although the binary variable relocation technique technically yields a suboptimal solution, focusing solely on the separation condition in this instance results in an optimal outcome. On the other hand, both the brute force strategy, which exhaustively explores all potential subregion configurations, and the parametric method also obtain the optimal solution, as expected, but with significantly more time than the binary variable relocation method with the parametric being the worse of the two, indicating that the parametric method, while promising, is not necessarily better than the brute force approach at least in this instance.

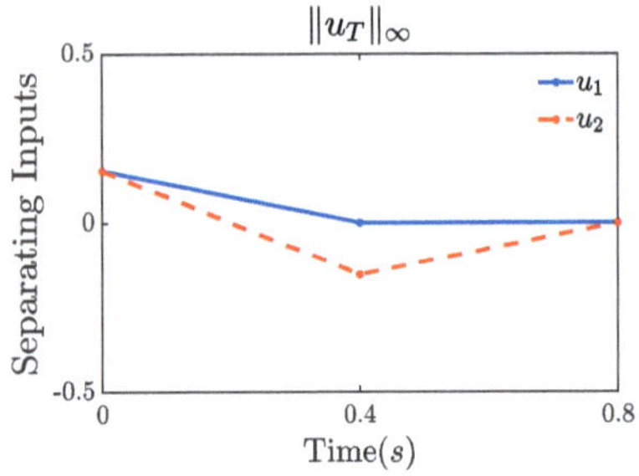

(a) Optimal inputs obtained by solving the proposed active model discrimination for intents, $\ell \in \{C, M\}$.

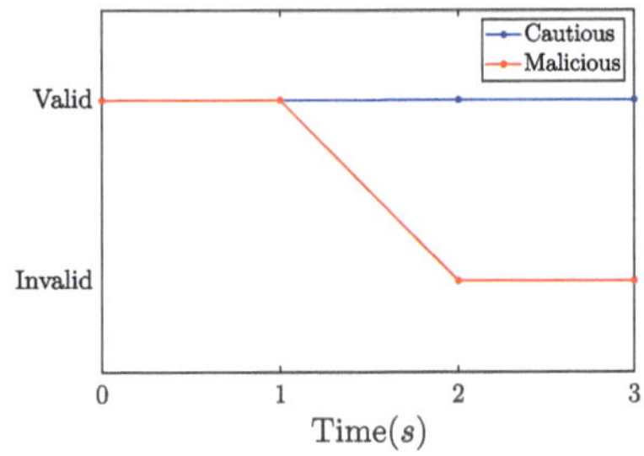

(b) Result of model invalidation algorithm with accurate determination made after $t = 2$.

Fig. 4 Optimal input obtained and model invalidation result

Finally, by using the designed inputs $u_1(k)$ and $u_2(k)$, along with the associated subregion sequencing and observed outputs (cf. Fig. 4a), the MI algorithm (cf. Theorem 3 and Algorithm 2) enables the differentiation of the two intents by the second time step, as shown in Fig. 4b.

7 Conclusions

The inevitable environmental and modeling uncertainties as well as the complex dynamics of CPS pose significant challenges in accurately characterizing their nominal behaviors and identifying faulty or malicious operations. To address these issues, this work introduces Active Model Discrimination (AMD) as a novel strategy for enhancing model differentiation and fault detection processes amidst uncertainties. Specifically, the AMD problem designs (perturbation) input signals that significantly facilitate the accuracy of model discrimination. This study specifically investigates the AMD problem for noisy switched piecewise affine inclusion systems, which are constrained by signal and metric temporal logic specifications. The presence of binary/integer variables due to the piecewise affine approximation of nonlinear systems and/or the encoding of STL/MTL specifications incurs substantial complexity in the bilevel optimization problems inherent to AMD. To navigate these complexities, we propose three distinct algorithmic solutions: the parametric approach, the binary variable relocation approach, and the brute force approach. Each method offers a unique strategy for simplifying and solving the mixed-integer bilevel optimization challenges presented by the AMD, with the overarching goal of recasting these problems into tractable forms that can be efficiently addressed. Moreover, this study complements these algorithmic strategies with a model invalidation algorithm. This additional tool is instrumental in excluding inconsistent models based on noisy observations, ensuring the accurate identification of the true model within a predefined detection time frame. The effectiveness and efficiency of the proposed algorithms are demonstrated in examples of parameter identification of a planetary assisted drive system and intent estimation in highway lane changing scenarios. These examples underscore the potential of AMD and the associated methodologies to significantly advance the safety and reliability of cyber-physical systems in a plethora of industrial and engineering applications.

Acknowledgements This work is supported in part by NSF grant CNS-2313814 and ONR grant N00014-23-1-2093.

Appendix

$$
\begin{aligned}
&M_\ell = \operatorname*{vec}_{i=1}^{q_\ell} \operatorname*{vec}_{j=1}^{|\Sigma|}\{M^j_{\ell,i}\}, &&\Omega^x_\ell = \operatorname*{vec}_{i=1}^{q_\ell} \operatorname*{vec}_{j=1}^{|\Sigma|}\{\Omega^{x,j}_{\ell,i}\}, &&F^\ell_{\underline{\tau},\overline{\tau}} = \operatorname*{vec}_{i=1}^{q_\ell} \operatorname*{vec}_{j=1}^{|\Sigma|}\{F^{\ell,i,j}_{\underline{\tau},\overline{\tau}}\},\\
&\beta^\ell_{\underline{\tau},\overline{\tau}} = \operatorname*{vec}_{i=1}^{q_\ell} \operatorname*{vec}_{j=1}^{|\Sigma|}\{\beta^{\ell,i,j}_{\underline{\tau},\overline{\tau}}\}, &&\alpha_\ell = \operatorname*{vec}_{i=1}^{q_\ell} \operatorname*{vec}_{j=1}^{|\Sigma|}\{\alpha^j_{\ell,i}\}, &&\zeta_\ell = \operatorname*{vec}_{i=1}^{q_\ell} \operatorname*{vec}_{j=1}^{|\Sigma|}\{\zeta^j_{\ell,i}\},\\
&E^\iota = \operatorname*{diag}_{\ell=\{\dagger,\ddagger\}}\{E_\ell\}, &&G^\iota = \operatorname*{vec}_{\ell=\{\dagger,\ddagger\}}\{G_\ell\}, &&F^u_\ell = \operatorname*{diag}_T\{D_{u,\ell}\},\\
&E_\ell = \operatorname*{diag}_T\{C_\ell\}, &&G_\ell = \operatorname*{vec}_T\{g_\ell\}, &&F^v_\ell = \operatorname*{diag}_T\{V_\ell\},
\end{aligned}
$$

$$
M^j_{\ell,i} = \begin{bmatrix}
\underline{A}^j_{\ell,i} & -\mathbb{I} & \mathbf{0} & \mathbf{0} & \cdots & \mathbf{0}\\
-\overline{A}^j_{\ell,i} & \mathbb{I} & \mathbf{0} & \mathbf{0} & \cdots & \mathbf{0}\\
\mathbf{0} & \underline{A}^j_{\ell,i} & -\mathbb{I} & \mathbf{0} & \cdots & \mathbf{0}\\
\mathbf{0} & -\overline{A}^j_{\ell,i} & \mathbb{I} & \mathbf{0} & \cdots & \mathbf{0}\\
\vdots & \vdots & \vdots & \vdots & \ddots & \vdots\\
\mathbf{0} & \mathbf{0} & \cdots & \cdots & \underline{A}^j_{\ell,i} & -\mathbb{I}\\
\mathbf{0} & \mathbf{0} & \cdots & \cdots & -\overline{A}^j_{\ell,i} & \mathbb{I}
\end{bmatrix},
\quad
M^\iota = \begin{bmatrix} \operatorname*{diag}_{\ell=\{\dagger,\ddagger\}}\{M_\ell\}\\ \operatorname*{diag}_{\ell=\{\dagger,\ddagger\}}\{\Omega_\ell\}\end{bmatrix},
\quad
F^\iota_T = \begin{bmatrix} \operatorname*{diag}_{\ell=\{\dagger,\ddagger\}}\{F_\ell\} + \operatorname*{diag}_{\ell=\{\dagger,\ddagger\}}\{\alpha_\ell\}\\ \operatorname*{diag}_{\ell=\{\dagger,\ddagger\}}\{\beta_\ell\} + \operatorname*{diag}_{\ell=\{\dagger,\ddagger\}}\{\zeta_\ell\}\end{bmatrix},
$$

$$
F^{\ell,i,j}_{\underline{\tau},\overline{\tau}} = \Big[{\underline{f}^j_{\ell,i}}^\top \; -{\overline{f}^j_{\ell,i}}^\top \; {\underline{f}^j_{\ell,i}}^\top \; -{\overline{f}^j_{\ell,i}}^\top \; \cdots \; {\underline{f}^j_{\ell,i}}^\top \; -{\overline{f}^j_{\ell,i}}^\top \Big]^\top_{2n(\overline{\tau}-\underline{\tau})\times 1},
$$

$$
\Omega^{x,j}_{\ell,i} = \Big[\operatorname*{diag}_{\overline{\tau}-\underline{\tau}}\{S^{x,j}_{\ell,i}\}\ \mathbf{0}\Big],\ \beta^{\ell,i,j}_{\underline{\tau},\overline{\tau}} = \mathrm{vec}_{T-1}\{\beta^j_{\ell,i}\},\ R^\iota_1 = \big[M^\iota\ \Gamma^\iota_d\ \Gamma^\iota_w\ \mathbf{0}\big],
$$

$$
r^\iota_1 = F^\iota_T,\ S^\iota_1 = \Gamma^\iota_u, \qquad r^\iota_2 = \begin{bmatrix} G_\dagger - G_\ddagger\\ G_\ddagger - G_\dagger\end{bmatrix}, \qquad S^\iota_2 = \begin{bmatrix} F_{u,\dagger} - F_{u,\ddagger}\\ F_{u,\ddagger} - F_{u,\dagger}\end{bmatrix}.
$$

$$
R^\iota_2 = \begin{bmatrix} E_\dagger & -E_\ddagger & F_{d,\dagger} & -F_{d,\ddagger} & \mathbf{0} & \mathbf{0} & F_{v,\dagger} & -F_{v,\ddagger}\\ -E_\dagger & E_\ddagger & -F_{d,\dagger} & F_{d,\ddagger} & \mathbf{0} & \mathbf{0} & -F_{v,\dagger} & F_{v,\ddagger}\end{bmatrix};
$$

$$
\text{For } \star = \{w,u,d\}: \begin{cases}
\Gamma^{\star,j}_{\ell,i} = \operatorname*{diag}_{\overline{\tau}-\underline{\tau}}\left\{\begin{bmatrix}\underline{B}^{\star,j}_{\ell,i}\\ -\overline{B}^{\star,j}_{\ell,i}\end{bmatrix}\right\}, & F^\iota_\star = \operatorname*{vec}_{\ell=\{\dagger,\ddagger\}}\{F^\star_\ell\},\\
\Gamma^\iota_\star = \begin{bmatrix} \operatorname*{diag}_{\ell=\{\dagger,\ddagger\}} \mathrm{vec}^{q_\ell}_{i=1}\,\mathrm{vec}^{|\Sigma|}_{j=1}\{\Gamma^{\star,j}_{\ell,i}\}\\ \operatorname*{diag}_{\ell=\{\dagger,\ddagger\}} \mathrm{vec}^{q_\ell}_{i=1}\,\mathrm{vec}^{|\Sigma|}_{j=1}\{\Omega^{\star,j}_{\ell,i}\}\end{bmatrix}, & \Omega^{\star,j}_{\ell,i} = \operatorname*{diag}_{\overline{\tau}-\underline{\tau}}\{S^{\star,j}_{\ell,i}\},\\
\Omega^\star_\ell = \mathrm{vec}^{q_\ell}_{i=1}\,\mathrm{vec}^{|\Sigma|}_{j=1}\{\Omega^{\star,j}_{\ell,i}\}, & \Gamma^\star_\ell = \mathrm{vec}^{q_\ell}_{i=1}\,\mathrm{vec}^{|\Sigma|}_{j=1}\{\Gamma^{\star,j}_{\ell,i}\},
\end{cases}
$$

$$\begin{aligned}
\overline{P}_x^\ell &= \begin{bmatrix} M_\ell \ \Gamma_\ell^d \ \Gamma_\ell^w \end{bmatrix}, \qquad \overline{P}_u^\ell = [-\Gamma_\ell^u], \quad \overline{p}^\ell = -[F_\ell + \alpha_\ell], \\
\overline{\overline{P}}_x^\ell &= \begin{bmatrix} \Omega_\ell^x & \Omega_\ell^d & \Omega_\ell^w \\ [\bar{P}_{x,\ell} \ \ \mathbf{0}] & \mathbf{0} & \mathbf{0} \end{bmatrix}, \overline{\overline{P}}_u^\ell = -\begin{bmatrix} \Omega_\ell^u \\ \mathbf{0} \end{bmatrix}, \overline{\overline{p}}^\ell = -\begin{bmatrix} \beta_\ell + \zeta_\ell \\ -\bar{p}_{x,\ell} \end{bmatrix}, \\
\overline{\overline{H}}^\ell &= \begin{bmatrix} \mathrm{diag}\{\overline{P}_0^\ell, Q_{d,\ell}, Q_{w,\ell}\}, \end{bmatrix}, \overline{\overline{h}}^\ell = \begin{bmatrix} \overline{p}_0^{\ell\top} \ \overline{p}_d^{\ell\top} \ \overline{q}_w^{\ell\top} \end{bmatrix}^\top.
\end{aligned}$$

References

Asarin E, Donzé A, Maler O, Nickovic D (2011) Parametric identification of temporal properties. In: International conference on runtime verification. Springer, pp 147–160

Azuma S, Imura J, Sugie T (2010) Lebesgue piecewise affine approximation of nonlinear systems. Nonlinear Anal Hybrid Syst 4(1):92–102

Belta C, Sadraddini S (2019) Formal methods for control synthesis: an optimization perspective. Annu Rev Control Robot Auton Syst 2(1):115–140

Cheong S, Manchester I (2015) Input design for discrimination between classes of LTI models. Automatica 53:103–110

Chiang LH, Russell EL, Braatz RD (2001) Fault detection and diagnosis in industrial systems. Springer, London

CPLEX II (2009) V12. 1: user's manual for CPLEX. Int Bus Mach Corp 46(53):157

Dang T, Maler O, Testylier R (2010) Accurate hybridization of nonlinear systems. In: ACM international conference on hybrid systems: computation and control, pp 11–20

Ding Y, Harirchi F, Yong SZ, Jacobsen E, Ozay N (2018) Optimal input design for affine model discrimination with applications in intention-aware vehicles. In: ACM/IEEE international conference on cyber-physical systems, arXiv:1702.01112

Dokhanchi A, Hoxha B, Fainekos G (2014) On-line monitoring for temporal logic robustness. In: International conference on runtime verification. Springer, pp 231–246

Dubins LE (1957) On curves of minimal length with a constraint on average curvature, and with prescribed initial and terminal positions and tangents. Am J Math 79:497–516

Grewal M, Glover K (1976) Identifiability of linear and nonlinear dynamical systems. IEEE Trans Aut Cont 21(6):833–837

Gurobi Optimization, Inc (2015) Gurobi optimizer reference manual. http://www.gurobi.com

Harirchi F, Yong SZ, Ozay N (2017) Guaranteed fault detection and isolation for switched affine models. In: IEEE conference on decision and control (CDC), pp 5161–5167

Harirchi F, Yong SZ, Ozay N (2018) Passive diagnosis of hidden-mode switched affine models with detection guarantees via model invalidation. In: Diagnosability, security and safety of hybrid dynamic and cyber-physical systems. Springer, pp 227–251

Herceg M, Kvasnica M, Jones CN, Morari M (2013) Multi-parametric toolbox 3.0. In: 2013 European control conference (ECC), IEEE, pp 502–510

Jin Z, Shen Q, Yong SZ (2019) Optimization-based approaches for affine abstraction and model discrimination of uncertain nonlinear systems. In: IEEE conference on decision and control, pp 7976–7981

Jin Z, Shen Q, Yong SZ (2020) Mesh-based piecewise affine abstraction with polytopic partitions for nonlinear systems. IEEE Control Syst Lett 5(5):1543–1548

Koymans R (1990) Specifying real-time properties with metric temporal logic. Real-Time Syst 2(4):255–299

Liu C, Shen Q, Niu R, Yong SZ (2021) Integrated passive-active model identification with tunable model discrimination for affine discrete-time systems. IEEE Control Syst Lett 6:1885–1890

Lou H, Si P (2009) The distinguishability of linear control systems. Nonlinear Anal Hybrid Syst 3(1):21–39

Maler O, Nickovic D (2004) Monitoring temporal properties of continuous signals. In: International symposium on formal techniques in real-time and fault-tolerant systems. Springer, pp 152–166
Marseglia GR, Raimondo DM (2017) Active fault diagnosis: a multi-parametric approach. Automatica 79:223–230
Nikoukhah R (1998) Guaranteed active failure detection and isolation for linear dynamical systems. Automatica 34(11):1345–1358
Nikoukhah R, Campbell S (2006) Auxiliary signal design for active failure detection in uncertain linear systems with a priori information. Automatica 42(2):219–228
Niu R, Shen Q, Yong SZ (2019) Partition-based parametric active model discrimination design with applications to driver intention estimation. In: European Control Conference, pp 3880–3885
Niu R, Hassaan SM, Yang L, Jin Z, Yong SZ (2021) Model discrimination of switched nonlinear systems with temporal logic-constrained switching. IEEE Control Syst Lett 6:151–156
Niu R, Hassaan SM, Yong SZ (2022) A multi-parametric method for active model discrimination of nonlinear systems with temporal logic-constrained switching. In: American control conference, pp 1652–1658
Niu R, Gah E, Yong SZ (2023) Active model discrimination for piecewise affine inclusion systems. IFAC-PapersOnLine 56(2):9448–9454
Raimondo DM, Braatz RD, Scott JK (2013) Active fault diagnosis using moving horizon input design. In: European control conference, pp 3131–3136
Scott JK, Findeison R, Braatz RD, Raimondo DM (2014) Input design for guaranteed fault diagnosis using zonotopes. Automatica 50(6):1580–1589
Shen Q, Yong SZ (2019) Robust optimization-based affine abstractions for uncertain affine dynamics. In: 2019 American control conference (ACC), IEEE, pp 2452–2457
Shen Q, Yong SZ (2020) Active model discrimination using partition-based output feedback. In: European control conference, pp 712–717
Shen Q, Niu R, Yong SZ (2022) Tractable model discrimination for safety-critical systems with disjunctive and coupled constraints. Nonlinear Anal Hybrid Syst 46:101217
Singh K, Ding Y, Ozay N, Yong SZ (2018a) Input design for nonlinear model discrimination via affine abstraction. IFAC PapersOnLine 51(16):175–180
Singh K, Shen Q, Yong SZ (2018b) Mesh-based affine abstraction of nonlinear systems with tighter bounds. In: IEEE conference on decision and control, pp 3056–3061
Smith RS, Doyle JC (1989) Model invalidation: a connection between robust control and identification. In: American control conference, IEEE, pp 1435–1440
Tabatabaeipour S (2015) Active fault detection and isolation of discrete-time linear time-varying systems: a set-membership approach. Int J Syst Sci 46(11):1917–1933
Vidal R, Chiuso A, Soatto S, Sastry S (2003) Observability of linear hybrid systems. In: International workshop on hybrid systems, computation and control. Springer, pp 526–539
Yang L, Ozay N (2019) Fault detectability analysis of switched affine systems with linear temporal logic constraints. In: IEEE conference on decision and control (CDC), pp 5779–5786

Optimal Separator for an Ellipse; Application to Localization

Luc Jaulin

Abstract This paper proposes a minimal contractor and a minimal separator for an ellipse in the plane. The task is facilitated using actions induced by the hyperoctahedral group of symmetries. An application related to the localization of an object using multiple sonars is proposed.

1 Introduction

Contractor programming (Chabert and Jaulin 2009) is an efficient tool to solve rigorously complex nonlinear problems involving bounded uncertainties (Cébério and Granvilliers 2001; Kreinovich et al. 1997; Rauh and Auer 2011). It is based on the notion of *contractor* which is an operator which shrinks an axis-aligned box $[\mathbf{x}]$ of $\mathbb{R}^n$ without removing any point of the subset $\mathbb{X}$ of $\mathbb{R}^n$ to which it is associated. The set $\mathbb{X}$ is assumed to be defined by equations or inequalities involving the components $x_1, \ldots, x_n$ of $\mathbf{x}$. As a result, combined with a paver (Sainudiin 2010) which bisects boxes, the contractor will allow us to build an outer approximation of the set $\mathbb{X}$. This outer approximation is represented as a list of boxes $\mathcal{L}$ with sides parallel to the coordinate axes. This list $\mathcal{L}$ contains as few elements not from $\mathbb{X}$ as possible. The resulting methodology can be applied in several domains of engineering such as identification (Ramdani and Poignet 2005), localization (Guyonneau et al. 2013), SLAM (Mustafa et al. 2018; Rohou et al. 2019), vision (Ehambram et al. 2021), control (Rauh and Auer 2009; Wan 2007), calibration (Dit Sandretto et al. 2014), motion planning (Crépon et al. 2018), etc.

Contractor programming relies on a catalog of efficient elementary contractors which are usually built using interval arithmetic (Moore et al. 2009). Basically, interval computation allows us to compute the range of a function the input of which are known to be inside intervals. Take for instance, $f(x) = x^2 - x$ where $x \in [x] = [x^-, x^+]$. We have $f(x) = a - x$, where $a = x^2$. To enclose $f(x)$, we first compute the smallest interval $[a]$ which contains all feasible $a = x^2$, $x \in [x]$.

L. Jaulin (✉)
ENSTA-Bretagne, Brest, France
e-mail: lucjaulin@gmail.com

T. N. Dinh et al. (eds.), *Set-Valued Approaches to Control and Estimation of Uncertain Systems*, Mathematical Engineering,
https://doi.org/10.1007/978-3-031-94239-6_6

Then, we compute the smallest interval $[y]$ which encloses the set of feasible $a - x$, assuming that $a \in [a]$ and $x \in [x]$. Interval computation can be used to build contractors for elementary equations. These contractors shrink intervals of feasible values for the variables, without removing a single feasible value. Note that straightforward (naive) interval arithmetic which simply replaces all arithmetic operations with corresponding operations of interval arithmetic leads to pessimistic contractors. As explained in Moore (1966) more accurate enclosure can be obtained using centered form, monotonicity checking, and other ideas now used in effective interval packages.

Combining all available elementary contractors, we can construct a more sophisticated one consistent with the solution set of the problem we want to solve. This operation introduces a pessimism which has to be balanced by additional bisections performed by the paver (van Hentenryck et al. 1997). For more efficiency, it is important to extend the catalog by adding some new specific contractors.

Note that contractors and separators use abstract domains that are boxes. Boxes are well suited to constraint propagation algorithms since the intersection (finite or infinite) of boxes is always a box. This is not the case for other types of containers such as ellipsoids (Durieu et al. 1996) or zonotopes (Combastel 2005). This is why we prefer in this paper to use boxes. However zonotopes bundles (Althoff and Krogh 2011) could probably provide a better approximation especially if a radial splitting is used.

In this paper, we propose a minimal contractor for an ellipse in the plane. We then extend the approach to build a minimal separator, which is a pair of two contractors: an inner contractor and an outer contractor. Consider the quadratic function

$$f(\mathbf{q}, \mathbf{x}) = q_0 + q_1 x_1 + q_2 x_2 + q_3 x_1^2 + q_4 x_1 x_2 + q_5 x_2^2 \tag{1}$$

where $\mathbf{q} = (q_0, \ldots, q_5)$ is the parameter vector and $\mathbf{x} = (x_1, x_2)$ is the vector of variables. Equivalently, we can write the function in a matrix form:

$$f(\mathbf{q}, \mathbf{x}) = \mathbf{x}^T \cdot \begin{pmatrix} q_3 & \frac{1}{2} q_4 \\ \frac{1}{2} q_4 & q_5 \end{pmatrix} \cdot \mathbf{x} + (q_1 \;\; q_2) \cdot \mathbf{x} + q_0. \tag{2}$$

The zeros of $f(\mathbf{q}, \mathbf{x})$ is a conic section (a circle or other ellipse, a parabola, or a hyperbola). We will assume here that $\mathbf{Q}$ has eigenvalues which are both positive. It means

$$\begin{aligned} q_3 &> 0 \\ q_5 &> 0 \\ 4q_3 q_5 - q_4^2 &> 0 \end{aligned} \tag{3}$$

Let us check this. The two first conditions are trivial. We only check the condition $4q_3 q_5 - q_4^2 > 0$. The characteristic polynomial of the matrix $\mathbf{Q}$ is

$$\begin{aligned} P(s) = & \;\; (s - q_3)(s - q_5) - \tfrac{1}{4} q_4^2 \\ & = s^2 - (q_3 + q_5)s + q_3 q_5 - \tfrac{1}{4} q_4^2 \end{aligned}$$

Its discriminant is

$$\begin{array}{rcl}\Delta &=& (q_3+q_5)^2 - 4q_3q_5 + q_4^2 \\ &=& q_3^2 + q_5^2 - 2q_3q_5 + q_4^2 \\ &=& (q_3-q_5)^2 + q_4^2\end{array} \tag{4}$$

which is always positive. Which means that the matrix $\mathbf{Q}$ has two real values (this is not a surprise since $\mathbf{Q}$ is symmetric). The solution are positive is the smallest one is positive. This means that

$$\begin{array}{rc} & q_3+q_5-\sqrt{\Delta}>0 \\ \Leftrightarrow & q_3+q_5>\sqrt{\Delta} \\ \Leftrightarrow & (q_3+q_5)^2>\Delta \\ \Leftrightarrow & (q_3+q_5)^2>(q_3-q_5)^2+q_4^2 \\ \Leftrightarrow & q_3^2+q_5^2+2q_3q_5>q_3^2+q_5^2-2q_3q_5+q_4^2 \\ \Leftrightarrow & 4q_3q_5-q_4^2>0 \\ \Leftrightarrow & \det\mathbf{Q}>0\end{array} \tag{5}$$

Define the set

$$\mathbb{X} = \{(x_1, x_2 | f(\mathbf{q}, \mathbf{x}) \leq 0\}. \tag{6}$$

The zeros of this quadratic function is, in general, a conic section (a circle or other ellipse, a parabola, or a hyperbola). Define the set

$$\mathbb{X} = \{(x_1, x_2 | f(\mathbf{q}, \mathbf{x}) \leq 0\}. \tag{7}$$

Since the square matrix involved in the matrix form has positive eigenvalues, the set $\mathbb{X}$ is an ellipse. In this paper, we propose an interval-based method (Moore 1979) to generate an optimal separator (Jaulin and Desrochers 2014) for the set $\mathbb{X}$. This separator will be used to generate an inner and an outer approximations for $\mathbb{X}$. As an application, we will consider the problem of the localization of an object using three sonars.

This paper is organized as follows. Section 2 introduces the notion of symmetries that will be used in the construction of the separators. Section 3 recalls the basic notions and contractors and separators. Section 4 builds the separator for the ellipse. Section 5 illustrates the use of the separator to approximate the set of position for an object using three sonars. Section 6 concludes the paper.

2 Symmetries

2.1 Hyperoctahedral Group

The hyperoctahedral group B_n is the group of symmetries of the hypercube $[-1, 1]^n$ (Coxeter 1999) of $\mathbb{R}^n$. It corresponds to the group of $n \times n$ orthogonal matrices

Fig. 1 Multiplication table

	σ_0	σ_1	σ_2	σ_3	σ_4	σ_5	σ_6	σ_7
σ_0	σ_0	σ_1	σ_2	σ_3	σ_4	σ_5	σ_6	σ_7
σ_1	σ_1	σ_0	σ_6	σ_4	σ_3	σ_7	σ_2	σ_5
σ_2	σ_2	σ_5	σ_0	σ_7	σ_6	σ_1	σ_4	σ_3
σ_3	σ_3	σ_4	σ_7	σ_0	σ_1	σ_6	σ_5	σ_2
σ_4	σ_4	σ_3	σ_5	σ_1	σ_0	σ_2	σ_7	σ_6
σ_5	σ_5	σ_2	σ_4	σ_6	σ_7	σ_3	σ_0	σ_1
σ_6	σ_6	σ_7	σ_1	σ_5	σ_2	σ_0	σ_3	σ_4
σ_7	σ_7	σ_6	σ_3	σ_2	σ_5	σ_4	σ_1	σ_0

whose entries are integers and contains $2^n \cdot n!$ elements. For instance, for $n = 2$, we have $2^2 \cdot 2! = 8$ elements. If we use the matrix form, the elements of B_2 are

$$\begin{array}{ll} \sigma_0 = \begin{pmatrix} 1 & 0 \\ 0 & 1 \end{pmatrix} & \sigma_1 = \begin{pmatrix} -1 & 0 \\ 0 & 1 \end{pmatrix} \\ \sigma_2 = \begin{pmatrix} 0 & 1 \\ 1 & 0 \end{pmatrix} & \sigma_3 = \begin{pmatrix} -1 & 0 \\ 0 & -1 \end{pmatrix} \\ \sigma_4 = \begin{pmatrix} 1 & 0 \\ 0 & -1 \end{pmatrix} & \sigma_5 = \begin{pmatrix} 0 & 1 \\ -1 & 0 \end{pmatrix} \\ \sigma_6 = \begin{pmatrix} 0 & -1 \\ 1 & 0 \end{pmatrix} & \sigma_7 = \begin{pmatrix} 0 & -1 \\ -1 & 0 \end{pmatrix} \end{array} \tag{8}$$

Matrices associated to B_n are called *hyperoctahedral matrices.* Each row and each column should contain one and only one non-zero entry which should be 1 or -1. It is trivial to check that B_n is a group with respect to the multiplication, or equivalently, for the composition $\circ$. In this paper, a symmetry σ will be seen both as a linear function from $\mathbb{R}^n$ to $\mathbb{R}^n$ or as a $n \times n$ matrix. For instance, we will write equivalently

$$\sigma_5 = \begin{pmatrix} 0 & 1 \\ -1 & 0 \end{pmatrix} \text{ or } \sigma_5 : \begin{cases} \mathbb{R}^2 & \mapsto & \mathbb{R}^2 \\ (x_1, x_2) & \mapsto & (x_2, -x_1) \end{cases} \tag{9}$$

Figure 1 corresponds to the multiplication table (known as Cayley table (Cayley 1854)).

Now, due to the fact that we have a group, the transitivity and the inversion property should be satisfied. As a consequence, the entries of the table are dependent. We

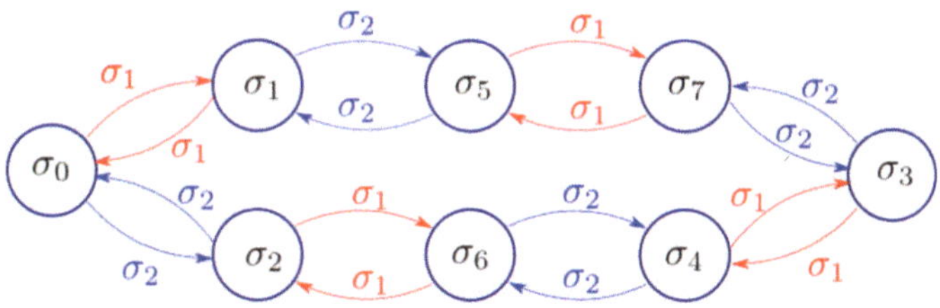

Fig. 2 Cayley graph associated to the hyperoctahedral group B_2

can easily check that from the second and the third lines, we can reconstruct the whole table. We say that the two elements σ_1, σ_2 are generators of the group B_2 or equivalently, we write $B_2 = \langle \sigma_1, \sigma_2 \rangle$.

Compute for instance $\sigma_6 \circ \sigma_7$ using these two lines (red and blue in the figure). We get:

$$\sigma_6 \circ \sigma_7 = \sigma_1 \circ \underbrace{\sigma_2 \circ \underbrace{\sigma_1 \circ \underbrace{\sigma_2 \circ \sigma_1}_{\sigma_5}}_{\sigma_7}}_{\sigma_3} = \sigma_4 \tag{10}$$

or equivalently

$$\begin{aligned} \sigma_2 \circ \sigma_1 &= \sigma_5 \\ \sigma_1 \circ \sigma_5 &= \sigma_7 \\ \sigma_2 \circ \sigma_7 &= \sigma_3 \\ \sigma_1 \circ \sigma_3 &= \sigma_4 \end{aligned} \tag{11}$$

To prove that $\{\sigma_1, \sigma_2\}$ is a generator pair of B_2, we build the Cayley graph (Coxeter 1935) (see Fig. 2). We first draw a node for each of the 8 symmetries. Then we draw the arcs by reading the lines of the multiplication table corresponding to σ_1, σ_2. For instance, since we read $\sigma_7 = \sigma_1 \circ \sigma_5$, we draw an arc labeled by σ_1 between σ_5 and σ_7. Since the graph is strongly connected, we conclude that all symmetries can be obtained by compositions of σ_1, σ_2.

An important property that will be used later for building optimal contractors is that any $\sigma \in B_n$ is *box-conservative* (Desrochers and Jaulin 2016), i.e., if for all $\mathbb{A} \subseteq \mathbb{R}^n$,

$$\sigma\left([\![\mathbb{A}]\!]\right) = [\![\sigma\left(\mathbb{A}\right)]\!] \tag{12}$$

where $[\![\mathbb{A}]\!]$ is the *hull* operator, i.e., the smallest box which encloses $\mathbb{A}$. An illustration is given by Fig. 3. On the left figure, $\sigma = \begin{pmatrix} 0 & 1 \\ 1 & 0 \end{pmatrix}$ is a permutation. The image of $\mathbb{A}$ is $\sigma(\mathbb{A})$. We easily check that $\sigma\left([\![\mathbb{A}]\!]\right) = [\![\sigma\left(\mathbb{A}\right)]\!]$. On the right figure, σ is a symmetry with respect to an axis which makes 40 degrees with the horizontal axis. This operator is not hyperoctahedral and not box-conservative.

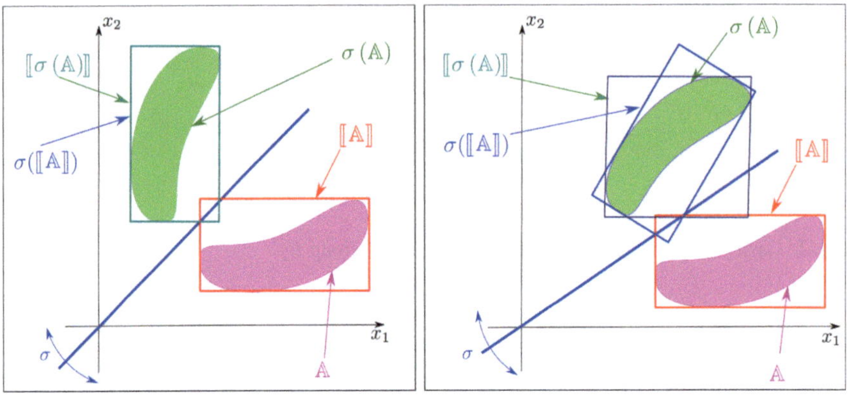

Fig. 3 (Left) σ is hyperoctahedral and the equality $\sigma([\![\mathbb{A}]\!]) = [\![\sigma(\mathbb{A})]\!]$ is satisfied. (Right) The equality is not anymore satisfied when σ is not hyperoctahedral

2.2 Symmetries in Equations

Define an equation of the form

$$f(\mathbf{q}, \mathbf{x}) = 0. \tag{13}$$

Two transformations ε and σ are conjugate with respect to f if

$$f(\varepsilon(\mathbf{q}), \sigma(\mathbf{x})) = 0 \Leftrightarrow f(\mathbf{q}, \mathbf{x}) = 0. \tag{14}$$

Transformations that will be considered are limited to the *hyperoctahedral group* B_n. Figure 4 shows different notations usually considered to represent a symmetry σ of B_5. We will prefer the Cauchy one line notation (Wussing 2007) which is shorter. We should understand the symmetry σ of the figure as the function:

Fig. 4 Different representations of an element σ of B_5. Left: graph; Top right: Matrix notation; Bottom right: Cauchy one line notation

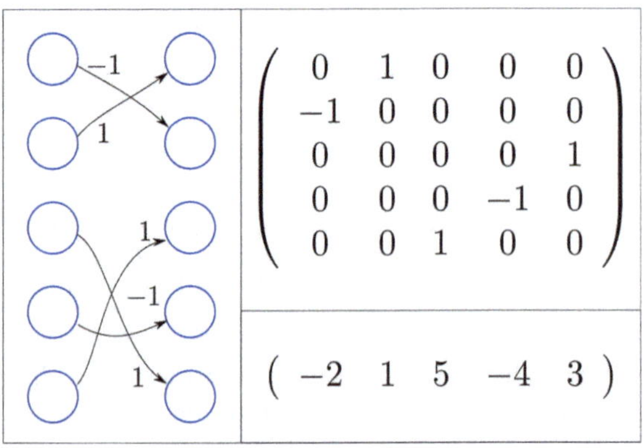

$$\sigma(x_1, x_2, x_3, x_4, x_5) = (-x_2, x_1, x_5, -x_4, x_3). \tag{15}$$

Even if the matrix representation looks more intuitive, for efficiency reasons, we use the Cauchy one line representation to compose the symmetries. Let us consider again the function

$$f(\mathbf{q}, \mathbf{x}) \stackrel{(1)}{=} q_0 + q_1x_1 + q_2x_2 + q_3x_1^2 + q_4x_1x_2 + q_5x_2^2. \tag{16}$$

Take the symmetry

$$\begin{pmatrix} x_1 \\ x_2 \end{pmatrix} \to \begin{pmatrix} \varepsilon_1 x_1 \\ \varepsilon_2 x_2 \end{pmatrix} \tag{17}$$

where $\varepsilon_i \in \{-1, 1\}$. With the Cauchy notation, this transformation is denoted by $\varepsilon = (\varepsilon_1, 2\varepsilon_2)$. The $\varepsilon_1, \varepsilon_2$ should be understood as parameters, i.e., for each different $\varepsilon_1, \varepsilon_2$, we get a different symmetry. We have

$$\begin{aligned} f(\mathbf{q}, \varepsilon\mathbf{x}) &= f(\varepsilon_1 x_1, \varepsilon_2 x_2) \\ &= q_0 + q_1(\varepsilon_1 x_1) + q_2(\varepsilon_2 x_2) + q_3(\varepsilon_1 x_1)^2 + q_4(\varepsilon_1 x_1)(\varepsilon_2 x_2) + q_5(\varepsilon_2 x_2)^2 \\ &= q_0 + (q_1\varepsilon_1)x_1 + (q_2\varepsilon_2)x_2 + (q_3)x_1^2 + (q_4\varepsilon_1\varepsilon_2)x_1x_2 + (q_5)x_2^2 \\ &= f(\sigma\mathbf{q}, \mathbf{x}) \end{aligned} \tag{18}$$

where

$$\sigma = (1, 2\varepsilon_1, 3\varepsilon_2, 4, 5\varepsilon_1\varepsilon_2, 6). \tag{19}$$

As a consequence, for each symmetry $\varepsilon = (\varepsilon_1, 2\varepsilon_2)$, the pair

$$((\varepsilon_1, 2\varepsilon_2), (1, 2\varepsilon_1, 3\varepsilon_2, 4, 5\varepsilon_1\varepsilon_2, 6)). \tag{20}$$

is conjugate.

In Jaulin (2023), we have defined the notion of choice function ψ. Given an equation $f(\mathbf{q}, \mathbf{x})$, to each symmetry ε, the choice function ψ yields a symmetry σ such that (ε, σ) is a conjugate pair. For our example, the choice function is defined by:

$$\psi(\varepsilon_1, 2\varepsilon_2) = (1, 2\varepsilon_1, 3\varepsilon_2, 4, 5\varepsilon_1\varepsilon_2, 6). \tag{21}$$

Equivalently, we have

$$\begin{aligned} \psi(1, 2) &= (1, 2, 3, 4, 5, 6) \\ \psi(-1, 2) &= (1, -2, 3, 4, -5, 6) \\ \psi(1, -2) &= (1, 2, -3, 4, -5, 6) \\ \psi(-1, -2) &= (1, -2, -3, 4, 5, 6) \end{aligned}$$

The fact that $\psi(1, -2) = (1, 2, -3, 4, -5, 6)$ for the equation $q_0 + q_1x_1 + q_2x_2 + q_3x_1^2 + q_4x_1x_2 + q_5x_2^2 = 0$ means the following: if x_2 is replaced by $-x_2$ (this is the symmetry $(1, -2)$) then q_2 should be replaced by $-q_2$ and q_4 by $-q_4$ (this is the symmetry $(1, 2, -3, 4, -5, 6)$) to maintain the equation valid.

3 Contractors and Separators

This section recalls the basic notions on intervals, contractors and separators that are needed to understand how the paver will approximate the solution set. A *contractor* $\mathcal{C}$ for the set $\mathbb{X} \subset \mathbb{R}^n$ is an operator $\mathbb{IR}^n \mapsto \mathbb{IR}^n$ which satisfies

$$\begin{array}{ll} \mathcal{C}([\mathbf{x}]) \subset [\mathbf{x}] & \text{(contractance)} \\ [\mathbf{x}] \subset [\mathbf{y}] \Rightarrow \mathcal{C}([\mathbf{x}]) \subset \mathcal{C}([\mathbf{y}]). & \text{(monotonicity)} \\ \mathcal{C}([\mathbf{x}]) \cap \mathbb{X} = [\mathbf{x}] \cap \mathbb{X} & \text{(consistency)} \end{array} \quad (22)$$

where $\mathbb{IR}^n$ is the set of axis-aligned boxes in $\mathbb{R}^n$. Figure 5a illustrates the notion of contractor. A contractor for $\mathbb{X}$ can be used inside a paver (an algorithm which bisects boxes and uses $\mathcal{C}$ to eliminate parts of the search space that are outside the solution set) to provide an outer approximation of $\mathbb{X}$. Figure 5b shows the paving generated by the paver. It corresponds to an outer approximation of $\mathbb{X}$. The blue boxes are outside $\mathbb{X}$.

The contractor $\mathcal{C}$ is *minimal* if $\mathcal{C}([\mathbf{x}])$ corresponds exactly to the smallest box that can be obtained by a contraction of $[\mathbf{x}]$ without removing a single point of $\mathbb{X}$.

If $\mathcal{C}_1$ and $\mathcal{C}_2$ are two contractors, we define the following operations on contractors (Chabert and Jaulin 2009):

$$(\mathcal{C}_1 \cap \mathcal{C}_2)([\mathbf{x}]) = \mathcal{C}_1([\mathbf{x}]) \cap \mathcal{C}_2([\mathbf{x}]) \quad (23)$$

$$(\mathcal{C}_1 \cup \mathcal{C}_2)([\mathbf{x}]) = \mathcal{C}_1([\mathbf{x}]) \sqcup \mathcal{C}_2([\mathbf{x}]) \quad (24)$$

$$(\mathcal{C}_1 \circ \mathcal{C}_2)([\mathbf{x}]) = \mathcal{C}_1 (\mathcal{C}_2([\mathbf{x}])) \quad (25)$$

where $\sqcup$ is the interval union hull, i.e., $[\mathbf{a}] \sqcup [\mathbf{b}]$ corresponds to the smallest box which encloses the two boxes $[\mathbf{a}]$ and $[\mathbf{b}]$. If $\sigma : \mathbb{R}^n \to \mathbb{R}^n$ is a symmetry, we define

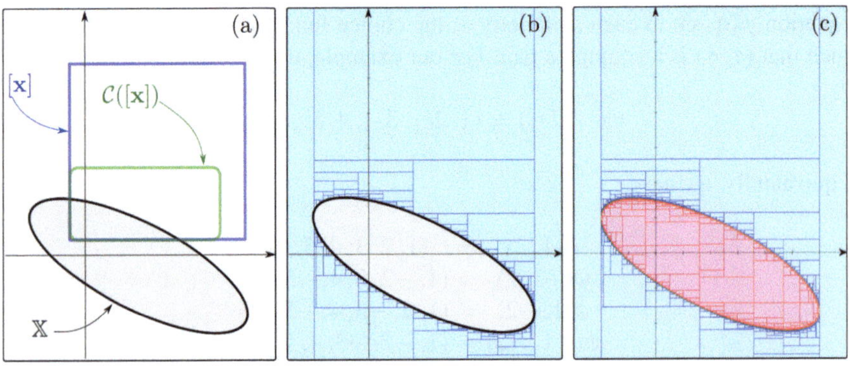

Fig. 5 **a** The box $[\mathbf{x}]$ is contracted by the contractor $\mathcal{C}$; **b** A paver uses the contractor $\mathcal{C}$ to get an outer approximation of $\mathbb{X}$; **c** A paver uses the separator $\mathcal{S}$ to get an outer and an inner approximations of $\mathbb{X}$

the *action* by σ of a contractor $\mathcal{C}$ for $\mathbb{X}$ as (Desrochers and Jaulin 2016):

$$\sigma\mathcal{C} = \sigma \circ \mathcal{C} \circ \sigma^{-1}. \tag{26}$$

It has been shown in (Desrochers and Jaulin 2016), Proposition 3 that $\sigma\mathcal{C}$ is a contractor for $\sigma\mathbb{X}$.

In order to characterize both an inner and outer approximation of the set $\mathbb{X}$, we need the notion of *separator*, as illustrated by Fig. 5c. A *separator* $\mathcal{S}$ for $\mathbb{X}$ is a pair of contractors $\{\mathcal{S}^{\text{in}}, \mathcal{S}^{\text{out}}\}$ such that, for all $[\mathbf{x}] \in \mathbb{IR}^n$, we have

$$\begin{array}{ll} \mathcal{S}^{\text{in}}([\mathbf{x}]) \cup \mathcal{S}^{\text{out}}([\mathbf{x}]) = [\mathbf{x}] & \text{(complementarity)} \\ \mathcal{S}^{\text{out}}([\mathbf{x}]) \cap \mathbb{X} = [\mathbf{x}] \cap \mathbb{X} & \text{(outer consistency)} \\ \mathcal{S}^{\text{in}}([\mathbf{x}]) \cap \mathbb{X}^C = [\mathbf{x}] \cap \mathbb{X}^C & \text{(inner consistency)} \end{array} \tag{27}$$

where $\mathbb{X}^C = \{\mathbf{x} \mid \mathbf{x} \notin \mathbb{X}\}$. We write $\mathcal{S} \sim \mathbb{X}$ if $\mathcal{S}$ is a separator for $\mathbb{X}$.

A separator $\mathcal{S}$ is *minimal* if its two contractors $\mathcal{S}^{\text{in}}$ and $\mathcal{S}^{\text{out}}$ are both minimal. We define the following operations

$$\begin{array}{llc} \mathcal{S}_1 \cap \mathcal{S}_2 = & \left\{\mathcal{S}_1^{\text{in}} \cup \mathcal{S}_2^{\text{in}}, \mathcal{S}_1^{\text{out}} \cap \mathcal{S}_2^{\text{out}}\right\} & \text{(intersection)} \\ \mathcal{S}_1 \cup \mathcal{S}_2 = & \left\{\mathcal{S}_1^{\text{in}} \cap \mathcal{S}_2^{\text{in}}, \mathcal{S}_1^{\text{out}} \cup \mathcal{S}_2^{\text{out}}\right\} & \text{(union)} \\ \mathcal{S}^C \quad = & \{\mathcal{S}^{\text{out}}, \mathcal{S}^{\text{in}}\} & \text{(complement)} \\ \sigma\mathcal{S} \quad = & \left\{\sigma \circ \mathcal{S}_{\mathbb{X}}^{\text{in}} \circ \sigma^{-1}, \sigma \circ \mathcal{S}_{\mathbb{X}}^{\text{out}} \circ \sigma^{-1}\right\} & \text{(action)} \end{array} \tag{28}$$

We have (Jaulin and Desrochers 2014):

$$\left\{\begin{array}{l} \mathcal{S}_1 \sim \mathbb{X}_1 \\ \mathcal{S}_2 \sim \mathbb{X}_2 \\ \mathcal{S} \sim \mathbb{X} \end{array}\right. \Rightarrow \left\{\begin{array}{ccc} \mathcal{S}_1 \cap \mathcal{S}_2 & \sim & \mathbb{X}_1 \cap \mathbb{X}_2 \\ \mathcal{S}_1 \cup \mathcal{S}_2 & \sim & \mathbb{X}_1 \cup \mathbb{X}_2 \\ \mathcal{S}^C & \sim & \mathbb{X}^C \\ \sigma\mathcal{S} & \sim & \sigma\mathbb{X} \end{array}\right. \tag{29}$$

As an illustration, consider the set

$$\mathbb{X} = \left\{\mathbf{x} \in \mathbb{R}^2, (x_1 - 2)^2 + (3x_2 + x_1 - 1)^2 \in [0, 4]\right\} \tag{30}$$

using a paver with a separator $\mathcal{S}$ for $\mathbb{X}$, we get Fig. 6a. The blue part has been eliminated by the outer contractor $\mathcal{S}^{out}$ of $\mathcal{S}$. Whereas the magenta part has been eliminated by the inner contractor $\mathcal{S}^{in}$. Figure 6b has been computed using the separator $\sigma\mathcal{S}$ where $\sigma = (2\,1)$ corresponds to the axial symmetry with respect the axis $x_1 - x_2 = 0$. We therefore get an approximation for $\sigma\mathbb{X}$. Figure 6c is obtained using the separator $\sigma\mathcal{S} \cup \mathcal{S}$ and corresponds to an approximation for $\sigma\mathbb{X} \cup \mathbb{X}$.

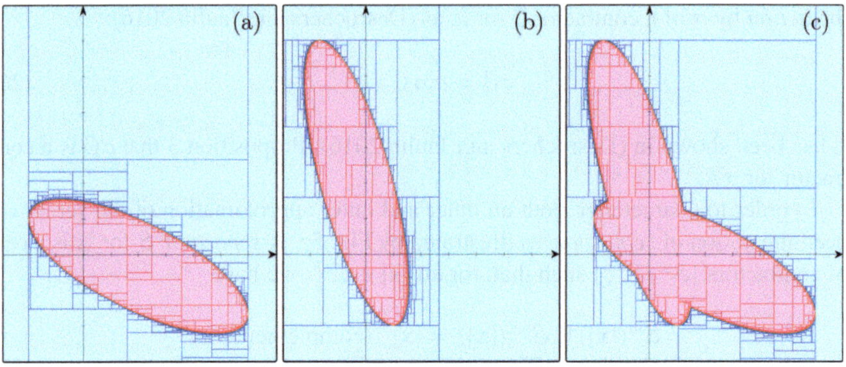

Fig. 6 A paver is used to compute an inner and an outer approximations of 3 sets: **a** an ellipse $\mathbb{X}$, **b** its symmetric $\sigma\mathbb{X}$ by σ, **c** the union of the two sets $\sigma\mathbb{X} \cup \mathbb{X}$. The frame box is $[-3, 7] \times [-3, 7]$

4 Minimal Separator for the Ellipse

This section proposes a minimal separator for an ellipse. To get the optimality, the principle is to solve the problem in a part of the plane (x_1, x_2) where the function $f(x_1, x_2)$ is monotonic. The symmetries will then allow us to extend the resolution to the whole plane. The minimal separator will be used later by a paver to compute boxes that are completely inside or outside the ellipse. In this section, we still consider the equation $f(\mathbf{q}, \mathbf{x}) = 0$, but $\mathbf{q}$ is considered as a parameter vector. The dependency with respect to $\mathbf{q}$ will not always be written.

4.1 Cardinal Points

We define the cardinal points as the points (x_1, x_2) which satisfy

$$\begin{cases} f(\mathbf{x}) = 0 \\ \frac{\partial f}{\partial x_1}(\mathbf{x}) = 0 \text{ or } \frac{\partial f}{\partial x_2}(\mathbf{x}) = 0 \end{cases} \tag{31}$$

Generically, there exist four cardinal points. The cardinal point, painted red in Fig. 7, at the top (left, bottom, right) is the North (West, South, East, respectively).

4.2 Contractor for the Positive Quadrant

The part of the ellipse between the North and the East is called the *positive arc* and is painted blue in Fig. 7. The smallest box which encloses this arc is the *positive*

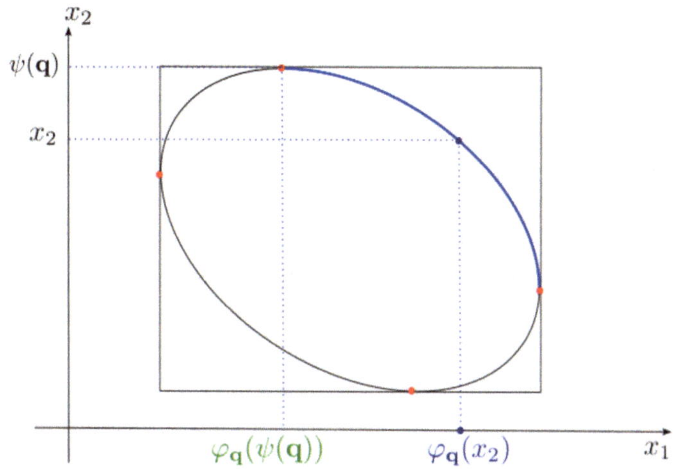

Fig. 7 Positive arc and the corresponding function $\varphi_{\mathbf{q}}(x_2)$

quadrant. The four red points are the cardinal points. The top (left, bottom, right, resp.) cardinal points are called the North, West, South, East.

Proposition 1 *Take a point $x = (x_1, x_2)$ such that $f(\mathbf{x}) = 0$ of the positive quadrant. We have*

$$\begin{aligned} x_1 &= \varphi_{\mathbf{q}}(x_2) \\ &= \frac{-(q_1+q_4x_2)+\sqrt{(q_1+q_4x_2)^2-4q_1(q_0+q_2x_2+q_5x_2^2)}}{2q_3} \end{aligned} \tag{32}$$

The largest feasible x_2 is

$$\begin{aligned} x_2 &= \psi(\mathbf{q}) \\ &= \frac{-2q_3q_2+q_1q_4+\sqrt{(2q_3q_2-q_1q_4)^2-(4q_3q_5-q_4^2)(4q_3q_0-q_1^2)}}{4q_3q_5-q_4^2} \end{aligned} \tag{33}$$

The North has the coordinates $(\varphi_{\mathbf{q}}(\psi(\mathbf{q})), \psi(\mathbf{q}))$.

Proof Assume that x_2 is known. Let us compute the possible values for x_1. Since

$$f(x_1, x_2) \overset{(1)}{=} q_3x_1^2 + (q_1 + q_4x_2)x_1 + q_2x_2 + q_0 + q_5x_2^2, \tag{34}$$

we get the following discriminant:

$$\Delta_1 = b_1^2 - 4a_1c_1 \tag{35}$$

where

$$a_1 = q_3,\ b_1 = q_1 + q_4x_2,\ c_1 = q_0 + q_2x_2 + q_5x_2^2. \tag{36}$$

The two solutions are

$$x_1 = \frac{-b_1 \pm \sqrt{\Delta_1}}{2a_1}. \tag{37}$$

We have thus proved (32).

A value for x_2 yields a feasible x_1 if $\Delta_1 \geq 0$, i.e.,

$$\begin{array}{ll} b_1^2 - 4a_1c_1 & \geq 0 \\ \Leftrightarrow -(q_1 + q_4x_2)^2 + 4q_3(q_0 + q_2x_2 + q_5x_2^2) & \geq 0 \\ \Leftrightarrow (4q_3q_5 - q_4^2)x_2^2 + (4q_3q_2 - 2q_1q_4)x_2 + 4q_3q_0 - q_1^2 & \leq 0 \end{array} \tag{38}$$

which is quadratic in x_2. The discriminant is

$$\Delta_2 = b_2^2 - 4a_2c_2 \tag{39}$$

where

$$\begin{array}{rl} a_2 = & 4q_3q_5 - q_4^2 \\ b_2 = & 4q_3q_2 - 2q_1q_4 \\ c_2 = & 4q_3q_0 - q_1^2 \end{array} \tag{40}$$

We thus get (33). □

Proposition 2 *Take a point $x = (x_1, x_2)$ such that $f(\mathbf{x}) = 0$ of the positive quadrant. We have*

$$x_2 = \varphi_{\sigma(\mathbf{q})}(x_1) \tag{41}$$

where $\sigma = (1, 3, 2, 6, 5, 4)$. The largest feasible x_1 is

$$x_1 = \psi(\sigma(\mathbf{q})). \tag{42}$$

The East has the coordinates $\left(\psi(\sigma(\mathbf{q})), \varphi_{\sigma(\mathbf{q})}(\psi(\sigma(\mathbf{q})))\right)$.

Proof The symmetry which permutes x_1, x_2 is $\sigma = (1, 3, 2, 6, 5, 4)$. Indeed:

$$\begin{array}{rl} f(x_2, x_1) \overset{(1)}{=} & q_0 + q_1x_2 + q_2x_1 + q_3x_2^2 + q_4x_1x_2 + q_5x_1^2 \\ = & q_0 + q_2x_1 + q_1x_2 + q_5x_1^2 + q_4x_1x_2 + q_3x_2^2 \end{array} \tag{43}$$

After application of the symmetry, Proposition 2 falls in the conditions of Proposition 1 (see Fig. 8). □

Proposition 3 *The smallest box which contains the North and the East is*

$$[\mathbf{a}] = [\varphi_{\mathbf{q}}(\psi(\mathbf{q})), \psi(\sigma(\mathbf{q}))] \times [\varphi_{\sigma(\mathbf{q})}(\psi(\sigma(\mathbf{q}))), \psi(\mathbf{q})]. \tag{44}$$

Proof The result can be read directly from Fig. 9. □

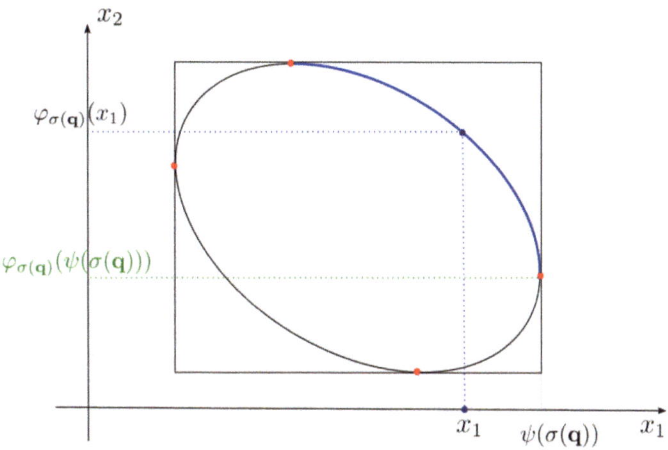

Fig. 8 Positive arc and the corresponding function $\varphi_{\sigma(\mathbf{q})}$ after permutation

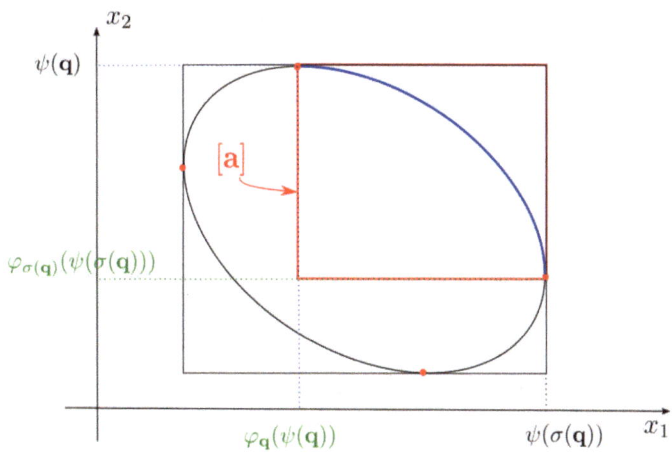

Fig. 9 Smallest box [**a**] which encloses the North and the East

Proposition 4 *The minimal contractor associated to the positive ellipse is*

$$C_0^{\mathbf{q}}([\mathbf{x}]) = [\mathbf{x}] \cap \begin{pmatrix} [\varphi_{\mathbf{q}}(b_2^+), \varphi_{\mathbf{q}}(b_2^-)] \\ [\varphi_{\sigma(\mathbf{q})}(b_1^+), \varphi_{\sigma(\mathbf{q})}(b_1^-)] \end{pmatrix} \tag{45}$$

with $[\mathbf{b}] = [\mathbf{x}] \cap [\mathbf{a}]$.

Proof This is a direct consequence of the monotonicity of the partial function $\varphi_{\mathbf{q}}$. □

If we apply this contractor in a paver with $\mathbf{q} = (-5, 1, 1, 3, 1, 2)$, we get Fig. 10. Note that we have here an outer approximation of the set

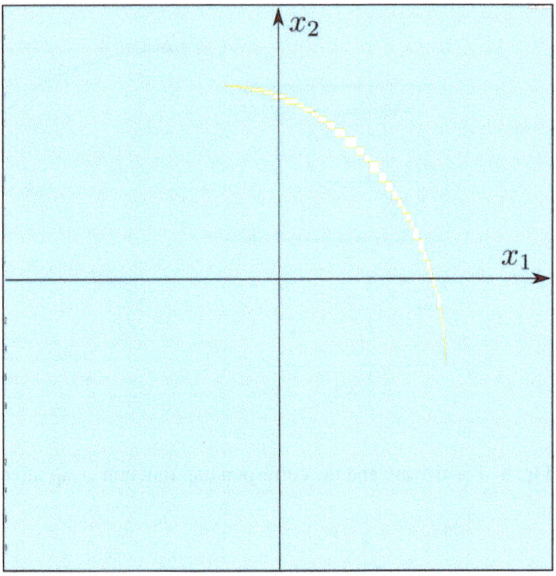

Fig. 10 Illustration of the application of the contractor for the ellipse on the positive quadrant.
The white boxes enclose all solutions. The blue boxes are proved to contain no solutions. The frame box is $[-2, 2] \times [-2, 2]$

$$-5 + x_1 + x_2 + 3x_1^2 + x_1x_2 + 2x_2^2 = 0 \tag{46}$$

only on its positive part. We understand that if we are able to obtain this approximation for one quadrant, the same methodology applies to the other quadrants. It is the role of symmetries to adapt the method working one quadrant to the others.

4.3 Contractor for the Ellipse Boundary

Section 4.2 has shown how to build a contractor for the North-East quadrant of the ellipse. Recall that $C_0^{\mathbf{q}}([\mathbf{x}])$ contracts the box $[\mathbf{x}]$ with respect to the positive quadrant of the ellipse. It depend on the parameter vector $\mathbf{q}$ of the ellipse. Using the notion of contractor action (Jaulin 2022), we show how we can extend this contractor $C_0^{\mathbf{q}}$ to other quadrants. We recall that the action of a symmetry ε to the contractor C is defined by

$$\varepsilon \bullet C([\mathbf{x}]) = \varepsilon \circ C \circ \varepsilon^{-1}([\mathbf{x}]). \tag{47}$$

This means that $\varepsilon \bullet C$ is a contractor that has been built from the contractor C as follows:

- Apply to the box $[\mathbf{x}]$ the symmetry ε^{-1}
- Apply the contractor C
- Apply to the resulting box $C \circ \varepsilon^{-1}([\mathbf{x}])$ the symmetry ε.

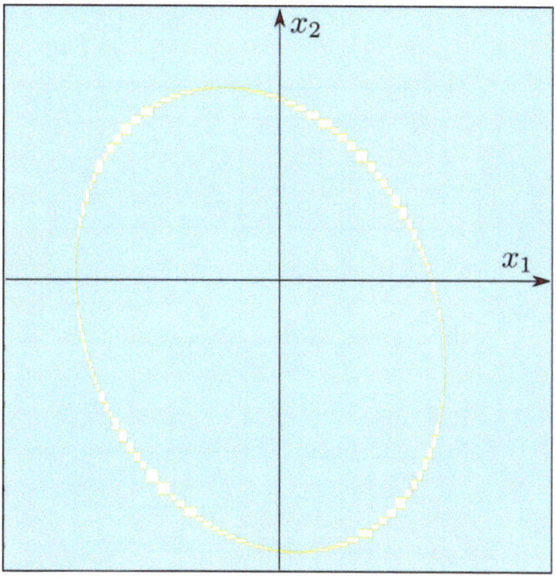

Fig. 11 Illustration of the application of the contractor for the ellipse on all quadrants. The frame box is $[-2, 2] \times [-2, 2]$

If we consider the pair (ε, σ) conjugate with respect to the ellipse, the contractor $\varepsilon \bullet C_0^{\sigma(\mathbf{q})}$ is associated to another quadrant of the ellipse. The selection of the symmetries (ε, σ) to be selected is made using the choice function (21). In the ellipse case, we clearly understand geometrically that 4 symmetries are needed since the ellipse has 4 quadrants (North-East, North-West, South-West, South-East).

The contractor for the ellipse boundary is thus

$$[\mathbf{x}] \mapsto \bigcup_{\varepsilon \in \{(1,2),(1,-2),(-1,2),(-1,-2)\}} (\varepsilon \bullet C_0^{\psi_\varepsilon(\mathbf{q})})([\mathbf{x}]). \tag{48}$$

The application of this contractor is illustrated by Fig. 11. Again, the white boxes enclose all solutions and the blue boxes are proved to contain no solutions.

4.4 Separator for the Ellipse

From a contractor on the boundary of a set $\mathbb{X}$ and a test for $\mathbb{X}$, we can obtain a separator. This is illustrated by Fig. 12. Subfigure (1) presents three boxes $[\mathbf{a}]$, $[\mathbf{b}]$ and $[\mathbf{c}]$. In Subfigure (2), the three boxes are contracted with respect to the boundary $\partial\mathbb{X}$ of $\mathbb{X}$ to get the yellow boxes. In Subfigure (3) we test one point in each connected component of the contracted zone (white). Since for the box $[\mathbf{b}]$, we have two connected components, we have to test two points. As shown in Subfigure (4), if the tested point is inside $\mathbb{X}$, the corresponding white polygon is fully inside $\mathbb{X}$ and if the tested point is outside $\mathbb{X}$, the corresponding white polygon is fully outside $\mathbb{X}$.

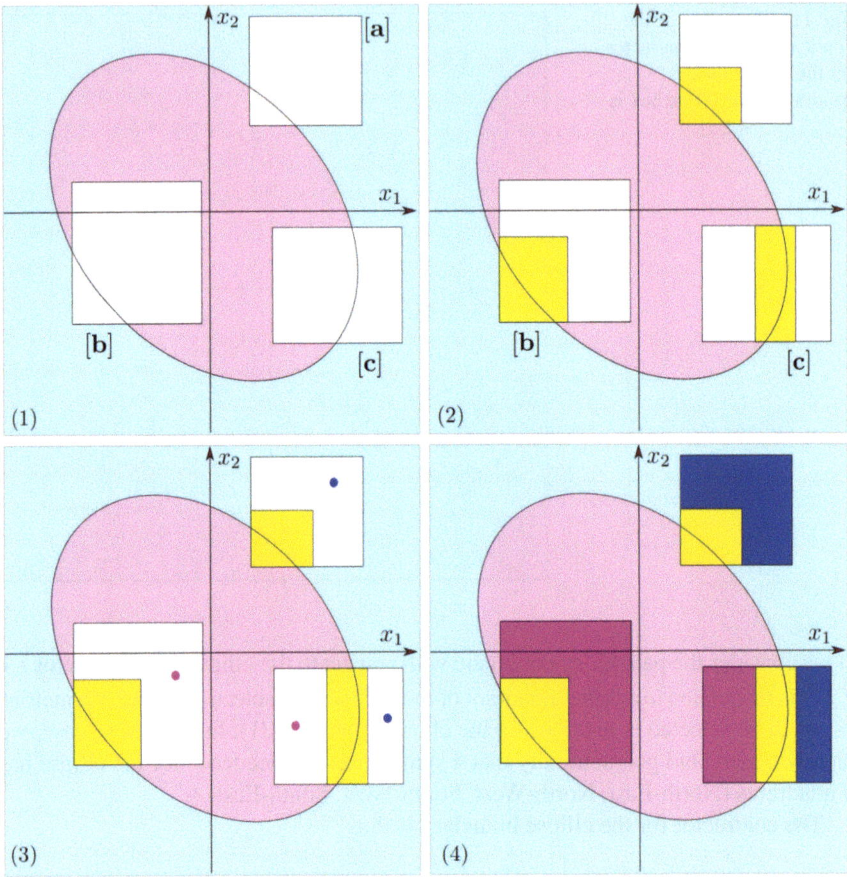

Fig. 12 With a contractor on the boundary $\partial\mathbb{X}$ and a test for $\mathbb{X}$, we build a separator for $\mathbb{X}$

As a consequence, we can get an inner and an outer approximations for $\mathbb{X}$ as illustrated by Fig. 13. Any point in a magenta box is a solution and no point in a blue box is a solution.

If we compare with a classical forward-backward contractor (Benhamou et al. 1999) (see Fig. 14) or other contractors such as Araya et al. (2012), we see that our contractor yields a more accurate approximation. This is consistent with the fact that our separator is minimal.

Remark. We have assumed that we had no uncertainties on $\mathbf{q}$. In case of interval uncertainty, the set to be characterized becomes

$$\mathbb{X} = \{\mathbf{x}|\exists\mathbf{q} \in [\mathbf{q}], q_0 + q_1x_1 + q_2x_2 + q_3x_1^2 + q_4x_1x_2 + q_5x_2^2 \leq 0\}. \quad (49)$$

The resolution is still possible as shown in Jaulin (2023).

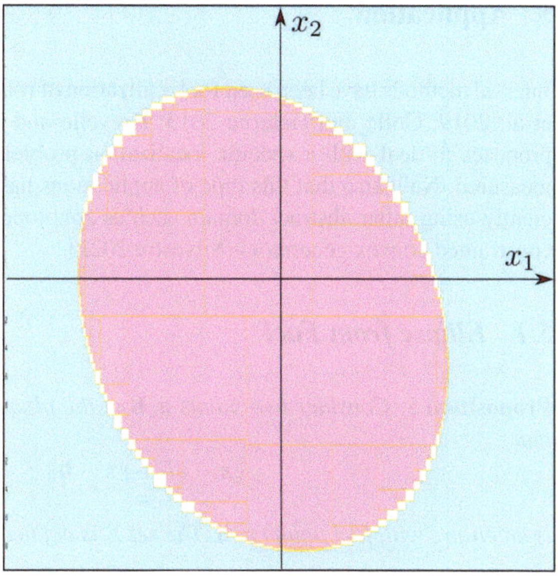

Fig. 13 Illustration of the application of the separator for the ellipse on all quadrants. The solution set contains the magenta zone and does not intersect the blue zone. The frame box is $[-2, 2] \times [-2, 2]$

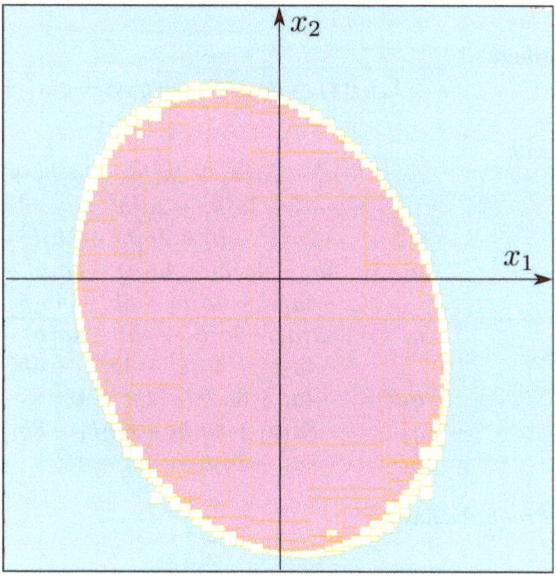

Fig. 14 Classical interval contractors require a larger number of bisections

5 Application

Interval methods have been used for localization of robots for several decades (Rohou et al. 2019; Colle and Galerne 2013; Drevelle and Bonnifait 2009). This section proposes to deal with a specific localization problem where sum of distances are measured. Note also that this type of applications has already been considered efficiently using other abstract domain such as zonotopes (Bhamidipati et al. 2022) or constrained convex generators (Silvestre 2022).

5.1 Ellipse from Foci

Proposition 5 *Consider two points* **a**, **b** *of the plane. The set* $\mathbb{X}$ *of all points such that*

$$\|\mathbf{x}-\mathbf{a}\|+\|\mathbf{x}-\mathbf{b}\| \leq \ell \tag{50}$$

is an ellipse with foci points **a**, **b**. *The set* $\mathbb{X}$ *is defined by the inequality*

$$\mathbf{f}_{\mathbf{a},\mathbf{b},\ell}(\mathbf{x}) \leq 0 \tag{51}$$

where

$$\mathbf{f}_{\mathbf{a},\mathbf{b},\ell}(\mathbf{x}) = q_0 + q_1x_1 + q_2x_2 + q_3x_1^2 + q_4x_1x_2 + q_5x_2^2 \tag{52}$$

with

$$\begin{array}{rcl} q_0 &=& -\,a_1^4 - 2a_1^2a_2^2 + 2a_1^2b_1^2 + 2a_1^2b_2^2 + 2a_1^2\ell^2 \\ && -a_2^4 + 2a_2^2b_1^2 + 2a_2^2b_2^2 \\ && +2a_2^2\ell^2 - b_1^4 - 2b_1^2b_2^2 + 2b_1^2\ell^2 - b_2^4 + 2b_2^2\ell^2 - \ell^4 \\ q_1 &=& 4a_1^3 - 4a_1^2b_1 + 4a_1a_2^2 - 4a_1b_1^2 - 4a_1b_2^2 \\ && -4a_1\ell^2 - 4a_2^2b_1 + 4b_1^3 + 4b_1b_2^2 - 4b_1\ell^2 \\ q_2 &=& 4a_1^2a_2 - 4a_1^2b_2 + 4a_2^3 - 4a_2^2b_2 - 4a_2b_1^2 \\ && -4a_2b_2^2 - 4a_2\ell^2 + 4b_1^2b_2 + 4b_2^3 - 4b_2\ell^2 \\ q_3 &=& -4a_1^2 + 8a_1b_1 - 4b_1^2 + 4\ell^2 \\ q_4 &=& -8a_1a_2 + 8a_1b_2 + 8a_2b_1 - 8b_1b_2 \\ q_5 &=& -4a_2^2 + 8a_2b_2 - 4b_2^2 + 4\ell^2 \end{array} \tag{53}$$

Proof We have

$$\begin{array}{l} \|\mathbf{x}-\mathbf{a}\|+\|\mathbf{x}-\mathbf{b}\| = \ell \\ \Leftrightarrow \sqrt{(x_1-a_1)^2+(x_2-a_2)^2} + \sqrt{(x_1-b_1)^2+(x_2-b_2)^2} = \ell \end{array} \tag{54}$$

After some trivial symbolic calculus, we get to get rid of the square root to get

$$\begin{array}{r} 4\left((x_1-a_1)^2+(x_2-a_2)^2\right)\left((x_1-b_1)^2+(x_2-b_2)^2\right) \\ -\left(\ell^2-(x_1-a_1)^2-(x_2-a_2)^2-(x_1-b_1)^2-(x_2-b_2)^2\right)^2 = 0 \end{array} \tag{55}$$

We can develop the expression to get the coefficients of the proposition. □

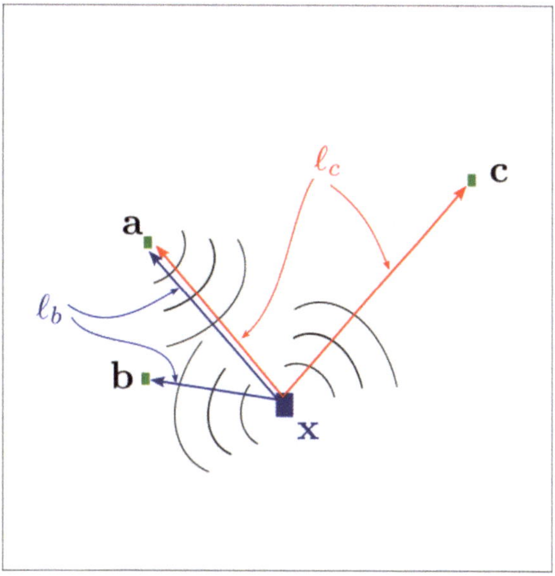

Fig. 15 The sonar system returns the two distances ℓ_b and ℓ_c

5.2 Localization

We consider an example related to localization which can be seen as special case of interval data fitting problem (Kreinovich and Shary 2016). Consider three sonars located at points $\mathbf{a} : (-2, 1)$, $\mathbf{b} : (-2, -1)$, $\mathbf{c} : (3, 2)$ of the plane. The emitter **a** sends a sound which is reflected by an object at position **x** received by **b** and **c** (see Fig. 15). We thus have two path for the sound: either $\mathbf{a} \to \mathbf{x} \to \mathbf{b}$ or $\mathbf{a} \to \mathbf{x} \to \mathbf{c}$. From the time of flight of the sound we want to estimate the position of the object.

We assume that we were able to collect two distance intervals such that $\ell_b \in [4, 6]$ and $\ell_c \in [7, 9]$. The solution set $\mathbb{X}$ is defined by

$$\begin{array}{ll} \text{(i)} & \|\mathbf{x}-\mathbf{a}\| + \|\mathbf{x}-\mathbf{b}\| = \ell_b \in [4, 6] \\ \text{(ii)} & \|\mathbf{x}-\mathbf{a}\| + \|\mathbf{x}-\mathbf{c}\| = \ell_c \in [7, 9] \end{array} \tag{56}$$

From Proposition 5, we get that $\mathbb{X}$ is defined by

$$\mathbb{X} : \begin{cases} \mathbf{f}_{\mathbf{a},\mathbf{b},6}(\mathbf{x}) \leq 0 \\ \mathbf{f}_{\mathbf{a},\mathbf{b},4}(\mathbf{x}) \geq 0 \\ \mathbf{f}_{\mathbf{a},\mathbf{c},9}(\mathbf{x}) \leq 0 \\ \mathbf{f}_{\mathbf{a},\mathbf{c},7}(\mathbf{x}) \geq 0 \end{cases} \tag{57}$$

Using a paver, we are thus able to get an inner and an outer approximations for the set of $\mathbb{X}$ (see Fig. 18). Figure 16 represents the inequality (56,i) and Fig. 17 corre-

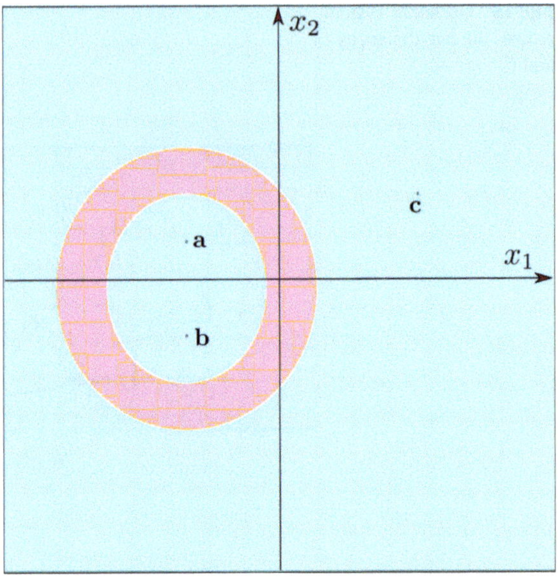

Fig. 16 Set of positions consistent with the path $\mathbf{a} \to \mathbf{x} \to \mathbf{b}$. The frame box is $[-7, 7] \times [-7, 7]$

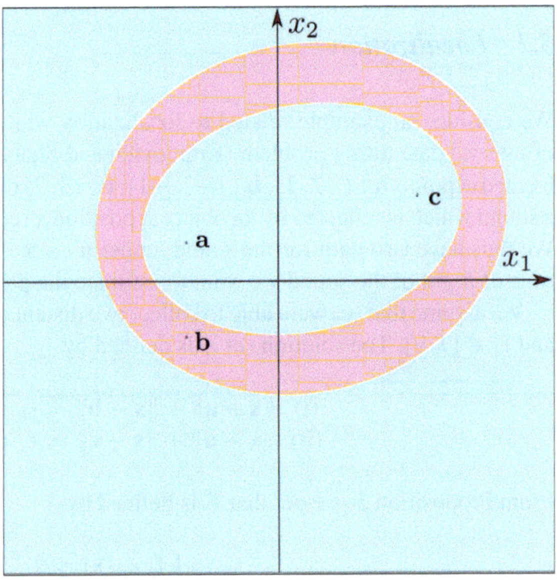

Fig. 17 Set of positions consistent with the path $\mathbf{a} \to \mathbf{x} \to \mathbf{c}$. The frame box is $[-7, 7] \times [-7, 7]$

sponds to the inequality (56,ii). All results are guaranteed since outward rounding is implemented (Revol et al. 2022).

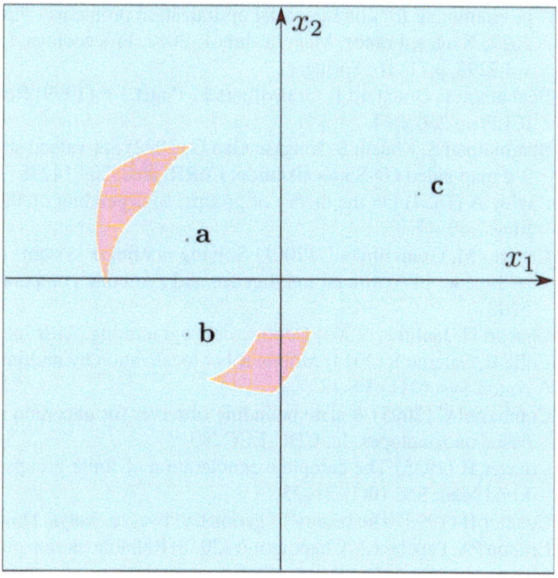

Fig. 18 Set of positions consistent with the path $\mathbf{a} \to \mathbf{x} \to \mathbf{b}$ and the path $\mathbf{a} \to \mathbf{x} \to \mathbf{c}$. The frame box is $[-7, 7] \times [-7, 7]$

6 Conclusion

This paper has proposed a minimal contractor and a minimal separator for an ellipse of the plane. The notion of actions derived from hyperoctahedral symmetries allowed us to limit the analysis in on part of the constraint where the monotonicity can be assumed. The symmetries was used to extend the analysis to the whole plane.

The goal of this paper was to provide a simple example which illustrates how to use the hyperoctahedral symmetries in order to build minimal separators. Now, as shown in Jaulin (2023), the use of these symmetries is more interesting when we deal with projection problems where quantifier elimination is needed. This type of projection problem is indeed much more difficult to solve with classical interval approaches (Hladík and Ratschan 2014).

The Python code based on Codac (http://codac.io/) is given at

www.ensta-bretagne.fr/jaulin/ctcellipse.html

References

Althoff M, Krogh B (2011) Zonotope bundles for the efficient computation of reachable sets. In: 2011 50th IEEE conference on decision and control and European control conference, pp 6814–6821

Araya I, Trombettoni G, Neveu B (2012) A contractor based on convex interval Taylor. In: Beldiceanu N, Jussien N, Pinson E (eds), Integration of AI and OR techniques in constraint programming for combinatorial optimization problems - 9th international conference, CPAIOR 2012, Nantes, France, May 28–June1, 2012. Proceedings, Lecture Notes in Computer Science, vol 7298, pp 1–16. Springer

Benhamou F, Goualard F, Granvilliers L, Puget J-F (1999) Revising hull and box consistency. In: ICLP, pp 230–244

Bhamidipati S, Kousik S, Xingxin Gao G (2022) Set-valued shadow matching using zonotopes for 3-d map-aided GNSS localization. CoRR, abs/2209.14238

Cayley A (1854) On the theory of groups, as depending on the symbolic equation $\theta^n = 1$. Philos Mag 7:40–47

Cébério M, Granvilliers L (2001) Solving nonlinear systems by constraint inversion and interval arithmetic. In: Artificial intelligence and symbolic computation, vol 1930, pp 127–141, LNCS 5202

Chabert G, Jaulin L (2009) Contractor programming. Artif Intell 173:1079–1100

Colle E, Galerne S (2013) Mobile robot localization by multiangulation using set inversion. Robot Auton Syst 61(1):39–48

Combastel C (2005) A state bounding observer for uncertain non-linear continuous-time systems based on zonotopes. In: CDC-ECC '05

Coxeter H (1935) The complete enumeration of finite groups of the form $r_i^2 = (r_i \cdot r_j)^{k_{ij}} = 1$. J Lond Math Soc 10(1):21–25

Coxeter H (1999) The beauty of geometry: twelve essays. Dover Books on Mathematics

Crépon PA, Panchea A, Chapoutot A (2018) Reliable motion plannning for a mobile robot. In: IEEE international conference on robotic computing

Desrochers B, Jaulin L (2016) A minimal contractor for the polar equation; application to robot localization. Eng Appl Artif Intell 55:83–92

Dit Sandretto JA, Trombettoni G, Daney D, Chabert G (2014) Certified calibration of a cable-driven robot using interval contractor programming. In: Thomas F, Pérez Gracia A (eds), Computational kinematics, mechanisms and machine science. Spinger

Drevelle V, Bonnifait P (2009) High integrity gnss location zone characterization using interval analysis. In: ION GNSS

Durieu C, Polyak B, Walter E (1996) Trace versus determinant in ellipsoidal outer bounding with application to state estimation. In: Proceedings of the 13th IFAC world congress, vol I, pp 43–48, San Francisco, CA

Ehambram A, Voges R, Wagner B (2021) Stereo-visual-lidar sensor fusion using set-membership methods. In: 17th IEEE international conference on automation science and engineering, CASE 2021, Lyon, France, Aug 23–27, 2021, pp 1132–1139. IEEE

Guyonneau R, Lagrange S, Hardouin L (2013) A visibility information for multi-robot localization. In: IEEE/RSJ international conference on intelligent robots and systems (IROS)

Hladík M, Ratschan S (2014) Efficient solution of a class of quantified constraints with quantifier prefix exists-forall. Math Comput Sci 8(3–4):329–340

Jaulin L (2022) Actions of the hyperoctahedral group to compute minimal contractors. Artif Intell 313:103790

Jaulin L (2023) Inner and outer characterization of the projection of polynomial equations using symmetries, quotients and intervals. Int J Approx Reason 159:108928

Jaulin L, Desrochers B (2014) Introduction to the algebra of separators with application to path planning. Eng Appl Artif Intell 33:141–147

Kreinovich V, Lakeyev AV, Rohn J, Kahl PT (1997) Computational complexity and feasibility of data processing and interval computations. Springer

Kreinovich V, Shary S (2016) Interval methods for data fitting under uncertainty: a probabilistic treatment. Reliab Comput 23:105–140

Moore R (1979) Methods and applications of interval analysis. Society for industrial and applied mathematics, Jan 1979

Moore RE (1966) Interval analysis. Prentice-Hall, Englewood Cliffs, NJ

Moore RE, Kearfott RB, Cloud MJ (2009) Introduction to interval analysis. SIAM, Philadelphia, PA

Mustafa M, Stancu A, Delanoue N, Codres E (2018) Guaranteed SLAM; An interval approach. Robot Auton Syst 100:160–170

Ramdani N, Poignet P (2005) Robust dynamic experimental identification of robots with set membership uncertainty. IEEE/ASME Trans Mechatron 10(2):253–256

Rauh A, Auer E (2009) Interval approaches to reliable control of dynamical systems. In: Brown BM, Kaltofen E, Oishi S, Rump SM (eds), Computer-assisted proofs - tools, methods and applications, 15.11. - 20.11.2009, Dagstuhl Seminar Proceedings, vol 09471. Schloss Dagstuhl - Leibniz-Zentrum für Informatik, Germany

Rauh A, Auer E (2011) Modeling, design, and simulation of systems with uncertainties. Springer

Revol N, Benet L, Ferranti L, Zhilin S (2022) Testing interval arithmetic libraries, including their ieee-1788 compliance. arXiv:2205.11837, math.NA

Rohou S, Jaulin L, Mihaylova L, Le Bars F, Veres S (2019) Reliable robot localization. Wiley

Sainudiin R (2010) Machine interval experiments: accounting for the physical limits on empirical and numerical resolutions. LAP Academic Publishers, Köln, Germany

Silvestre D (2022) Constrained convex generators: a tool suitable for set-based estimation with range and bearing measurements. IEEE Control Syst Lett 6:1610–1615

van Hentenryck P, Deville Y, Michel L (1997) Numerica: a modeling language for global optimization. MIT Press, Boston, MA

Wan J (2007) Computationally reliable approaches of contractive model predictive control for discrete-time systems. PhD dissertation, Universitat de Girona, Girona, Spain

Wussing H (2007) The genesis of the abstract group concept: a contribution to the history of the origin of abstract group theory. Dover Publications

Zonotopic Velocity Sensor Fault Detection for Autonomous Underwater Vehicle: An LMI Approach

Jitao Li, Yuxi Liu, Yushi Zhang, Mingjun Zhang, and Zhenhua Wang

Abstract In this paper, a novel zonotopic velocity sensor fault detection approach for autonomous underwater vehicles is presented. To make the generated residual robust against disturbance, we develop a fault detection observer based on H_∞ performance index. Sufficient design conditions are formulated in terms of linear matrix inequalities. Fault detection is accomplished by checking whether the zero point lies outside the residual zonotope. The performance of the presented approach is analyzed via a minimum detectable fault set. Simulation examples are given to validate the effectiveness and feasibility of the presented approach.

1 Introduction

With the gradual depletion of non-renewable resources on land, the development of marine resources has become a focal point of public attention. To promote the exploitation of marine resources, it is imperative to provide the necessary support relying on advanced equipment and technology (Liu et al. 2022). Compared

This work was supported by the National Natural Science Foundation of China (62303128), the Natural Science Foundation for Excellent Youth Scholars in Heilongjiang Province, China (YQ2022F003), and the Heilongjiang Postdoctoral Fund (LBH-Z24114).

J. Li (✉) · Y. Liu · Y. Zhang · M. Zhang
College of Mechanical and Electrical Engineering, Harbin Engineering University, Harbin, China
e-mail: jitaoli@hrbeu.edu.cn

Y. Liu
e-mail: jitaoli@hrbeu.edu.cn

Y. Zhang
e-mail: jitaoli@hrbeu.edu.cn

M. Zhang
e-mail: jitaoli@hrbeu.edu.cn

Z. Wang
School of Astronautics, Harbin Institute of Technology, Harbin, China
e-mail: zhenhua.wang@hit.edu.cn

T. N. Dinh et al. (eds.), *Set-Valued Approaches to Control and Estimation of Uncertain Systems*, Mathematical Engineering,
https://doi.org/10.1007/978-3-031-94239-6_7

to human-occupied vehicles, Autonomous Underwater Vehicles (AUVs) offer safe, efficient, and cost-effective means for numerous deep-sea missions, e.g., environmental monitoring (Ji et al. 2021), submarine cable/pipeline inspection (Liu et al. 2020), oceanographic and biological surveys (Fang et al. 2023), and underwater resource exploration and exploitation (Chu et al. 2024), etc. Safety is of significance for AUV during its unmanned and cableless operations in the marine environment (Xia et al. 2021). With the prolonged operation in complex marine environments, faults are inevitable and can seriously affect the performance of AUVs (Gao et al. 2021). Therefore, such faults need to be detected when they occur and appropriate measures need to be taken to mitigate their influence. As such, many works have been reported on the fault detection of AUVs (Li et al. 2024).

AUV fault detection is usually achieved using two types of methods: model-based fault detection and data-driven fault detection (Jiang et al. 2020). The data-driven fault detection method is achieved through signal processing. This processing is based on a large accumulation of historical process data collected during the AUV's operation (Chu et al. 2020). The model-based fault detection method involves establishing a mathematical model of the normal and faulty AUV system. A residual signal between the actual and estimated states is then generated for fault detection (Xia et al. 2022). In recent decades, the model-based fault detection method has attracted much attention due to its relatively mature applications (Wang and Shen 2022). However, for practical systems, model uncertainty, disturbances and measurement noise are inevitable (Wang et al. 2022; Dinh and Tran 2024; Shao et al. 2023). To minimize the effects of these uncertainties, robust fault detection has been intensively developed and widely applied in many fields (Xu et al. 2019; Zheng et al. 2021). For example, in Xu et al. (2019), a robust fault detection approach using the unknown input observer was proposed. Recently, in Wang et al. (2025), set-membership has been successfully applied for fault detection and has yielded significant results. Therefore, we aim to achieve fault detection via the set-membership method in this paper.

Under a mild assumption that uncertainties are unknown but bounded, the set-membership method aims to enclose these uncertainties using known compact sets (Ma et al. 2025a). These sets are represented by some simple geometrical forms such as interval (Zhu et al. 2023), polytope (Wan et al. 2018), ellipsoid (Tang et al. 2022), and zonotope (Wang et al. 2020a). During the past decades, zonotopes have become popular due to their flexibility, less complexity and high computation efficiency (Le et al. 2013). Scott et al. (2016) applied zonotopes to state estimation and fault detection. In Zhou et al. (2019), a novel interval fault estimation approach using the zonotope technique for discrete-time linear parameter-varying systems was proposed. Wang et al. (2020b) studied the problem of zonotopic fault detection for 2-D systems under an event-triggered mechanism, which utilized a novel zonotope-based dynamic threshold for residual evaluation. In Zhang et al. (2023), zonotopes were used to achieve active fault diagnosis for uncertain LPV systems. Li et al. (2023a) proposed an attack detection approach for a class of cyber-physical systems with bounded disturbances, which is based on zonotopic reachability analysis. In this paper, zonotope is used to enclose model uncertainties, disturbances, and measurement noise to achieve fault detection. The sufficient design condition of the zonotopic fault detection observer is given in terms of linear matrix inequalities (LMIs).

LMIs are widely used to solve some problems in systems and control (Zhou et al. 2025). As shown in reference Bisoffi et al. (2022), a wide variety of problems arising in systems and control theory can be reduced to a few standard convex or quasiconvex optimization problems involving LMIs. Ma et al. (2025b) focused on the observer design problem via LMIs for discrete-time descriptor systems. In Guerra et al. (2017), an H_∞ observer for continuous-time nonlinear systems was designed, where sufficient conditions for the H_∞ observer were formulated in terms of LMIs. In Fridman and Shaikhet (2017), the static output-feedback stabilization problem for the n-th order vector differential equations by using artificial multiple delays was considered, wherein conditions for the stability analysis of the closed-loop system were presented in terms of LMIs. Haj Brahim et al. (2017) addressed the problem of robust sensor fault-tolerant control for T-S fuzzy descriptor systems, where sufficient conditions for the design of a robust fuzzy descriptor observer and fault-tolerant controller were formulated in terms of LMIs. Given the success of these works, we will provide the design conditions of the observer in the form of LMIs.

Based on the above discussions, this paper aims to develop a novel zonotopic velocity sensor fault detection approach for AUVs. The main contributions of this paper are as follows:

1. A velocity sensor fault detection problem based on zonotope for AUVs is solved. Fault detection is accomplished by checking whether the zero point lies outside the residual zonotope.
2. Compared with the interval observer in Li et al. (2022), The proposed method achieves higher estimation accuracy and provides more reliable fault detection.
3. The H_∞ performance is applied in the zonotopic observer design to achieve disturbance robustness. The observer gain is obtained by solving an optimization problem formulated as an LMI.
4. A so-called minimum detectable fault set is considered to analyze the fault detection performance of the proposed method.

The remainder of this paper is organized as follows: In Sect. 2, we present some necessary preliminaries and the problem formulation. Section 3 provides the main results of the proposed fault detection method for AUV. A simulation example is given to verify the effectiveness of the proposed approach in Sect. 4. Finally, this research is concluded in Sect. 5.

2 Preliminaries and Problem Formulation

In this section, we will introduce some preliminaries and problem formulation.

2.1 Preliminaries

The following notations used in this paper: The set of real numbers is denoted by $\mathbb{R}$, $\mathbb{R}^n$ and $\mathbb{R}^{n\times m}$ are the set composed of n dimensional Euclidean space and $n \times m$ dimensional matrix. The set of nonnegative integers is denoted by $\mathbb{N}_+$. For a matrix A, A^T denotes its transposition, and $A^\dagger$ denotes its Moore-Penrose inverse. Given matrices A_1 and A_2 of adequate dimensions, we use $\begin{bmatrix} A_1 & A_2 \end{bmatrix}$ and $\begin{bmatrix} A_1 \\ A_2 \end{bmatrix}$ to denote the matrices obtained by stacking A_1 and A_2 horizontally and vertically. I denotes the identity matrix. diag(·) denotes the block-diagonal matrix. $\star$ denotes the term introduced by symmetry. A matrix P is considered positive definite if the scalar $x^T Px$ is positive for any non-zero column vector x of real numbers, indicated as $P \succ 0$. Conversely, the negative definite matrix is indicated as $P \prec 0$. The comparison operators $>$, $\geq$, $<$, and $\leq$ are understood elementwise. We use $\oplus$ to denote the Minkowski sum of two sets $\mathcal{S}_1$ and $\mathcal{S}_2$, i.e.,

$$\mathcal{S}_1 \oplus \mathcal{S}_2 = \{s_1 + s_2 : s_1 \in \mathcal{S}_1, s_2 \in \mathcal{S}_2\}.$$

In this paper, the subsequent definitions and properties are fundamental.

Definition 1 (*Zonotope*) Althoff (2010): Given a vector $p \in \mathbb{R}^n$ and segment generators $h_i \in \mathbb{R}^n$, $i = 1, 2, \ldots, m$, a zonotope is defined as:

$$\mathcal{Z} = \left\{ x \in \mathbb{R}^n : x = p + \sum_{i=1}^{m} h_i b_i, b_i \in \mathbb{B} \right\},$$

where $\mathbb{B} = \{b \in \mathbb{R} : -1 \leq b \leq 1\}$ is the unitary interval. The geometrical interpretation of $\mathcal{Z}$ is the Minkowski sum of m line segments $\tilde{h}_i = h_i\mathbb{B}$, $i = 1, \ldots, m$. The order of the zonotope $\mathcal{Z}$ is $\frac{m}{n}$. For the sake of concise notation, the center and generators of a zonotope are written in a compact form as $\mathcal{Z} = \langle p, H\rangle$, where $H = \begin{bmatrix} h_1 & h_2 & \cdots & h_m \end{bmatrix}$. The construction of a zonotope step by step can be found in Fig. 1.

Property 1 (Minkowski sum) *Tang et al. (2020): The Minkowski sum of two zonotopes $\mathcal{Z}_1 = \langle p_1, H_1\rangle$ and $\mathcal{Z}_2 = \langle p_2, H_2\rangle$ is also a zonotope, and the subsequent equality is satisfied:*

$$\langle p_1, H_1\rangle \oplus \langle p_1, H_1\rangle = \langle p_1 + p_2, [H_1, H_2]\rangle$$

Property 2 (Linear transformation) *Tang et al. (2020): Given a zonotope $\mathcal{Z} = \langle p, H\rangle$, its linear transformation induced by a matrix K is*

$$K\langle p, H\rangle = \langle Kp, KH\rangle$$

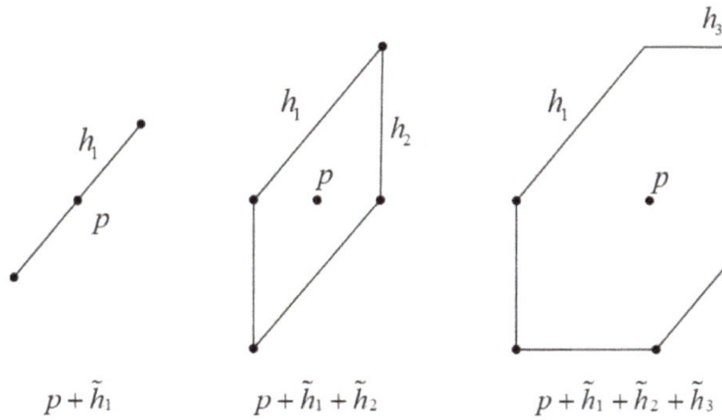

Fig. 1 Construction of a zonotope

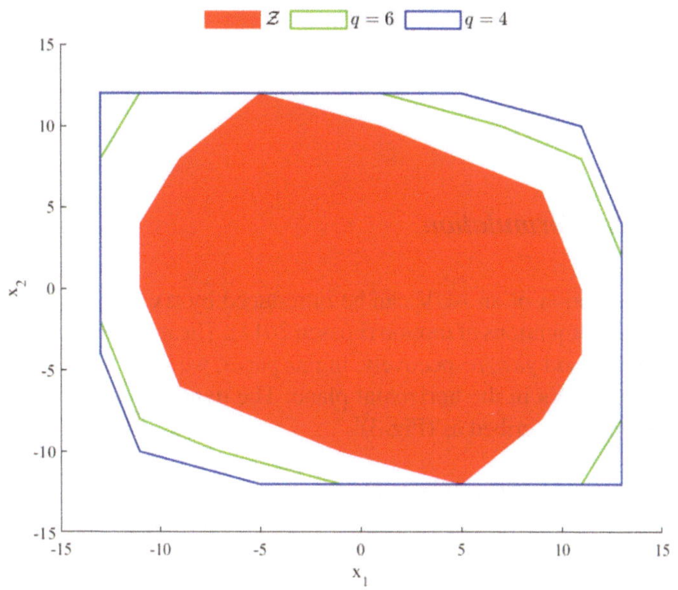

Fig. 2 Illustration of order reduction with different q values

Property 3 (Order reduction) *Alamo et al. (2005): Given a zonotope* $\mathcal{Z} = p \oplus HB^m \subset R^n$ *and a fixed integer* $l(n < l < m)$, *show* $\tilde{H}$ *as the matrix which rearrange* H *in decreasing Euclidean norm, we have* $\mathcal{Z} \subseteq p \oplus \downarrow_l (H)B^l$, *where* $\downarrow_l (H) = [H_> \; rs(H_<)]$, $H_>$ *is the first* $l - n$ *columns of* $\tilde{H}$ *and* $H_<$ *is consisted of the rest of the columns of* $\tilde{H}$.

For ease of understanding, Fig. 2 provides the illustration of order reduction with different q values.

Lemma 1 (Schur complements) *Boyd et al. (1994): For* $Q = Q^T$ *and* $R = R^T$, *the following descriptions are equivalent,*

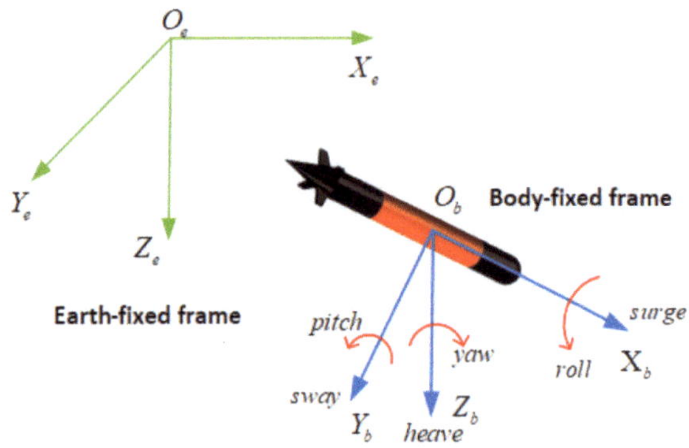

Fig. 3 Earth-fixed frame and body-fixed frame of AUV

1. $R \prec 0,\ Q - SR^{-1}S^T \prec 0,$
2. $\begin{bmatrix} Q & S \\ S^T & R \end{bmatrix} \prec 0.$

2.2 Problem Formulation

Generally, the motion of an AUV can be introduced by six degrees of freedom (6-DOF) differential equations of motion (Fossen 2011). These motions are surge, sway, heave, roll, pitch, and yaw, respectively. In this paper, we study the surge, sway, and yaw motion of AUVs in the horizontal plane. The dynamic model of AUV in the horizontal plane is described as (Fig. 3)

$$M\dot{\mathbf{v}} + C_{RB}(\mathbf{v})\mathbf{v} + C_A(\mathbf{v})\mathbf{v} + D(\mathbf{v})\mathbf{v} = \tau + \mathbf{w}, \tag{1}$$

where

$$M = \begin{bmatrix} m + X_{\dot{\rho}} & 0 & 0 \\ 0 & m + Y_{\dot{v}} & 0 \\ 0 & 0 & I_{zz} + N_{\dot{r}} \end{bmatrix}, C_{RB}(\mathbf{v}) = \begin{bmatrix} 0 & 0 & -mv \\ 0 & 0 & m\rho \\ mv & -m\rho & 0 \end{bmatrix},$$
$$C_A(\mathbf{v}) = \begin{bmatrix} 0 & 0 & Y_{\dot{v}}v \\ 0 & 0 & -X_{\dot{\rho}}\rho \\ -Y_{\dot{v}}v & X_{\dot{\rho}}\rho & 0 \end{bmatrix}, D(\mathbf{v}) = \begin{bmatrix} X_\rho & 0 & 0 \\ 0 & Y_v & 0 \\ 0 & 0 & N_r \end{bmatrix},$$

herein, $\mathbf{v} = \begin{bmatrix} \rho & v & r \end{bmatrix}^T$ is the linear and angular velocity vector of the vehicle with ρ, v, and r denoting the surge velocity, sway velocity, and yaw angular velocity, respectively, $\tau = \begin{bmatrix} \tau_X & \tau_Y & \tau_N \end{bmatrix}^T$ is the vector of thruster input, $\mathbf{w} = \begin{bmatrix} \mathbf{w}_X & \mathbf{w}_Y & \mathbf{w}_N \end{bmatrix}^T$ is the vector of wind-induced and wave-induced disturbances, m is the mass of the vehicle, I_{zz} is the inertia moment about the z-axis, and $X_{\dot{\rho}}$, $Y_{\dot{v}}$, $N_{\dot{r}}$, X_ρ, Y_v, N_r are hydrodynamic coefficients.

Assuming that the vehicle is cruising at a constant speed, we have that $\rho = \rho_0$ with ρ_0 is a constant. Then, the dynamic Eq. (1) can be reformulated as

$$M_1\dot{\mathbf{v}}_1 + C_{RB}\mathbf{v}_1 + C_A\mathbf{v}_1 + D\mathbf{v}_1 = \tau_1 + \mathbf{w}_1, \tag{2}$$

where

$$\begin{aligned}
\mathbf{v}_1 =& [v, r]^T, \quad \tau_1 = [\tau_Y, \tau_N]^T, \quad \mathbf{w}_1 = [\mathbf{w}_Y, \mathbf{w}_N]^T, \\
M_1 =& \begin{bmatrix} m + Y_{\dot{v}} & 0 \\ 0 & I_{zz} + N_{\dot{r}} \end{bmatrix}, D = \begin{bmatrix} Y_v & 0 \\ 0 & N_r \end{bmatrix}, \\
C_{RB} =& \begin{bmatrix} 0 & m\rho_0 \\ -m\rho_0 & 0 \end{bmatrix}, C_A = \begin{bmatrix} 0 & -X_{\dot{\rho}}\rho_0 \\ X_{\dot{\rho}}\rho_0 & 0 \end{bmatrix}.
\end{aligned}$$

Define

$$x(t) = \begin{bmatrix} v & r \end{bmatrix}^T, u(t) = \begin{bmatrix} \tau_Y & \tau_N \end{bmatrix}^T, w(t) = \begin{bmatrix} \mathbf{w}_Y & \mathbf{w}_N \end{bmatrix}^T,$$

with $x(t) \in \mathbb{R}^{n_x}$, $u(t) \in \mathbb{R}^{n_u}$, and $w(t) \in \mathbb{R}^{n_w}$, the AUV system can be described as

$$\dot{x}(t) = A_t x(t) + B_t u(t) + D_t w(t), \tag{3}$$

where

$$\begin{aligned}
A_t &= -M_1^{-1}(C_{RB} + C_A + D), \\
B_t &= M_1^{-1}, \\
D_t &= M_1^{-1}.
\end{aligned}$$

Using Euler's discretization with Δt, the AUV system (3) is expressed as:

$$x(k+1) = Ax(k) + Bu(k) + D_w w(k), \tag{4}$$

where

$$\begin{aligned}
A &= A_t \Delta t + I, \\
B &= B_t \Delta t, \\
D_w &= D_t \Delta t.
\end{aligned}$$

The system output could be expressed as:

$$y(k) = Cx(k) + D_\eta \eta(k) + Ef(k), \tag{5}$$

where $y(k) \in \mathbb{R}^{n_y}$ is the measurement output, $\eta(k) \in \mathbb{R}^{n_\eta}$ is the measurement noise, $f(k) \in \mathbb{R}^{n_f}$ is the sensor fault, C, D_η, and E are known matrices with compatible dimensions.

The following assumption is essential.

Assumption 1 The process disturbance $w(k)$, the measurement noise $\eta(k)$ and the initial state $x(0)$ are unknown but bounded, i.e.

$$|x(0) - p_0| \leq \bar{x}_0, |w(k)| \leq \bar{w}, |\eta(k)| \leq \bar{\eta},$$

where $p_0 \in R^{n_x}$, $\bar{x}_0 \in R^{n_x}$, $\bar{w} \in R^{n_w}$, and $\bar{\eta} \in R^{n_\eta}$ are known vectors.

According to Definition 1, Assumption 1 could be reformulated as:

$$\begin{aligned} x(0) &\in \mathcal{X}_0 = \langle p_0, H_0 \rangle, \\ w(k) &\in \mathcal{Z}_w = \langle 0, H_w \rangle, \\ \eta(k) &\in \mathcal{Z}_\eta = \langle 0, H_\eta \rangle, \end{aligned}$$

where $H_0 = diag(\bar{x}_0)$, $H_w = diag(\bar{w})$, and $H_\eta = diag(\bar{\eta})$.
A fault detection observer is designed as follows:

$$\begin{cases} \hat{x}(k+1) = A\hat{x}(k) + Bu(k) + L(y(k) - C\hat{x}(k)) \\ \delta(k) = y(k) - C\hat{x}(k) \end{cases} \tag{6}$$

where $\hat{x}(k) \in R^{n_x}$ is the state estimation vector, $\delta(k) \in R^{n_y}$ is residual signal, and $L \in R^{n_x \times n_y}$ is the gain matrix of the observer.
Define the state estimation error vector as

$$e(k) = x(k) - \hat{x}(k). \tag{7}$$

Combining equations (4), (5) and (6), the dynamic error system is obtained as

$$\begin{cases} e(k+1) = (A - LC)e(k) + D_w w(k) - LD_\eta \eta(k) - LEf(k), \\ \delta(k) = Ce(k) + D_\eta \eta(k) + Ef(k). \end{cases} \tag{8}$$

Since $x(0) \in \mathcal{X}_0 = \langle p_0, H_0 \rangle$, we have $e(0) \in \mathcal{Z}_{e(0)} = \langle 0, H_0 \rangle$ provided $\hat{x}(0) = p_0$. Consider $e(k) \in \mathcal{Z}_e(k) = \langle p_e(k), H_e(k) \rangle$, according to the dynamic error system (8), $e(k+1)$ is governed by

$$\begin{aligned} e(k+1) &\in \mathcal{Z}_e(k+1) \\ &= (A - LC) \odot \mathcal{Z}_e(k) \oplus D_w \odot \mathcal{Z}_w \oplus (-LD_\eta) \odot \mathcal{Z}_\eta \oplus Ef(k), \end{aligned} \tag{9}$$

Based on the Property 1 and Property 2, the center $p_e(k+1)$ and segment matrix $H_e(k+1)$ of $\mathcal{Z}_e(k+1)$ are expressed as follows:

$$p_e(k+1) = (A - LC)p_e(k) - LEf(k), \tag{10a}$$

$$H_e(k+1) = [(A - LC) \downarrow_l H_e(k) \quad D_w H_w \quad -LD_\eta H_\eta]. \tag{10b}$$

Similarly, based on the dynamic error system (8), $\delta(k)$ is governed by

$$\begin{aligned}\delta(k) &\in \mathcal{Z}_\delta(k)\\ &= C \odot \mathcal{Z}_e(k) \oplus D_\eta \odot \mathcal{Z}_\eta \oplus Ef(k),\end{aligned} \tag{11}$$

where the center $p_\delta(k)$ and segment matrix $H_\delta(k)$ of $\mathcal{Z}_\delta(k+1)$ are expressed as follows:

$$p_\delta(k) = Cp_e(k) + Ef(k), \tag{12a}$$

$$H_\delta(k) = [CH_e(k) \quad D_\eta H_\eta]. \tag{12b}$$

To derive the disturbance robustness conditions for designing fault detection observers, we divide the dynamic error system (8) into two subsystems as follows:

$$S_1 : \begin{cases} e_d(k+1) = (A - LC)e_d(k) + D_x d(k), \\ \delta_d(k) = Ce_d(k) + D_y d(k), \end{cases} \tag{13}$$

$$S_2 : \begin{cases} e_f(k+1) = (A - LC)e_f(k) - LEf(k), \\ \delta_f(k) = Ce_f(k) + Ef(k). \end{cases} \tag{14}$$

where

$$e(k) = e_d(k) + e_f(k),\ D_x = \begin{bmatrix} D_w & -LD_\eta \end{bmatrix},\ D_y = \begin{bmatrix} 0 & D_\eta \end{bmatrix},\ d(k) = \begin{bmatrix} w(k) \\ \eta(k) \end{bmatrix}.$$

herein, the term $e_d(k)$ represents the influence of disturbance on the system error, whereas $e_f(k)$ reflects the impact of fault on the system error.

The following definition is needed.

Definition 2 The dynamic error system (8) has an H_∞ performance indicator γ, if the system (8) satisfies the following inequality:

$$\sum_{k=0}^{\infty} \delta^T(k)\delta(k) \le \gamma^2 \sum_{k=0}^{\infty} d^T(k)d(k) \tag{15}$$

Based on the above discussions, now, we are ready to present the main objective of this paper as follows.

Objective: In this paper, we aim to propose a sensor fault detection method for AUV, which mainly includes three parts:

1. How to design the fault detection observer?
2. How to determine the occurrence of a fault?
3. How to analyze fault detection performance?

3 Main Results

In this section, the design of the fault detection observer will be detailed. One step further, the fault detection logic and the performance of the proposed method will be analyzed.

3.1 Observer Gain Matrix

Theorem 1 *For fault detection observer* (6)*, given a scalar* $\gamma > 0$*, the* H_∞ *performance index in inequality* (15) *is satisfied if there exist matrices* $P = P^T \succ 0 \in \mathbb{R}^{n_x \times n_x}$ *and* $N \in \mathbb{R}^{n_x \times n_y}$*, such that*

$$\begin{bmatrix} C^T C - P & \star & \star & \star \\ 0 & -\gamma^2 I & \star & \star \\ D_\eta^T C & 0 & D_\eta^T D_\eta - \gamma^2 I & \star \\ PA - NC & PD_w & -ND_\eta & -P \end{bmatrix} \prec 0. \tag{16}$$

Proof The Lyapunov function is defined as

$$V_d(k) = e_d^T(k) P e_d(k), \tag{17}$$

it follows that

$$\begin{aligned} \Delta V_d(k) &= V_d(k+1) - V_d(k) \\ &= e_d^T(k+1) P e_d(k+1) - e_d^T(k) P e_d(k) \\ &= [(A - LC)e_d(k) + D_x d(k)]^T P[(A - LC)e_d(k) \\ &\quad + D_x d(k)] - e_d^T(k) P e_d(k). \end{aligned}$$

According to Definition 2, the following inequality should be applied to satisfy the H_∞ performance:

$$\sum_{k=0}^{\infty} [\delta^T(k)\delta(k) - \gamma^2 d^T(k) d(k) + \Delta V_d(k)] \prec 0. \tag{18}$$

Then, the following inequality can be obtained from inequality (18):

$$\begin{bmatrix} \Omega_{11} & \Omega_{12} \\ \star & \Omega_{22} \end{bmatrix} \prec 0, \tag{19}$$

where

$$\Omega_{11} = (A - LC)^T P(A - LC) + C^T C - P,$$
$$\Omega_{12} = (A - LC)PD_x + C^T D_y,$$
$$\Omega_{22} = D_x^T PD_x + D_y^T D_y - \gamma^2 I.$$

According to Lemma 1, inequality (19) could be rewritten as follows:

$$\begin{bmatrix} C^T C - P & \star & \star \\ D_y^T C & D_y^T D_y - \gamma^2 I & \star \\ P(A - LC) & PD_x & -P \end{bmatrix} \prec 0. \tag{20}$$

Substitute D_x and D_y into inequality (20) to obtain that

$$\begin{bmatrix} C^T C - P & \star & \star & \star \\ 0 & -\gamma^2 I & \star & \star \\ D_\eta^T C & 0 & D_\eta^T D_\eta - \gamma^2 I & \star \\ P(A - LC) & PD_w & -PLD_\eta & -P \end{bmatrix} \prec 0. \tag{21}$$

Letting $N = PL$, the inequality (21) can be rewritten as inequality (16). □

Then, the matrix L in the observer (6) can be obtained by

$$L = P^{-1} N. \tag{22}$$

Remark 1 According to the interval estimation based on reachability analysis algorithm of Tang et al. (2019), if the observer gain is computed using the H_∞ performance to minimize the effect of disturbances, the size of the zonotope remains bounded as well. For more details, please refer to Tang et al. (2019).

3.2 Fault Detection Logic

We use the following methods to determine whether the fault occurs

$$\begin{cases} 0 \in \mathcal{Z}_\delta(k), & f_s = 0 \text{ (fault-free)} \\ 0 \notin \mathcal{Z}_\delta(k), & f_s = 1 \text{ (fault-alarm)} \end{cases} \tag{23}$$

When there is no fault, the residual zonotope $\mathcal{Z}_\delta(k)$ will always contain zero, and the size of the residual zonotope serves as an indicator for the extent of uncertainty influence. When a fault occurs, $\mathcal{Z}_\delta(k)$ shifts and eventually no longer contains zero, triggering an alarm. For clarity, we summarize the zonotopic sensor fault detection approach for AUVs as Algorithm 1, with its graphical interpretation depicted in Fig. 4. In the figure, the green geometry represents the residual zonotope, and the solid black dot represents zero. When zero lies within the zonotope (left), the sensor is functioning normally. When zero is outside the zonotope (right), a fault has occurred.

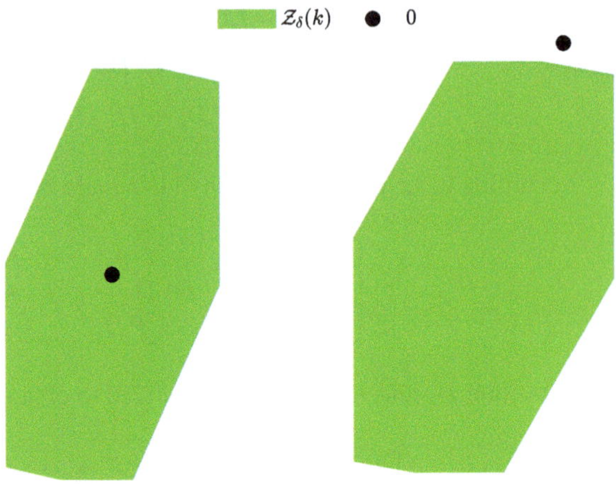

Fig. 4 Left: fault-free case. Right: fault-alarm case

Algorithm 1 Sensor fault detection for AUV based on zonotope

Input: AUV model matrices M_1, $C_{RB}(\mathbf{v}_1)$, $C_A(\mathbf{v}_1)$, and $D(\mathbf{v}_1)$
Output: Fault occurrence f_s
initialization:
Using Euler discretization with Δt to obtain A, B, D_w, C, D_η, E.
Solving the optimization problem (16)
Compute the gain matrix L using (22)
Initialize $k = 0$ and $x(0) \in \mathcal{X}_0 = \langle p_0, H_0 \rangle$
while $k \geq 0$ **do**
  Run the dynamic error system (8)
  Compute the residual zonotope $\mathcal{Z}_\delta(k)$ in (11) using (12a) and (12b)
  if $0 \in \mathcal{Z}_\delta(k)$ **then**
    $f_s = 0$ (fault-free)
    else
    $f_s = 1$ (fault-alarm)
  end if
end while
return: Fault occurrence f_s

3.3 *Fault Detection Performance*

We will analyze the performance of the presented approach using a minimum detectable fault set.

Before proceeding, the subsequent definition is presented.

Definition 3 (*Minimum detectable fault set*) Li et al. (2023b): For a given fault detection approach and a time instant k, a set $\mathcal{F}_k$ is called the minimum detectable fault set if any fault outside $\mathcal{F}_k$ will be detected.

The following theorem is proposed for the calculation of the minimum detectable fault set $\mathcal{F}_k$ to the fault detection approach given above.

Theorem 2 *Consider the system (4) and* (5), *a fault* $f(k) \in \mathbb{R}^{n_f}$ *will be detected via the above method if there exists a time instant* $k \in \mathbb{N}_+$ *such that* $f(k) \notin \mathcal{F}_k = \langle p_f(k), H_f(k) \rangle$, *where*

$$p_f(k) = -E^{\dagger} C p_e(k), \tag{24a}$$

$$H_f(k) = \begin{bmatrix} -E^{\dagger} C H_e(k) & -E^{\dagger} D_{\eta} H_{\eta} \end{bmatrix}. \tag{24b}$$

Proof According to the fault detection logic (23), we know that a fault is detected if $0 \notin \mathcal{Z}_{\delta}(k)$. Note that $\mathcal{Z}_{\delta}(k) = \langle p_{\delta}(k), H_{\delta}(k) \rangle$. Thus,

$$0 \notin \langle p_{\delta}(k), H_{\delta}(k) \rangle. \tag{25}$$

Take (12) into account, (25) becomes

$$0 \notin \langle C p_e(k) + E f(k), \begin{bmatrix} C H_e(k) & D_{\eta} H_{\eta} \end{bmatrix} \rangle. \tag{26}$$

Then, we have

$$- E f(k) \notin \langle C p_e(k), \begin{bmatrix} C H_e(k) & D_{\eta} H_{\eta} \end{bmatrix} \rangle. \tag{27}$$

Assume that E is full column rank, we have

$$f(k) \notin -E^{\dagger} \langle C p_e(k), \begin{bmatrix} C H_e(k) & D_{\eta} H_{\eta} \end{bmatrix} \rangle. \tag{28}$$

Based on Property 2, $\mathcal{F}_k = \langle p_f(k), H_f(k) \rangle$ holds. □

4 Simulation Example

In this section, the "UVIC-I" AUV model developed by the laboratory is adopted to demonstrate the effectiveness of the proposed method. The structure of the "UVIC-I" AUV is shown in Fig. 5. This AUV adopts a streamlined design. It utilizes Doppler to measure velocity and electronic compasses to measure pose. It is equipped with a propulsion system comprising two main, two side and four vertical thrusters. Additionally, the "UVIC-I" AUV is outfitted with a depth gauge, a hydraulic system, and a machine arm, enabling it to perform a wide range of underwater tasks and operations. The parameters of UVIC-I AUV are shown in Table 1. Then, the coefficient matrix of the "UVIC-I" AUV model in (2) is represented as

$$M_1 = \begin{bmatrix} 249.5328 & 0 \\ 0 & 5021.3 \end{bmatrix}, D_1(\mathbf{v}) = \begin{bmatrix} -7.6610 & 0 \\ 0 & -3.5160 \end{bmatrix},$$

$$C_{RB1}(\mathbf{v}) = \begin{bmatrix} 0 & 102 \\ -102 & 0 \end{bmatrix}, C_{A1}(\mathbf{v}) = \begin{bmatrix} 0 & -12.7612 \\ 12.7612 & 0 \end{bmatrix}.$$

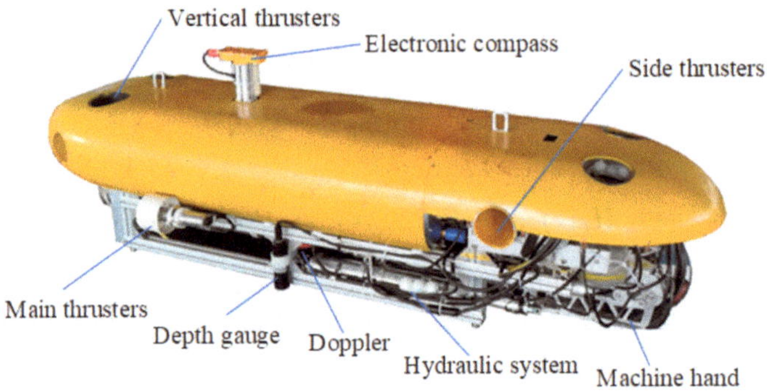

Fig. 5 "UVIC-I" AUV

Table 1 Parameters of UVIC-I AUV

Variable	Value	Unit	Comments
m	204	kg	Mass of "UVIC-I" AUV
ρ_0	2	m/s	Cruising speed of "UVIC-I" AUV
I_{zz}	4840	$kg \cdot m^2$	Inertia moment about the z-axis
$X_{\dot{\rho}}$	25.5224	kg	Hydrodynamic added mass
$Y_{\dot{v}}$	45.5328	kg	Hydrodynamic added mass
$N_{\dot{r}}$	181.254	$kg \cdot m^2$	Hydrodynamic added mass
Y_v	-7.661	kg/s	Linear damping coefficient
N_r	-3.516	$kg \cdot m^2/(s \cdot rad)$	Linear damping coefficient

We select a sampling time of 0.1 s, the form of (4) can be obtained with

$$A = \begin{bmatrix} 0.9969 & 0.0358 \\ -0.0018 & 0.9999 \end{bmatrix}, B = \begin{bmatrix} 0.00040075 & 0 \\ 0 & 0.000019915 \end{bmatrix},$$
$$D_w = \begin{bmatrix} 0.00040075 & 0 \\ 0 & 0.000019915 \end{bmatrix}.$$

In this paper, we suppose that both sway velocity and yaw angular velocity can be measured via two sensors. In this case,

$$C = \begin{bmatrix} 1 & 0 \\ 0 & 1 \end{bmatrix}, D_\eta = \begin{bmatrix} 1 & 0 \\ 0 & 1 \end{bmatrix}, E = \begin{bmatrix} 1 & 0 \\ 0 & 1 \end{bmatrix}.$$

In the simulation, we consider that the disturbance w_k and measurement noise η_k are unknown but bounded as

$$|w_k| \leq \left[0.2\ 0.2\right]^T, |\eta_k| \leq \left[0.1\ 0.1\right]^T,$$

and they are randomly generated with uniform distributions corresponding to the assumed ranges. The control input is $u(k) = \begin{bmatrix} 0.1 & 0.1 \end{bmatrix}^T$ The initial state is $x(0) = \begin{bmatrix} 1 & 1 \end{bmatrix}^T$, so we set the initial state zonotope is $\mathcal{X}_0 = \langle \begin{bmatrix} 0 & 0 \end{bmatrix}^T \, 0.1I \rangle$. The maximal zonotope order is chosen as 20. We set the H_∞ index as $\gamma = 2$, then the observer gain matrix is obtained as:

$$L = \begin{bmatrix} 0.3975 & 0.0081 \\ 0.0081 & 0.3988 \end{bmatrix}.$$

To verify the effectiveness of the proposed method, two fault cases are considered as follows:
Case 1: abrupt fault

$$f(k) = \begin{cases} \begin{bmatrix} 0 \\ 0 \end{bmatrix} & k < 20, \\ \begin{bmatrix} 0.35 \\ 0.3 \end{bmatrix} & k \geq 20, \end{cases}$$

Case 2: time-varying fault

$$f(k) = \begin{cases} \begin{bmatrix} 0 \\ 0 \end{bmatrix} & k < 20, \\ \begin{bmatrix} 0.35 - 0.1\sin(0.1\pi k - 5) \\ 0.3 - 0.05\sin(0.1\pi k - 5) \end{bmatrix} & k \geq 20. \end{cases}$$

The residual zonotopes are obtained by applying the proposed method as shown in Figs. 6 and 7, respectively. In Figs. 6 and 7, the green geometry is the residual zonotope, and the red circles represent zero points. It can be seen from Fig. 6 and Fig. 7 that, when $k < 20$, the residual zonotopes contain zero points. According to the fault detection logic (23), this infers that there is no fault occurs in the velocity sensors. When $k \geq 20$, the residual zonotopes no longer contain zero points, indicates that a fault is occurred during these time. Besides, from Figs. 6 and 7, it can be seen that, for the considered fault, it is detected at 21 time instant. This almost immediately after the fault occurs, which demonstrates practical applicability and effectiveness of our proposed method. The fault detection result is shown in Fig. 8. Obviously, we can see that the occurrence of both abrupt and time-varying faults can be checked. Thus, velocity sensor faults for "UVIC-I" AUV can be detected by using the proposed method.

The detection performance of the present approach is analyzed via the minimum detectable fault set. The simulation result of the minimum detectable fault set is shown in Fig. 9. The shape of the minimum detectable fault set is a zonotope. The boundary in the direction of sway velocity measurement is 0.3 and 0.4, while the boundary of yaw angular velocity measurement is 0.15 and 0.25. According to Definition 3, these indicate that any fault outside this set can be detected.

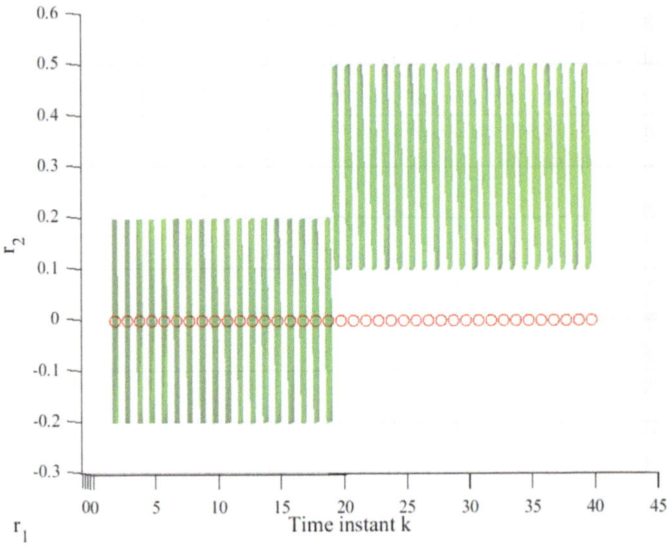

Fig. 6 Residual zonotopes for abrupt fault

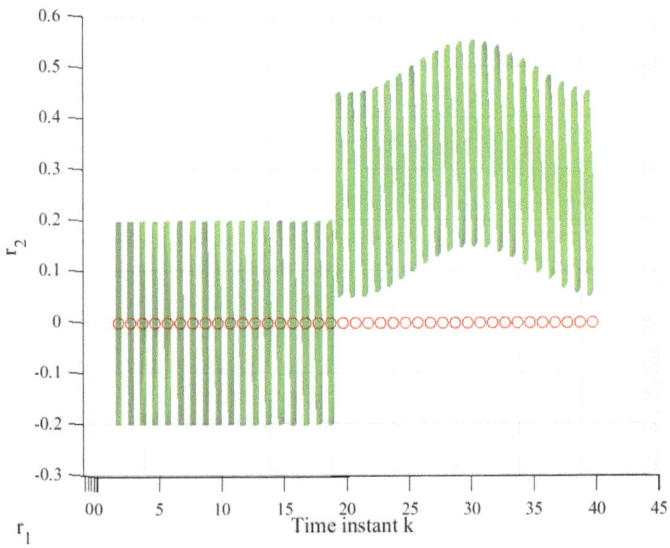

Fig. 7 Residual zonotopes for time-varying fault

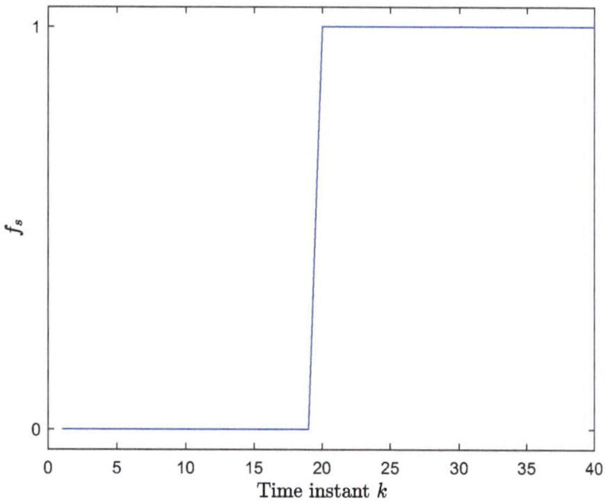

Fig. 8 Fault detection result

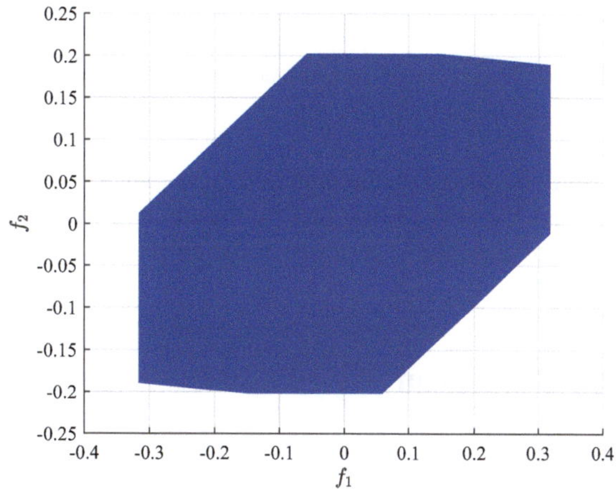

Fig. 9 The minimum detectable fault set of the proposed method

To demonstrate the superiority of the proposed method, we compare it with the interval observer method in Li et al. (2022). Li et al. (2022) used an interval observer for state estimation, which can also be employed for fault detection through residual evaluation. The comparison between the proposed method and the interval observer approach in Li et al. (2022) is presented in Fig. 10. In Fig. 10, the green geometry represents the residual zonotope, which is the same as Fig. 6, the red rectangle denotes the residual interval obtained from the interval observer method in Li et al. (2022), and the red circles indicate zero points. From Fig. 10, it can be seen that the residual zonotope by the proposed method is smaller than that by the interval observer in Li et al. (2022), which implies that the proposed method has higher estimation accuracy. This is reasonable, since the interval observer suffers from wrapping effects (Tang et al. 2019). Moreover, after the fault occurs, the residual zonotope by the proposed method no longer contains zero, while that by the interval observer in Li et al. (2022)

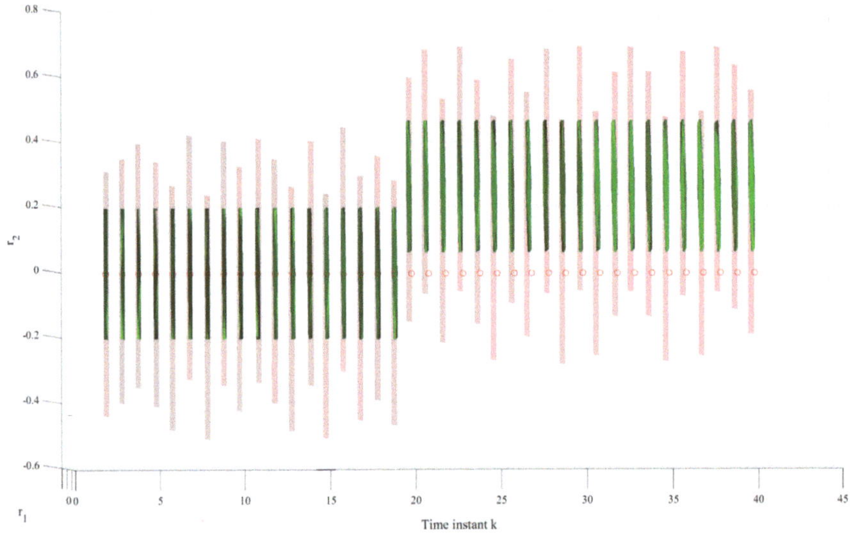

Fig. 10 The residual comparison obtained by different methods

still does. Based on the fault detection logic (23), the proposed method successfully detects the fault, while that by the interval observer in Li et al. (2022) fails. This demonstrates the superiority of the proposed method.

5 Conclusion

In this paper, a novel zonotopic velocity sensor fault detection approach for AUVs has been proposed. To attenuate the effect of disturbances and noises, an observer design method based on H_∞ performance index has been developed. The design conditions have been derived as LMIs. The fault has been detected by checking whether the zero point has lain outside the residual zonotope. The fault detection performance of the proposed method has been evaluated via the minimum detectable fault set. Finally, simulation results have demonstrated the effectiveness of the proposed method. In the future, fault detection for nonlinear models of AUVs will be considered. The application of the proposed method on a real AUV will also be our future work.

References

Alamo T, Bravo J, Camacho E (2005) Guaranteed state estimation by zonotopes. Automatica 41(6):1035–1043

Althoff M (2010) Reachability analysis and its application to the safety assessment of autonomous cars. Technische Universitt Mnchen
Bisoffi A, De Persis C, Tesi P (2022) Learning controllers for performance through LMI regions. IEEE Trans Autom Control 68(7):4351–4358
Boyd S, El Ghaoui L, Feron E, Balakrishnan V (1994) Linear matrix inequalities in system and control theory. SIAM
Chu Z, Chen Y, Zhu D, Zhang M (2020) Observer-based fault detection for magnetic coupling underwater thrusters with applications in Jiaolong HOV. Ocean Eng 210:107570
Chu Z, Gu Z, Chen Y, Zhu D, Tang J (2024) A fault diagnostic approach for underwater thrusters based on generative adversarial network. IEEE Trans Instrum Meas 73:3524614. https://doi.org/10.1109/TIM.2024.3417591
Dinh TN, Tran GQB (2024) A unified KKL-based interval observer for nonlinear discrete-time systems. IEEE Control Syst Lett 8:430–435
Fang X, Xie L, Li X (2023) Distributed localization in dynamic networks via complex Laplacian. Automatica 151:110915
Fossen TI (2011) Handbook of marine craft hydrodynamics and motion control. Wiley
Fridman E, Shaikhet L (2017) Stabilization by using artificial delays: an LMI approach. Automatica 81:429–437
Gao S, He B, Yu F, Zhang X, Yan T, Feng C (2021) An abnormal motion condition monitoring method based on the dynamic model and complex network for AUV. Ocean Eng 237:109472
Guerra TM, Márquez R, Kruszewski A, Bernal M (2017) H_∞ LMI-based observer design for nonlinear systems via takagi–sugeno models with unmeasured premise variables. IEEE Trans Fuzzy Syst 26(3):1498–1509
Haj Brahim I, Chaabane M, Mehdi D (2017) Fault-tolerant control for T-S fuzzy descriptor systems with sensor faults: an lmi approach. Int J Fuzzy Syst 19(2):516–527
Ji D, Yao X, Li S, Tang Y, Tian Y (2021) Model-free fault diagnosis for autonomous underwater vehicles using sequence convolutional neural network. Ocean Eng 232:108874
Jiang Y, Yin S, Kaynak O (2020) Optimized design of parity relation-based residual generator for fault detection: data-driven approaches. IEEE Trans Ind Inf 17(2):1449–1458
Le VTH, Stoica C, Alamo T, Camacho EF, Dumur D (2013) Zonotopes: from guaranteed state-estimation to control. Wiley
Li J, Wang Z, Wang Y, Zhang M, Raïssi T (2022) Interval velocity estimation for unmanned underwater vehicles. IEEE Control Syst Lett 7:715–720
Li J, Wang Z, Shen Y, Xie L (2023) Attack detection for cyber-physical systems: a zonotopic approach. IEEE Trans Autom Control 8(11):6828–6835
Li J, Zhang Y, Yao F, Liu X, Zhang M, Wang Z, Wang X (2023) A zonotopic velocity sensor fault detection approach for autonomous underwater vehicle. IFAC-PapersOnLine 56(2):8470–8475
Li J, Zhang Y, Wang Z, Qiu J, Zhang M (2024) Towards sensor fault detection for autonomous underwater vehicles: a zonotopic approach. IEEE Trans Fuzzy Syst 32(12):6646–6657
Liu F, Tang H, Qin Y, Duan C, Luo J, Pu H (2022) Review on fault diagnosis of unmanned underwater vehicles. Ocean Eng 243:110290
Liu X, Zhang M, Wang S (2020) Adaptive region tracking control with prescribed transient performance for autonomous underwater vehicle with thruster fault. Ocean Eng 196:106804
Ma Y, Wang Z, Kinnaert M, Shen Y (2025) A zonotope-based interval estimation method for systems with nonlinear equality constraints. IEEE Trans Autom Control 70(2):1358–1365
Ma Y, Wang Z, Meslem N, Raïssi T (2025) Functional interval estimation for continuous-time linear systems with time-invariant uncertainties. Automatica 172:112017
Scott JK, Raimondo DM, Marseglia GR, Braatz RD (2016) Constrained zonotopes: a new tool for set-based estimation and fault detection. Automatica 69:126–136
Shao X, Pustina P, Stölzle M, Sun G, De Luca A, Wu L, Della Santina C (2023) Model-based control for soft robots with system uncertainties and input saturation. IEEE Trans Ind Electron
Tang W, Wang Z, Wang Y, Raïssi T, Shen Y (2019) Interval estimation methods for discrete-time linear time-invariant systems. IEEE Trans Autom Control 64(11):4717–4724

Tang W, Wang Z, Zhang Q, Shen Y (2020) Set-membership estimation for linear time-varying descriptor systems. Automatica 115:108867

Tang W, Zhang Q, Wang Z, Shen Y (2022) Set-membership estimation based on ellipsoid bundles for discrete-time LPV descriptor systems. Automatica 145:110580

Wan J, Sharma S, Sutton R (2018) Guaranteed state estimation for nonlinear discrete-time systems via indirectly implemented polytopic set computation. IEEE Trans Autom Control 63(12):4317–4322

Wang X, Fei Z, Qiu J, Gao H (2020) Zonotopic fault detection for fuzzy systems with event-triggered transmission. IEEE Trans Fuzzy Syst 29(12):3734–3742

Wang X, Fei Z, Shi P, Yu J (2020) Zonotopic fault detection for 2-D systems under event-triggered mechanism. IEEE Trans Cybern 52(5):3510–3518

Wang Z, Shen Y (2022) Model-based fault diagnosis: methods for state-space systems, vol 221. Springer Nature

Wang Z, Dinh TN, Zhang Q, Raissi T, Shen Y (2022) Fast interval estimation for discrete-time linear systems: an L1 optimization method. Automatica 137:110029

Wang Z, Lian D, Puig V, Shen Y (2025) Fault detection for brushless direct-current motor using descriptor system-based set-membership estimation. IEEE Trans Control Syst Technol

Xia S, Zhou X, Shi H, Li S, Xu C (2021) A fault diagnosis method based on attention mechanism with application in Qianlong-2 autonomous underwater vehicle. Ocean Eng 233:109049

Xia S, Zhou X, Shi H, Li S, Xu C (2022) A fault diagnosis method with multi-source data fusion based on hierarchical attention for AUV. Ocean Eng 266:112595

Xu F, Tan J, Wang Y, Wang X, Liang B, Yuan B (2019) Robust fault detection and set-theoretic UIO for discrete-time LPV systems with state and output equations scheduled by inexact scheduling variables. IEEE Trans Autom Control 64(12):4982–4997

Zhang Z, He X, Zhou D (2023) Active fault diagnosis for uncertain LPV systems: a zonotopic set-membership approach. IEEE Trans Autom Sci Eng

Zheng X, Zhang H, Wang Z, Zhang C, Yan H (2021) Finite-time dynamic event-triggered distributed H_∞ filtering for TS fuzzy systems. IEEE Trans Fuzzy Syst 30(7):2476–2486

Zhou M, Cao Z, Zhou M, Wang J, Wang Z (2019) Zonotoptic fault estimation for discrete-time LPV systems with bounded parametric uncertainty. IEEE Trans Intell Transp Syst 21(2):690–700

Zhou M, Li H, Xue T, Wang J, Raïssi T (2025) Event-triggered state interval estimation and fault-tolerant control for multi-agent systems under deception attacks. IEEE Trans Autom Control 70(1):689–696

Zhu F, Fu Y, Dinh TN (2023) Asymptotic convergence unknown input observer design via interval observer. Automatica 147:110744

Robust State Feedback Control of LPV Systems Based on Zonotopic Kalman Filter

Shuang Zhang, Sara Ifqir, and Vicenç Puig

Abstract In this chapter, we develop a robust state feedback control scheme based on a Zonotopic Kalman Filter (ZKF) for Linear Parameter Varying (LPV) systems subject to disturbances and measurement noises, not yet considered in the literature. For this purpose, first, a ZKF with offline filter gains design is proposed so that the system states can be obtained. Then, a Linear Quadratic Zonotopic (LQZ) control is proposed to tackle the synthesis of the ZKF and the state feedback controller. The LQZ control scheme is performed by means of zonotopes to find the set of gains that are robust against uncertainties and that satisfy predefined performances. Based on the separation principle, the computation of the optimal controller and estimator are done separately by finding the solutions to a finite number of Linear Matrix Inequalities (LMIs). At last, a reference path tracking controller of the vehicle lateral dynamics is designed to demonstrate the validity and performance of the proposed method.

1 Introduction

Over the past few decades, Linear Parameter-Varying (LPV) technique has been demonstrated as an effective way to represent nonlinear systems. Within the LPV framework, results concerning the design of observers (Efimov et al. 2013; Zhang et al. 2023a), state feedback controllers (Rotondo et al. 2014; Liu et al. 2017) and H_∞ controllers (Hang et al. 2019) have been widely studied. On the other hand,

S. Zhang (✉) · V. Puig
Institut de Robòtica i informàtica Industrial, (CSIC-UPC), Universitat Politècnica de Catalunya-BarcelonaTech, Barcelona, Spain
e-mail: shuang.zhang@upc.edu

V. Puig
e-mail: vicenc.puig@upc.edu

S. Ifqir
CRIStAL, Centrale Lille Institut, Villeneuve d'Ascq, France
e-mail: sara.ifqir@centralelille.fr

T. N. Dinh et al. (eds.), *Set-Valued Approaches to Control and Estimation of Uncertain Systems*, Mathematical Engineering,
https://doi.org/10.1007/978-3-031-94239-6_8

LPV systems and control theory have been used in several practical applications, for example, autonomous ground vehicles (Alcalá et al. 2019; Li et al. 2021), Unmanned Aircraft Vehicle (UAV) (Cavanini et al. 2021) and energy production system (Yang et al. 2020a). The goal of the chapter is to investigate a state feedback/observer design strategy for discrete-time LPV systems affected by additive uncertainties, such as unknown parameters, process disturbances or measurement noises.

Among the common robust state feedback control methods (e.g., H_∞ control (Hu and Duan 2019), sliding mode control (Ren et al. 2019) or Linear Quadratic Gaussian (LQG) control (Ostertag 2011)), LQG control can provide an optimal solution in terms of minimizing a quadratic cost function that includes both state errors and control efforts. Furthermore, this control approach can handle stochastic disturbances and noises by means of Kalman Filter (KF). However, the characterization of uncertainties requires the assumption that they follow Gaussian probability distributions, which may fail in many practical situations (Combastel 2015). Hence, in the remaining of the paper, set-based approaches (Puig et al. 2003) are considered, in which the modelling uncertainties are assumed to be unknown but bounded by a priori known bound. The bound can be described by several types of sets, such as intervals (Wang et al. 2024), polytopes (Blesa et al. 2012), ellipsoids (Zhang et al. 2023b), and zonotopes (Wang et al. 2023). The effectiveness of these set-based techniques has been proved in state estimation (Chen and Patton 2011; Rego et al. 2020; Zhang et al. 2023a), fault diagnosis (Ifqir et al. 2019a; Yang et al. 2020b) and fault tolerant control (Puig 2010; Xu et al. 2017) schemes.

Regarding the research on set-based control, an interval observer-based feedback control framework was first proposed by Efimov et al. (2013) for solving the problem of output stabilization of a class of nonlinear and LPV systems subject to uncertainties. After that, the interval observer-based control scheme was extended to nonlinear switched systems (He and Xie 2016) and switched LPV systems (Lv et al. 2023). Meanwhile, the interval Luenberger observer-based state feedback control has been applied to several practical scenarios, e.g., multi-DOF micropositioner (Hammouche et al. 2018) and solid oxide fuel cells stacks (Ifqir et al. 2019b). Furthermore, recent studies have applied interval observers to the active Fault Tolerant Control (FTC) scheme (Wang et al. 2021; Lamouchi et al. 2022; Shan and Zhu 2022), where the interval observer was used to perform fault estimation. However, the condition of designing an interval observer could be very restrictive, as the designed observer gain is required to ensure not only that the observer system is robustly stable, but also that the estimation error system matrix is Metzler (Ifqir et al. 2019a). This motivates the focus of this paper on zonotopes, that provide efficient computation (Yang and Scott 2018), a much more compact representation (Althoff et al. 2021) and effectively mitigate wrapping effect (Kühn 1998). To this end, the zonotopic observer/estimator has been widely applied to state estimation (Combastel 2015; Rego et al. 2020) and fault diagnosis (Puig 2010; Pourasghar et al. 2019; Wang et al. 2023). In particular, Trapiello and Puig (2022) proposed a Linear-Quadratic Regulator (LQR) controller that operates based on the estimates of a Zonotopic Kalman Filter (ZKF) for Linear Time Invariant (LTI) system. This paper extends these results to the LPV systems proving the separation principle in design of controller and observer. To the best of our knowledge, there is no literature on zonotopic estimator-based state feedback control for LPV systems. This motivates the work of the present paper.

Inspired by the aforementioned discussions, the aim of this paper is to address the state feedback control problem using zonotopes for uncertain LPV systems. To this end, this chapter develops a zonotopic counterpart of the LQG scheme, in which the feedback loop is closed through the optimal estimates of the ZKF, yielding the Linear Quadratic Zonotopic (LQZ) control scheme. Moreover, the separation principle of LQZ control is proved, which allows to design the optimal controller gain and the optimal filter gain separately. The main contributions of this paper are summarized as follows: (1) A Linear Quadratic Regulator (LQR) controller that operates based on the estimates of a ZKF is proposed for LPV systems, subject to disturbances and noises. The control scheme considers that the uncertainties are described by zonotopic sets instead of Gaussian probability distributions. In this context, it can be regarded as a zonotopic analogous case of the LQG scheme. (2) Different from the ZKF proposed by Combastel (2015), which provided an explicit solution to compute the optimal filter gain online, this chapter redesigns ZKF for LPV systems by calculating the optimal filter gain offline. This is achieved by solving an LMI optimization problem that minimizes the Frobenius norm. (3) The proposed LQZ control scheme is applied to vehicle reference path-tracking control and the effectiveness is validated through a simulation test.

This paper is organized as: Some preliminary results including definitions, properties and lemmas that will be used in this paper are addressed in Sect. 2.1. The problem statement is formulated in Sect. 2.2. In Sect. 3, the ZKF for LPV systems and its optimal polytopic filter gain computation are presented. Section 4 presents the LQZ control scheme, which is the synthesis of LQR-based state feedback control and ZKF. Section 5 illustrates the effectiveness of the proposed method through a reference tracking control case of vehicle lateral dynamics. Finally, some conclusions are drawn in Sect. 6.

Notation: In the following, $\mathbb{R}^n$ denotes the set of n-dimensional real numbers and $\oplus$ denotes the Minkowski sum. The matrices are written using capital letter, e.g., A, the calligraphic notation is used for denoting sets, e.g., $\mathcal{X}$, $[\underline{x}, \bar{x}]$ is an interval with lower bound $\underline{x}$ and upper bound $\overline{x}$. For a vector $s \in \mathbb{R}^n$, $\|s\|$ denotes Euclidean vector norm, $\|s\|_P = \sqrt{s^T P s}$ denotes weighted Euclidean vector norm, where $P > 0$.

2 Preliminaries and Problem Statement

2.1 Preliminaries

Definition 1 A unitary interval is a vector denoted by $\boldsymbol{B} = [-1, 1]$. A unitary box in $\mathbb{R}^m$, denoted by $\boldsymbol{B}^m$, is a box composed of m unitary intervals. Given a box $M = ([a_1, b_1], \ldots, [a_n, b_n])^T$, $\text{mid}(M)$ denotes its center and $\text{diam}(M) = (b_1 - a_1, \ldots, b_n - a_n)^T$.

Definition 2 (*Zonotope* Le et al. (2013)) A zonotope of order m in $\mathbb{R}^n$ is the translation by the center $c \in \mathbb{R}^n$ of the image of an unitary hypercube of dimension m in $\mathbb{R}^n$ under a linear transformation $R \in \mathbb{R}^{n\times m}$, the zonotope $\mathcal{X}$ is defined by:

$$\mathcal{X} = \langle c, R\rangle = c \oplus R\boldsymbol{B}^m = \left\{p + Rz : z \in \boldsymbol{B}^m\right\}$$

Besides, the following properties related to zonotopes hold:

$$\langle c_1, R_1\rangle \oplus \langle c_2, R_2\rangle = \langle c_1 + c_2, [R_1, R_2]\rangle\,,$$
$$L \odot \langle c, R\rangle = \langle Lc, LR\rangle,$$

where L is an arbitrary matrix of appropriate dimension.

Definition 3 (F_W-Radius). Let $W \in \mathbb{R}^{n\times n}$ be a Symmetric Positive Definite (SPD) matrix: $W = W^T \succ 0$. The weighted Frobenius radius (F_W-radius) of the zonotope $\langle c, R\rangle \subset \mathbb{R}^n$ is the weighted Frobenius norm of R : $\|\langle c, R\rangle\|_{F,W} = \|R\|_{F,W}$.

Definition 4 (*Zonotopic quadratic performance* Trapiello and Puig (2022)): Given an SPD matrix $S \in \mathbb{R}^{n\times n}$ and the unknown but zonotopically bounded vector $x \in \langle c, H\rangle \subset \mathbb{R}^n$, the performance of x is evaluated according to

$$Q\left[x^T Sx\right] = c^T Sc + \|H\|_{F,S}^2 = c^T Sc + \mathrm{Tr}[SP]$$

with $P = HH^T$.

Property 1 (Zonotope reduction Combastel (2015)). *A reduction operator, denoted $\downarrow_{q,W}$, allows to reduce the number of generators of a zonotope $\mathcal{X}$ to a fixed number $n \le q < m$, such that $\mathcal{X} = \langle c, R\rangle \subset \langle p, \downarrow_{q,W} R\rangle$. A common procedure for implementing the operator $\downarrow_{q,W}$ is summarized as follows: Sort the column of segment matrix $R \in \mathbb{R}^{n\times m}$ in decreasing weighted vector norm $\|\cdot\|_W$, $R = \left[r_1, \ldots, r_j, \ldots, r_m\right]$, $\left\|r_j\right\|_W^2 \ge \left\|r_{j+1}\right\|_W^2$; Enclose the set $R_>$generated by the $m - q + n$ smaller columns into a box (i.e., interval hull):*
If $m \le q$, then $\downarrow_{q,W} R = R$, Else

$$\downarrow_{q,W} R = [R_>, \mathrm{rs}\,(R_<)] \in \mathbb{R}^{n\times q},\ R_> = \left[r_1, \ldots, r_{q-n}\right],\ R_< = \left[r_{q-n+1}, \ldots, r_m\right]$$

where $r\,(R_<)$ is a diagonal matrix with diagonal elements of $\mathrm{rs}\,(R_<)_{i,i} = \sum_{j=1}^m |R_<|_{i,j}$, $i = 1, \ldots, n$.

Property 2 (Distributive property Trapiello and Puig (2022))*: Given the variables $x \in \mathbb{R}^n$ and $y \in \mathbb{R}^m$ such that $x \in \langle c_x, R_x\rangle \subset \mathbb{R}^n$ and $y \in \langle c_y, R_y\rangle \subset \mathbb{R}^m$, then*

$$Q\left[x^T S_x x + y^T S_y y\right] = Q\left[x^T S_x x\right] + Q\left[y^T S_y y\right]$$

with S_x and S_y SPD matrices of appropriate dimensions.

Lemma 1 *Consider the following double-sum condition*

$$\Pi\left(\rho_k, \rho_k\right) = \sum_{i=1}^{r} \sum_{j=1}^{r} h_i\left(\rho_k\right) h_j\left(\rho_k\right) \Pi_{i,j} > 0.$$

Then, the above condition is fulfilled provided that the following conditions hold:

$$\Pi_{i,i} > 0, i = 1, \ldots, r$$
$$\frac{2}{r-1}\Pi_{i,i} + \Pi_{i,j} + \Pi_{j,i} \geq 0, \quad , 1 \leq i < j \leq r.$$

2.2 Problem Statement

Consider the following discrete-time LPV system subject to disturbances and noises

$$\begin{cases} x(k+1) = A\left(\rho_k\right) x(k) + B\left(\rho_k\right) u(k) + E_w w(k) \\ \quad y(k) \quad = Cx(k) + E_v v(k) \end{cases} \tag{1}$$

where $u_k \in \mathbb{R}^{n_u}$, $y_k \in \mathbb{R}^{n_y}$, $x_k \in \mathbb{R}^{n_x}$, $w_k \in \mathbb{R}^{n_x}$, $v_k \in \mathbb{R}^{n_y}$ are the input, output, state, process disturbance and measurement noise vectors, respectively. Moreover, the disturbance and measurement noises are assumed to be unknown but bounded. Besides, $A\left(\rho_k\right) \in \mathbb{R}^{n_x \times n_x}$, $B\left(\rho_k\right) \in \mathbb{R}^{n_u \times n_u}$, $C \in \mathbb{R}^{n_y \times n_x}$ are the state-space matrices, where $\rho_k \in \mathbb{R}^{n_\theta}$ is the vector of time-varying parameters belonging to a known set Θ. $E_w \in \mathbb{R}^{n_x \times n_w}$, $E_v \in \mathbb{R}^{n_y \times n_v}$ and $C \in \mathbb{R}^{n_y \times n_x}$ are constant matrices. $w(k) \in \mathbb{R}^{n_w}$ and $v(k) \in \mathbb{R}^{n_v}$ are assumed to be unknown but bounded by zonotopes, i.e., $w(k) \in \mathcal{W} = \langle 0, I_{n_w} \rangle$, $v(k) \in \mathcal{V} = \langle 0, I_{n_v} \rangle$, where $I_{n_w} \in \mathbb{R}^{n_w \times n_w}$ and $I_{n_v} \in \mathbb{R}^{n_v \times n_v}$ denote the identity matrices. Considering that the variable ρ_k is online measurable and setting it as the scheduling variable leads to the following polytopic form of system (1):

$$\begin{cases} x(k+1) = \sum_{j=1}^{J} h_j\left(\rho_k\right)\left(A_j x(k) + B_j u(k)\right) + E_w w(k) \\ \quad y(k) \quad = Cx(k) + E_v v(k) \end{cases} \tag{2}$$

where A_j, B_j for $j = 1, \ldots, J$ are known constant matrices and $h_j\left(\rho(k)\right)$ are the weighting function depending on the time-varying parameter $\rho(k)$ that fulfills the following properties

$$\sum_{j=1}^{J} h_j\left(\rho_k\right) = 1, \quad 0 \leq h_j\left(\rho_k\right) \leq 1, \quad \forall j \in \mathcal{J} = \{1, \ldots, J\} \tag{3}$$

3 Zonotopic Kalman Filter for LPV Systems

For the LPV system (1), we would like to estimate and bound the uncertain system states $x(k), \forall k \in N$ in a zonotopic set. An optimal filter gain λ is used to reduce the state estimation error with a measured output $y(k)$.

Theorem 1 (Estimation-type Zonotopic Kalman Filter) *Consider the discrete-time LPV system (1), let $\hat{\mathcal{X}}(k) = \langle c(k), R(k)\rangle \in \mathbb{R}^{n_x}$ be the zonotopic estimated state, where $c(k) \in \mathbb{R}^{n_x}$ and $R(k) \in \mathbb{R}^{n_x \times n_r}$ represent the center and shape matrix, respectively. $\downarrow_{q,W} R(k)$ is the reduced generator matrix that is computed based on the Property 1. Assume that the initial state $x(k)$ belongs to the set $\hat{\mathcal{X}}(0) = \langle c(0), R(0)\rangle$, the estimated state can be propagated as follows:*

Predict:

$$\bar{c}(k+1) = A(\rho_k)\, c(k) + B(\rho_k)\, u(k) \tag{4a}$$

$$\bar{R}(k+1) = \left[A(\rho_k) \downarrow_{q,W} R(k) \quad E_w\right] \tag{4b}$$

Update:

$$c(k+1) = \bar{c}(k+1) + \lambda(\rho_k)\,(y(k+1) - C\bar{c}(k)) \tag{5a}$$

$$R(k+1) = \left[\left(I_{n_x} - \lambda(\rho_k)\, C\right)\bar{R}(k+1) \quad \lambda(\rho_k)\, E_v\right] \tag{5b}$$

where $\langle \bar{c}(k), \bar{R}(k)\rangle \in \mathbb{R}^{n_x}$ is the zonotopic predicted state, $\bar{c}(k) \in \mathbb{R}^{n_x}$ and $\bar{R}(k) \in \mathbb{R}^{n_x \times n_{\bar{r}}}$ represent the predicted center and shape matrix, respectively.

Proof Considering the inclusion $x(k) \in \langle c(k), R(k)\rangle$ and assuming that the $\downarrow_{q,W}$ preserves inclusion $x(k) \in \langle c(k), \downarrow_{q,W} R(k)\rangle$, the zonotopic predicted set can be computed as $\langle \bar{c}(k+1), \bar{R}(k+1)\rangle = A(\rho_k)\langle c(k), \downarrow_{q,W} R(k)\rangle \oplus \langle B(\rho_k)\, u(k), 0\rangle \oplus \langle 0, E_w\rangle$. Thus, the predicted state set (4a) and (4b) is derived. Then, the estimated state is updated as $\langle c(k+1), R(k+1)\rangle = \langle \bar{c}(k+1), \bar{R}(k+1)\rangle \oplus \lambda(\rho_k)\langle y(k+1), E_v\rangle \oplus \lambda(\rho_k)\, C\langle -\bar{c}(k+1), -\bar{R}(k+1)\rangle$. Therefore, the estimated state set (5a) and (5b) is derived. □

Remark 1 According to Combastel (2015), a time-varying zonotopic Kalman optimal filter gain can be obtained by minimizing F_W-radius of the state bounding zonotope $\langle c(k), R(k)\rangle$ and given by an online explicit solution. By minimizing the same radius, this paper aims to compute the parameter-dependent optimal filter gain through offline solving an LMI-based optimization problem. Furthermore, the proposed offline LMI-based solution can guarantee the state bounding zonotope has non-increasing size in time.

Profiting from the polytopic form (2) of the LPV system, this paper proposes an offline solution in the following Theorem 2 to compute the optimal filter gain λ_j for each vertex. Then, the parameter-dependent filter gain $\lambda(\rho_k)$ can be obtained by interpolation as $\lambda(\rho_k) = \sum_{j=1}^{J} h_j(\rho_k)\lambda_j$.

Theorem 2 *Given the polytopic LPV system (2), $\forall i, j \in \mathcal{J}$, the optimal filter gain $\lambda(\rho_k) = \sum_{j=1}^{J} h_j(\rho_k)\lambda_j$ can be computed if there exists a matrix $W_j \in \mathbb{R}^{n_x \times n_y}$, a positive definite matrix $\Gamma \in \mathbb{R}^{n_x \times n_x}$ and a positive scalar γ that are obtained by solving the following LMI optimization problem*

$$\min \gamma, \tag{6a}$$

$$s.t., \begin{bmatrix} \gamma I & I \\ I & \Gamma \end{bmatrix} > 0 \tag{6b}$$

$$\Pi_{i,i} < 0, \quad i = j \tag{6c}$$

$$\frac{2}{J-1}\Pi_{i,i} + \Pi_{i,j} + \Pi_{j,i} \leq 0, \quad i < j \tag{6d}$$

where

$$\Pi_{i,j} = \begin{bmatrix} -\Gamma & * & * & * \\ A_i^T(\Gamma - W_j C)^T & -\Gamma & * & * \\ E_w^T(\Gamma - W_j C)^T & 0 & -I & * \\ W_j^T & 0 & 0 & -F^{-1} \end{bmatrix}, \lambda_j = \Gamma^{-1}W_j, F = E_v E_v^T. \tag{7}$$

Proof As mentioned before, the optimal filter gain aims to minimize the F_W-radius of the state bounding zonotope, i.e.,

$$\min J_f = \|R(k)\|_{F,W}^2 \tag{8}$$

According to Combastel (2015), the optimal observer gain is independent of the weighting matrix W. The optimal observer gain is equivalently obtained through minimising the following cost function

$$J_f = \|R(k)\|_F^2 = tr(P(k)) = \sum_{i=1}^{n} \gamma_i \tag{9}$$

where $P(k) = R(k)R^T(k)$, $\gamma_1, \gamma_2, \ldots, \gamma_n$ represent the eigenvalues of P. Therefore, a possible approach to minimize the cost function J is to minimize a given scalar γ, s.t., $\gamma_i < \gamma$. It follows that

$$P(k) - \gamma I < 0 \tag{10}$$

Recalling that

$$R(k+1) = [(I - \lambda(\rho_k) C)A(\rho_k) R(k) \quad (I - \lambda(\rho_k) C)E_w \quad \lambda(\rho_k) E_v], \tag{11}$$

it follows that

$$\begin{aligned} P(k+1) = {} & (I - \lambda(\rho_k) C)A(\rho_k) P(k)((I - \lambda(\rho_k) C)A(\rho_k))^T \\ & + (I - \lambda(\rho_k) C)E_w E_w^T(I - \lambda(\rho_k) C)^T + \lambda(\rho_k) E_v E_v^T \lambda(\rho_k)^T \end{aligned} \tag{12}$$

As LMI provides the solution for steady state, then, $P(k+1) = P(k) = P_{ARE}$ satisfies the following Algebraic Riccati Equation (ARE)

$$\begin{aligned}&(I-\lambda(\rho_k)C)A(\rho_k)P_{ARE}((I-\lambda(\rho_k)C)A(\rho_k))^T\\&-P_{ARE}+(I-\lambda(\rho_k)C)E_wE_w^T(I-\lambda(\rho_k)C)^T+\lambda(\rho_k)E_vE_v^T\lambda(\rho_k)^T=0\end{aligned} \tag{13}$$

According to Boyd et al. (1994), the optimal value of the following inequality equals the solution to the above ARE,

$$\begin{aligned}&(I-\lambda(\rho_k)C)A(\rho_k)P((I-\lambda(\rho_k)C)A(\rho_k))^T-P\\&+(I-\lambda(\rho_k)C)E_wE_w^T(I-\lambda(\rho_k)C)^T+\lambda(\rho_k)E_vE_v^T\lambda(\rho_k)^T<0\end{aligned} \tag{14}$$

Let $\Gamma = P^{-1}$, $W(\rho_k) = \Gamma\lambda(\rho_k)$, and multiplying by Γ on the left and right of Eq. (14) yields to

$$\begin{aligned}&(\Gamma-W(\rho_k)C)A(\rho_k)\Gamma^{-1}A^T(\rho_k)(\Gamma-W(\rho_k)C)^T-\Gamma\\&+(\Gamma-W(\rho_k)C)E_wE_w^T(\Gamma-W(\rho_k)C)^T+W(\rho_k)E_vE_v^TW^T(\rho_k)<0\end{aligned} \tag{15}$$

which is equivalent to

$$-\Gamma-\Lambda\begin{bmatrix}-\Gamma^{-1} & 0 & 0\\ 0 & -I & 0\\ 0 & 0 & -E_vE_v^T\end{bmatrix}\Lambda^T<0. \tag{16}$$

where $\Lambda = [(\Gamma-W(\rho_k)C)A(\rho_k) \quad (\Gamma-W(\rho_k)C)E_w \quad W(\rho_k)]$.

Applying Schur complement to (16) and denoting $F = E_vE_v^T$, it follows that

$$\Pi(\rho_k,\rho_k)=\begin{bmatrix}-\Gamma & * & * & *\\ A^T(\rho_k)(\Gamma-W(\rho_k)C)^T & -\Gamma & * & *\\ E_w^T(\Gamma-W(\rho_k)C)^T & 0 & -I & *\\ W(\rho_k)^T & 0 & 0 & -F^{-1}\end{bmatrix}<0 \tag{17}$$

Considering the polytopic form for $A(\rho_k)$ and $W(\rho_k)$ in (17), it yields $\Pi(\rho_k,\rho_k) = \sum_{i=1}^{J}\sum_{j=1}^{J}h_i(\rho_k)h_j(\rho_k)\Pi_{i,j}$. Then, (6c) and (6d) can be derived by applying Lemma (1). Considering $\Gamma = P^{-1}$, the previous condition (10) becomes $\gamma I - \Gamma^{-1} > 0$, i.e.,

$$\begin{bmatrix}\gamma I & I\\ I & \Gamma\end{bmatrix}>0$$

Then, the optimization problem (6) should be solved by minimizing the value of γ. Hence, we complete the proof. □

4 Linear Quadratic Zonotopic Control Design

This section deals with the optimal control of the uncertain LPV system (1), in which the feedback loop is closed using the optimal state estimator ZKF. As we know, the Linear Quadratic Gaussian (LQG) control, a combination of a Kalman Filter (KF) together with a Linear Quadratic Regulator (LQR), applies to the system driven by additive Gaussian noises. Since the considered system (1) is affected by zonotopic uncertainties, a zonotopic counterpart of LQG control is proposed, i.e., Linear Quadratic Zonotopic (LQZ) control. In order to assess the system operation, the zonotopic performance criterion is introduced in Definition (4).

The LQZ control scheme is based on the optimal feedback of the estimated bounding state $\hat{x}(k) \in \langle c(k), R(k) \rangle$ to the system input via the optimal gain K, as depicted by Fig. 1. Now, a zonotopic quadratic cost function is given for the proposed control scheme

$$J = Q\left[\sum_{k=0}^{\infty} \left(x^T(k) Q_x x(k) + u^T(k) Q_u u(k)\right)\right]. \tag{18}$$

Usually, the weighting matrices for the system state and input vector, Q_x and Q_u, are implemented as diagonal matrices.

4.1 Separation Principle

In the following, it is proved that the proposed filter gain λ and controller gain K can be designed separately for the following closed-loop system:

$$\begin{cases} x(k+1) = A(\rho_k)\, x(k) + B(\rho_k)\, u(k) + E_w w(k), \\ y(k) = Cx(k) + E_v v(k), \\ c(k+1) = (I - \lambda(\rho_k)C)A(\rho_k)c(k) + (I - \lambda(\rho_k)C)B(\rho_k)u(k) + \lambda(\rho_k)y(k+1), \\ R(k+1) = [(I - \lambda(\rho_k)\, C)A(\rho_k)\, R(k) \quad (I - \lambda(\rho_k)\, C)E_w \quad \lambda(\rho_k)\, E_v], \\ u(k) = -K(\rho_k)c(k) \end{cases} \tag{19}$$

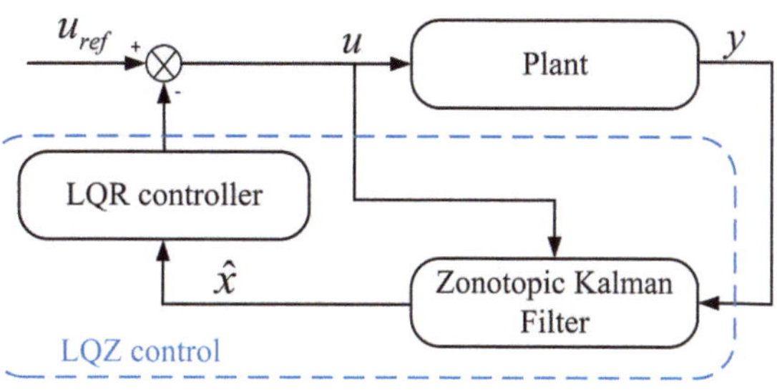

Fig. 1 Diagram of linear quadratic zonotopic control

In this regard, the LQZ design is decomposed into two independent problems:

- the optimization of the LQR control of the nominal plant;
- the optimization of the estimated state $\hat{x}$ from the measurements y.

Theorem 3 *The optimal control of the system (1), with the cost functional (18), is obtained by taking the optimal control law* $u(k) = -K(\rho_k)x_c(k)$*, calculated for the corresponding nominal system,*

$$x_c(k+1) = A(\rho_k)\,x_c(k) + B(\rho_k)\,u(k), \tag{20}$$

with the cost function

$$J_c = \sum_{k=0}^{\infty} \left(x_c^T(k)Q_x x_c(k) + u^T(k)Q_u u(k)\right). \tag{21}$$

The nominal plant state $x_c(k)$ *is then replaced by its optimal estimate* $c(k)$*, obtained by the zonotopic Kalman filter (4-5) with the optimal filter gain* $\lambda(\rho_k)$ *calculated using the minimization of the cost function*

$$J_f = \|R(k)\|_{F,W}^2 \tag{22}$$

Proof Denote the estimation error as $e(k+1) = x(k+1) - c(k+1)$, considering the estimated center

$$c(k+1) = (I - \lambda(\rho_k)\,C)A(\rho_k)\,c(k) + (I - \lambda(\rho_k)\,C)Bu(k) + \lambda(\rho_k)\,y(k+1),$$

the error dynamics is given by

$$e(k+1) = (I - \lambda(\rho_k)\,C)A(\rho_k)\,e(k) + (I - \lambda(\rho_k)\,C)E_w w(k) - \lambda(\rho_k)\,E_v v(k).$$

Replacing the input by $u(k) = -K(\rho_k)\,c(k)$, the system state equation becomes

$$\begin{aligned} x(k+1) &= A(\rho_k)\,x(k) - B(\rho_k)\,K(\rho_k)\,c(k) + E_w w(k) \\ &= (A(\rho_k) - B(\rho_k)\,K(\rho_k))x(k) + B(\rho_k)\,K(\rho_k)\,e(k) + E_w w(k). \end{aligned} \tag{23}$$

Therefore, the composite system (controller+filter) is described as follows.

$$\begin{aligned} \begin{bmatrix} x(k+1) \\ e(k+1) \end{bmatrix} &= \begin{bmatrix} A(\rho_k) - B(\rho_k)\,K(\rho_k) & B(\rho_k)\,K(\rho_k) \\ 0 & (I - \lambda(\rho_k)\,C)A(\rho_k) \end{bmatrix} \begin{bmatrix} x(k) \\ e(k) \end{bmatrix} \\ &+ \begin{bmatrix} E_w & 0 \\ (I - \lambda(\rho_k)\,C)E_w & -\lambda(\rho_k)\,E_v \end{bmatrix} \begin{bmatrix} w(k) \\ v(k) \end{bmatrix} \end{aligned} \tag{24}$$

It reveals that the dynamics of the estimation error is completely decoupled from that of the state. It is notable that $x(k) = c(k) + e(k) \in \langle c(k), R(k)\rangle$. According to Definition 4, the cost function (18) can be rewritten as

$$J = Q\left[\sum_{k=0}^{\infty}\left(x^T(k)Q_x x(k) + u^T(k)Q_u u(k)\right)\right]$$
$$= \sum_{k=0}^{\infty}\left(c^T(k)Q_x c(k) + u^T(k)Q_u u(k)\right) + \sum_{k=0}^{\infty}\|R(k)\|_{F,Q_x}^2 \quad (25)$$

The first part in expression (25) yields the part which will be minimized by a suitable choice of the sequence of control signals $u(k)$. The second part will be minimized as well when a suitable filter gain $\lambda(\rho_k)$ is selected to get the optimal estimate of $x(k)$. □

Remark 2 As the system state is not measurable, the estimated state is utilized to replace $x_c(k)$ for the feedback. To guarantee that the magnitude of $c(k)$ is as close as possible to $x_c(k)$, the estimation error is optimally minimized by minimizing its Frobenius norm $\|R(k)\|_{F,Q_x}^2$. It is notable that the estimation error is bounded by $\langle 0, R(k)\rangle$.

Remark 3 It is worth stating that the weighting matrix Q_x is not related to the calculation of the optimal filter gain (Combastel 2015). Therefore, Theorem 2 is directly applicable to obtain the optimal filter gain of the LQZ control scheme.

4.2 Optimal Controller Gain Design

The following Theorem 4 is proposed for computing the optimal robust controller $u(k) = -K(\rho_k)x_c(k)$ through minimizing the cost function (21).

Theorem 4 *Given the nominal LPV system (20), $\forall i, j \in \mathcal{J}$, the optimal controller gain $K(\rho_k) = \sum_{j=1}^{J} h_j(\rho_k)K_j$ can be calculated if there exists a matrix $W_i \in \mathbb{R}^{n_x \times n_u}$, a positive definite matrix $\Gamma \in \mathbb{R}^{n_x \times n_x}$ and a positive scalar γ that are obtained by solving the following optimization problem*

$$\min \gamma$$
$$s.t., \begin{bmatrix} \gamma I & I \\ I & \Gamma \end{bmatrix} > 0 \quad (26a)$$
$$\Upsilon_{i,i} < 0, \quad i = j \quad (26b)$$
$$\frac{2}{J-1}\Upsilon_{i,i} + \Upsilon_{i,j} + \Upsilon_{j,i} \le 0, \quad i < j \quad (26c)$$

where

$$\Upsilon_{i,j} = \begin{bmatrix} -\Gamma & * & * & * \\ A_j\Gamma - B_j W_i & -\Gamma & * & * \\ H\Gamma & 0 & -I & * \\ W_j & 0 & 0 & -Q_u^{-1} \end{bmatrix} \tag{27}$$

with $K_i = W_i\Gamma^{-1}$, $H^T H = Q_x$.

Proof Let us choose a Lyapunov function: $V(x_c(k)) = x_c^T(k)Px_c(k))$, $P = P^T > 0$, satisfying the following conditions:

$$V(x_c(0)) < \gamma \tag{28}$$

$$V(x_c(k+1)) - V(x_c(k)) + [(x_c^T(k)Q_x x_c(k) + u^T(k)Q_u u(k))] < 0 \tag{29}$$

Therefore,

$$\sum_{k=0}^{\infty} \left(x_c^T(k)Q_x x_c(k) + u^T(k)Q_u u(k)\right) + V(x_c(\infty)) - V(x_c(0)) < 0, \tag{30}$$

which we can rewrite as

$$J_c < V(x_c(0)) < \gamma. \tag{31}$$

In this context, the cost function J is minimized along with the objective scalar γ.

Recalling that

$$\begin{aligned} V(x_c(k+1)) &= x_c^T(k+1)Px_c(k+1) \\ &= ((A(\rho_k) - B(\rho_k)K(\rho_k))x_c(k))^T P((A(\rho_k) - B(\rho_k)K(\rho_k))x_c(k)), \end{aligned}$$

inequality (29) is equivalent to

$$(A(\rho_k) - B(\rho_k)K(\rho_k))^T P(A(\rho_k) - B(\rho_k)K(\rho_k)) - P + Q_x + K^T(\rho_k)Q_u K(\rho_k) < 0 \tag{32}$$

Applying Schur complementation, (32) can be rewritten as the following inequality:

$$\Upsilon(\rho_k, \rho_k) = \begin{bmatrix} -\Gamma & * & * & * \\ A(\rho_k)\Gamma - B(\rho_k)W(\rho_k) & -\Gamma & * & * \\ H\Gamma & 0 & -I & * \\ W(\rho_k) & 0 & 0 & -Q_u^{-1} \end{bmatrix} < 0 \tag{33}$$

where $\Gamma = P^{-1}$, $W(\rho_k) = K(\rho_k)\Gamma$, $Q_x = H^T H$. Considering the polytopic form of $A(\rho_k) = \sum_{j=1}^{J} h_j(\rho_k)A_j$, $B(\rho_k) = \sum_{j=1}^{J} h_j(\rho_k)B_j$, $K(\rho_k) = \sum_{j=1}^{J} h_j(\rho_k)K_j$, it yields $\Upsilon(\rho_k, \rho_k) = \sum_{i=1}^{J}\sum_{j=1}^{J} h_i(\rho_k)h_j(\rho_k)\Upsilon_{i,j}$, which is valid for all vertices

$\forall i, j \in J$. Then, (26b) and (26c) can be obtained according to Lemma 1. Furthermore, the condition (28) can be rewritten as $\gamma I - \Gamma^{-1} > 0$, i.e., $\begin{bmatrix} \gamma I & I \\ I & \Gamma \end{bmatrix} > 0$. Hence, we complete the proof. □

5 Simulation Results

5.1 Vehicle Lateral Dynamics

Vehicle lateral dynamics may be modeled using a two degree of freedom model known as the "bicycle model" (see Fig. 2) that describes the lateral and yaw motions. The equations of this model are as follows

$$mv_x(\dot{\beta} + \dot{\psi}) = F_{yf} + F_{yr} \quad I_z\ddot{\psi} = l_f F_{yf} - l_r F_{yr} \tag{34}$$

where m and I_z, are the mass and the yaw moment, v_x is the longitudinal velocity, y is the vehicle lateral position, $\dot{\psi}$ is vehicle yaw rate, l_f and l_r are the distances from front and rear axle to the center of gravity. F_{yf} and F_{yr} are lateral tire force of front and rear tires are modelled as below, with the assumption of small sideslip angle variations:

$$F_{yf} = c_f\left(\delta_f - \beta - \frac{l_f}{v_x}\dot{\psi}\right), \; F_{yr} = c_r\left(-\beta + \frac{l_r}{v_x}\dot{\psi}\right) \tag{35}$$

where δ_f is the steering angle, c_f, c_r are the cornering stiffness of front and rear tires. Combining Eqs. (34) and (35), the following state-space model can be derived as follows:

$$\frac{d}{dt}\begin{bmatrix} \beta \\ \dot{\psi} \end{bmatrix} = \begin{bmatrix} -\frac{c_f+c_r}{mv_x} & \frac{c_r l_r - c_f l_f}{mv_x^2} - 1 \\ \frac{c_r l_r - c_f l_f}{I_z} & -\frac{c_r l_r^2 + c_f l_f^2}{I_z v_x} \end{bmatrix}\begin{bmatrix} \beta \\ \dot{\psi} \end{bmatrix} + \begin{bmatrix} \frac{c_f}{mv_x} \\ \frac{c_f l_f}{Iz} \end{bmatrix}\delta_f \tag{36}$$

The detailed meaning of each parameter is depicted in Fig. 2. System parameters and nominal values are described in Table 1.

Note that the yaw rate $\dot{\psi}$ and longitudinal speed v_x can be measured online. In this chapter, we perform the controller and observer design in the discrete-time domain for real-time implementation. To this end, Euler's discretization method, with the sampling time T= 0.01s, is used to obtain the discrete-time model of system (36), as follows:

$$\begin{aligned} x(k+1) &= A(\rho_k)\,x(k) + B(\rho_k)\,u(k) + E_w w(k) \\ y(k) &= Cx(k) + E_v v(k) \end{aligned} \tag{37}$$

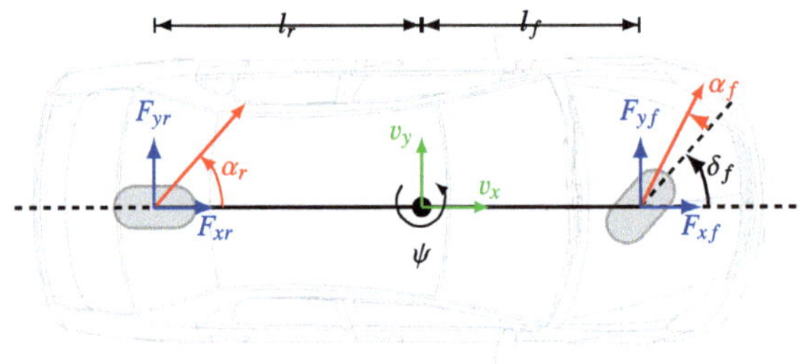

Fig. 2 The 2-DOF bicycle vehicle model

Table 1 System parameters

Parameter	Description	Value	Unit
m	Mass	1600	kg
I_z	Yaw moment	2454	$\text{kg} \cdot \text{m}^2$
l_f	Distance from front axle to the center of gravity	1	m
l_r	Distance from rear axle to the center of gravity	1.44	m
c_f	Cornering stiffness of front tires	35468	N/rad
c_r	Cornering stiffness of rear tires	40057	N/rad

where $x = [\beta \quad \dot{\psi}]^T$, $u = \delta_f$, $\rho_k = [\frac{1}{v_x} \quad \frac{1}{v_x^2}]^T$, and

$$A(\rho_k) = T\begin{bmatrix} -\frac{c_f+c_r}{mv_x} & \frac{c_r l_r - c_f l_f}{mv_x^2} - 1 \\ \frac{c_r l_r - c_f l_f}{I_z} & -\frac{c_r l_r^2 + c_f l_f^2}{I_z v_x} \end{bmatrix} + I,\ B(\rho_k) = T\begin{bmatrix} \frac{c_f}{mv_x} \\ \frac{c_f l_f}{Iz} \end{bmatrix},\ C = [0 \quad 1].$$

$E_w = \begin{bmatrix} 0.001 & 0 \\ 0 & 0.02 \end{bmatrix}$, $E_v = 0.03$, and the system disturbances w and measurement noise v are random white noise bounded using zonotopes: $w \in \mathcal{W} = \langle 0, I_2 \rangle$, $v \in \mathcal{V} = \langle 0, I_1 \rangle$.

Considering the physical limitations during normal driving conditions, the longitudinal speed is bounded by a compact set as $v_x \in [\underline{v}_x, \ \overline{v}_x]$, where $\underline{v}_x = 6, \overline{v}_x = 25$. Using the sector nonlinearity method with the scheduling variable $\rho_k = \left[\frac{1}{v_x}, \ \frac{1}{v_x^2}\right]^T$, the polytopic model (2) of system (37) can be obtained,

$$\begin{cases} x(k+1) = \sum_{j=1}^{4} h_j\big(\rho(k)\big)\big(A_j x(k) + B_j u(k)\big) + E_w w(k) \\ \quad y(k) \quad = Cx(k) + E_v v(k) \end{cases} \tag{38}$$

where

$$A_1 = A\left(\frac{1}{\underline{v}_x}, \frac{1}{\underline{v}_x^2}\right), A_2 = A\left(\frac{1}{\underline{v}_x}, \frac{1}{\overline{v}_x^2}\right), A_3 = A\left(\frac{1}{\overline{v}_x}, \frac{1}{\underline{v}_x^2}\right), A_4 = A\left(\frac{1}{\overline{v}_x}, \frac{1}{\overline{v}_x^2}\right),$$

$$B_1 = B_2 = B\left(\frac{1}{\underline{v}_x}\right), B_3 = B_4 = B\left(\frac{1}{\overline{v}_x}\right),$$

and

$$h_1(\rho) = \frac{\left(\frac{1}{\underline{v}_x} - \frac{1}{v_x}\right)\left(\frac{1}{\underline{v}_x^2} - \frac{1}{v_x^2}\right)}{\left(\frac{1}{\underline{v}_x} - \frac{1}{\overline{v}_x}\right)\left(\frac{1}{\underline{v}_x^2} - \frac{1}{\overline{v}_x^2}\right)}, h_2(\rho) = \frac{\left(\frac{1}{\underline{v}_x} - \frac{1}{v_x}\right)\left(\frac{1}{v_x^2} - \frac{1}{\overline{v}_x^2}\right)}{\left(\frac{1}{\underline{v}_x} - \frac{1}{\overline{v}_x}\right)\left(\frac{1}{\underline{v}_x^2} - \frac{1}{\overline{v}_x^2}\right)},$$

$$h_3(\rho) = \frac{\left(\frac{1}{v_x} - \frac{1}{\overline{v}_x}\right)\left(\frac{1}{\underline{v}_x^2} - \frac{1}{v_x^2}\right)}{\left(\frac{1}{\underline{v}_x} - \frac{1}{\overline{v}_x}\right)\left(\frac{1}{\underline{v}_x^2} - \frac{1}{\overline{v}_x^2}\right)}, h_4(\rho) = \frac{\left(\frac{1}{v_x} - \frac{1}{\overline{v}_x}\right)\left(\frac{1}{v_x^2} - \frac{1}{\overline{v}_x^2}\right)}{\left(\frac{1}{\underline{v}_x} - \frac{1}{\overline{v}_x}\right)\left(\frac{1}{\underline{v}_x^2} - \frac{1}{\overline{v}_x^2}\right)},$$

5.2 *Reference Tracking Control*

As the objective is to design a reference tracking control of the vehicle lateral dynamics using the proposed LQZ control, Fig. 3 depicts the planning-control-filter diagram designed for this vehicle application. The goal of this controller is to ensure that the vehicle tracks a desired reference trajectory by regulating the steering angle δ_f. The desired sideslip angle and yaw rate are given as follows

$$\beta_{\text{ref}} = \frac{l_r - \frac{l_f m v_x^2}{C_r(l_f+l_r)}}{\ell_f + \ell_r + \frac{m v_x^2(\ell_r C_r - \ell_f C_f)}{C_f C_r(\ell_f+\ell_r)}}\delta_{ref} \tag{39}$$

$$\dot{\psi}_{\text{ref}} = \frac{v_x}{\ell_f + \ell_r + \frac{m v_x^2(\ell_r C_r - \ell_f C_f)}{C_f C_r(\ell_f+\ell_r)}}\delta_{ref} \tag{40}$$

Remark 4 The position of the vehicle in the global (world) coordinates (X, Y) can be computed from the body-fixed coordinates $(\dot{x}, \dot{y})$ as

$$\begin{cases} X = \int_0^t (v_x \cos(\psi) - \dot{y}\sin(\psi))\, dt \\ Y = \int_0^t (v_x \sin(\psi) + \dot{y}\cos(\psi))\, dt \end{cases} \tag{41}$$

where $\psi = \psi_0 + \int_0^t \dot{\psi} dt$

As the trajectory planner has full-state information, a full-state feedback controller can be designed. As depicted by Fig. 3, the controller aims to reduce the state error.

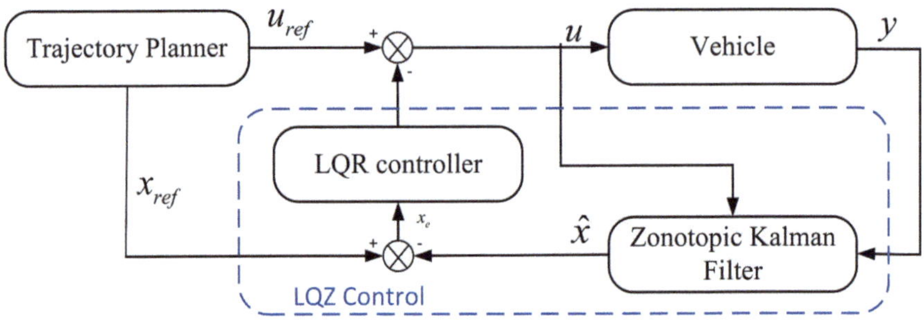

Fig. 3 Reference tracking control diagram for vehicle lateral dynamics

Thus, a reference tracking model is designed based on the lateral dynamics model using offsets of the desired path. By denoting the difference between the desired sideslip angle, yaw rate and steering angle $(\beta_{ref}, \dot{\psi}_{ref}, \delta_{ref})$ and the real vehicle sideslip angle, yaw rate and steering angle $(\beta, \dot{\psi}, \delta_f)$ as $x_e = \begin{bmatrix} \beta_{ref} \\ \dot{\psi}_{ref} \end{bmatrix} - \begin{bmatrix} \beta \\ \dot{\psi} \end{bmatrix}$, $u_e = \delta_{ref} - u$, the state-space reference tracking model can be obtained:

$$x_e(k+1) = A(\rho_k)x_e(k) + B(\rho_k)u_e(k) \tag{42}$$

Then, the real input of the vehicle is represented as $u(k) = K(\rho_k)x_e(k) + u_{ref}(k)$, where $K(\rho_k)$ is the controller gain to be designed. As the system states are not fully available, a zonotopic Kalman filter is employed to obtain the optimal estimated states bounded by a zonotope set $\hat{x}(k) \in \langle c(k), R(k) \rangle$, as well as to deal with the uncertainty problem. According to Theorem 3, the controller and filter gain can be designed separately. Now, consider the following optimal state feedback control (43) for the reference tracking model (42),

$$u_e(k) = -K(\rho_k)x_e(k). \tag{43}$$

$$J = \sum_{k=0}^{\infty} \left(x_e^T(k) Q_x x_e(k) + u_e^T(k) Q_u u_e(k) \right) \tag{44}$$

The cost function to be minimized is described in (44). It is possible to determine the appropriate cost by setting the ratio of state and control gain through the selection of Q_x an Q_u. Then, the optimal controller gain $K(\rho_k) = \sum_{j=1}^{J} h_j(\rho_k) K_j$ can be obtained by solving the LMI optimization problem (26). Meanwhile, by minimizing the cost function (8), the optimal filter gain $\lambda(\rho_k) = \sum_{j=1}^{J} h_j(\rho_k) \lambda_j$ can be obtained by solving the LMI optimization problem (6).

By assigning $Q_x = \begin{bmatrix} 20 & 0 \\ 0 & 60 \end{bmatrix}$, $Q_u = 1$, and solving the LMI optimization problem (26) and (6) by means of YALMIP toolbox, we have

$$K_1 = [2.1803 \quad 5.8860]\,,\; K_2 = [2.1464 \quad 5.9086]\,,$$

$$K3 = [1.0381 \quad 6.2211]\,,\; K_4 = [0.9358 \quad 6.0077]\,,$$

$$\lambda_1 = [0.0017 \quad 0.4566]^T\,,\; \lambda_2 = [0.0007 \quad 0.4567]^T\,,$$

$$\lambda_3 = [0.0022 \quad 0.4741]^T\,,\; \lambda_4 = [0.0010 \quad 0.4739]^T\,.$$

The results are depicted in Figs. 4, 5 and 6. In particular, Fig. 4 reveals a comparison between the reference and the vehicle trajectories in the world coordinates. From this figure, it is evident that the proposed algorithm allows to achieve good performance in terms of trajectory tracking. This result is also further detailed in Fig. 5, in which a comparison between the referenced states and real/estimated states, and the lateral and heading offsets are presented. Furthermore, Fig. 6 shows the equivalence between the proposed offline ZKF and the online ZKF proposed in Combastel (2015) in terms of estimation precision during steady state. In comparison to the approach presented in Combastel (2015), the proposed method demonstrates good performance with a slight improvement in state estimation precision. Moreover, it is worth noting that the proposed design not only outperforms in terms of accuracy but also offers a more attractive solution from a computational point of view. The optimal filter gain design of our method is conducted offline, minimizing the need for real-time computations. In contrast, the methodology employed by Combastel (2015) involves online computation of filter gain, which is time-consuming. This distinction underscores the efficiency and practicality of our approach, making it a compelling choice for vehicle applications where computational resources are a critical consideration.

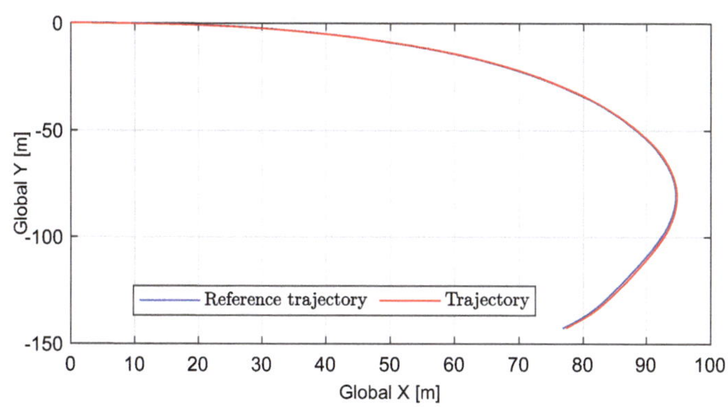

Fig. 4 Vehicle trajectory

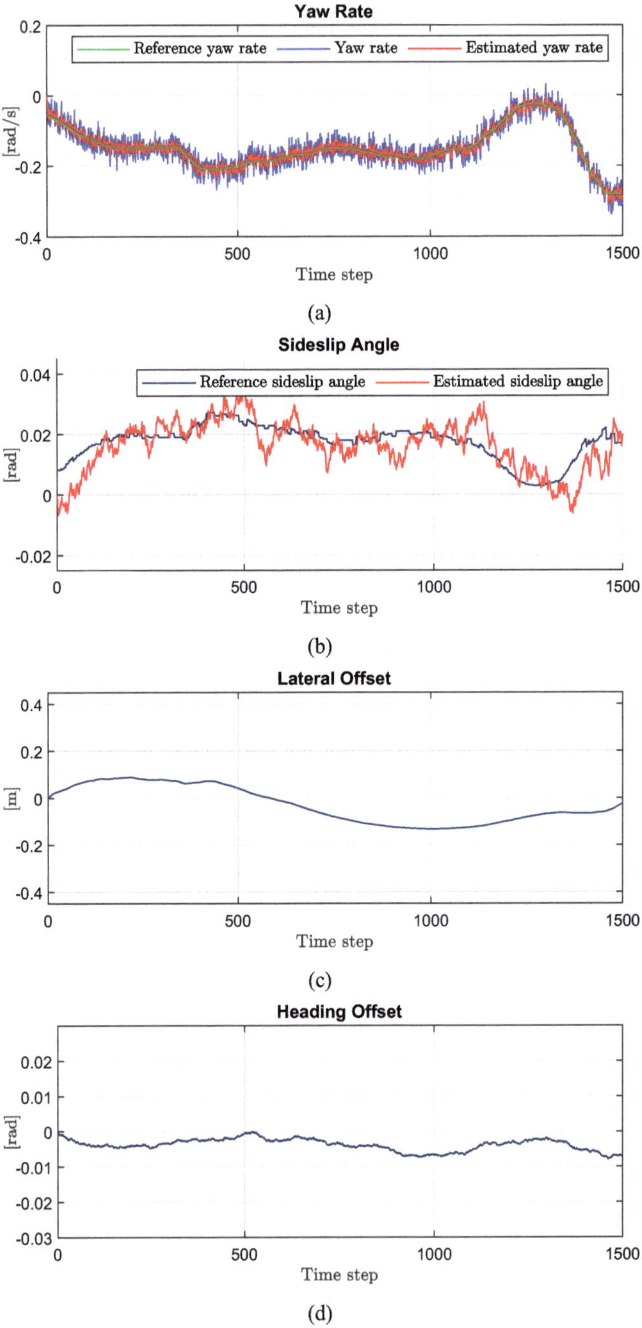

Fig. 5 Tracking control performance simulation results with the proposed LQZ control scheme. **a** Yaw rate, **b** Sideslip angle, **c** Lateral offset, **d** Heading offset

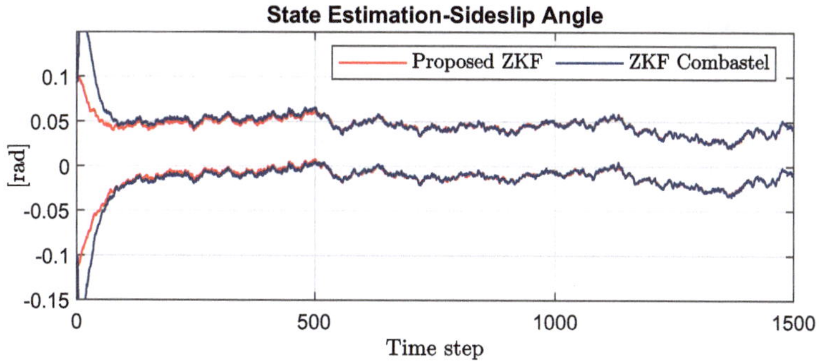

(a)

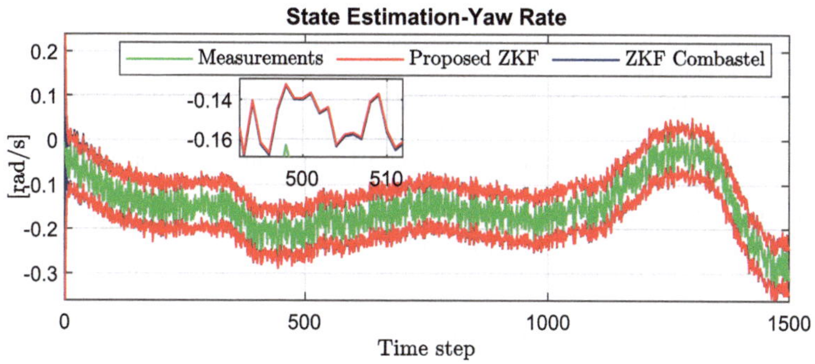

(b)

Fig. 6 Comparison of state estimation performance with the proposed ZKF and Combasetl ZKF. **a** Estimated sideslip angle, **b** Estimated yaw rate

6 Conclusion

In this paper, a robust state feedback control scheme based on a ZKF is proposed for LPV systems subject to disturbances and measurement noises. In order to obtain the unmeasurable states, the ZKF has been redesigned for LPV systems in terms of calculating the optimal filter gain offline by finding the solutions to a finite number of LMIs. Then, a LQZ-based control synthesis has been proposed as a zonotopic counterpart of LQG, in which the separation principle has been proved. Therefore, the optimal controller gain and optimal filter gain has been designed separately. Finally, the proposed method has been tested in simulation via a reference tracking control of vehicle lateral dynamics. From the simulation results of controller performance and filter performance, it revealed the feasibility and effectiveness of the proposed control scheme.

Acknowledgements This work has been co-financed by the Spanish State Research Agency (AEI) and the European Regional Development Fund (ERFD) through the project SaCoAV (ref. MINECO PID2020-114244RB-I00), by the European Regional Development Fund of the European Union in the framework of the ERDF Operational Program of Catalonia 2014–2020 (ref. 001-P-001643 Looming Factory), by the DGR of Generalitat de Catalunya (SAC group ref. 2017/SGR/482) and by the Chinese Scholarship Council (CSC) under grant (202006120006).

References

Alcalá E, Puig V, Quevedo J (2019) LPV-MPC control for autonomous vehicles. IFAC-PapersOnLine 52(28):106–113

Althoff M, Frehse G, Girard A (2021) Set propagation techniques for reachability analysis. Annu Rev Control Robot Auton Syst 4:369–395

Blesa J, Puig V, Saludes J (2012) Robust fault detection using polytope-based set-membership consistency test. IET Control Theory Appl 6(12):1767–1777

Boyd S, El Ghaoui L, Feron E, Balakrishnan V (1994) Linear matrix inequalities in system and control theory. SIAM

Cavanini L, Ippoliti G, Camacho EF (2021) Model predictive control for a linear parameter varying model of an UAV. J Intell Robot Syst 101:1–18

Chen L, Patton RJ (2011) Polytope lPV estimation for non-linear flight control. IFAC Proc Vol 44(1):6680–6685

Combastel C (2015) Zonotopes and Kalman observers: gain optimality under distinct uncertainty paradigms and robust convergence. Automatica 55:265–273

Efimov D, Raissi T, Zolghadri A (2013) Control of nonlinear and lPV systems: interval observer-based framework. IEEE Trans Autom Control 58(3):773–778

Hammouche M, Mohand-Ousaid A, Lutz P, Rakotondrabe M (2018) Robust interval Luenberger observer-based state feedback control: Application to a multi-DOF micropositioner. IEEE Trans Control Syst Technol 27(6):2672–2679

Hang P, Chen X, Luo F (2019) LPV/H? controller design for path tracking of autonomous ground vehicles through four-wheel steering and direct yaw-moment control. Int J Automot Technol 20:679–691

He Z, Xie W (2016) Control of non-linear switched systems with average dwell time: interval observer-based framework. IET Control Theory Appl 10(1):10–16

Hu Y, Duan G (2019) H? finite-time control for lPV systems with parameter-varying time delays and external disturbance via observer-based state feedback. J Franklin Inst 356(12):6303–6327

Ifqir S, Dalil I, Naïma AO, Saïd M (2019a) Adaptive threshold generation for vehicle fault detection using switched T-S interval observers. IEEE Trans Ind Electron 67(6):5030–5040

Ifqir S, Rauh A, Kersten J, Ichalal D, Ait-Oufroukh N, Mammar S (2019b) Interval observer-based controller design for systems with state constraints: application to solid oxide fuel cells stacks. In: 2019 24th International conference on methods and models in automation and robotics (MMAR). IEEE, pp 372–377

Kühn W (1998) Rigorously computed orbits of dynamical systems without the wrapping effect. Computing 61:47–67

Lamouchi R, Raissi T, Amairi M, Aoun M (2022) On interval observer design for active fault tolerant control of linear parameter-varying systems. Syst Control Lett 164:105218

Le VTH, Stoica C, Alamo T, Camacho EF, Dumur D (2013) Zonotopes: from guaranteed state-estimation to control. Wiley

Li P, Nguyen AT, Du H, Wang Y, Zhang H (2021) Polytopic LPV approaches for intelligent automotive systems: State of the art and future challenges. Mech Syst Signal Process 161:107931

Liu Z, Theilliol D, Gu F, He Y, Yang L, Han J (2017) State feedback controller design for affine parameter-dependent LPV systems. IFAC-PapersOnLine 50(1):9760–9765
Lv J, Wang YE, Wu D (2023) Observer-based control for continuous-time switched linear parameter-varying systems. Circuits Syst Signal Process 42(4):2049–2064
Ostertag E (2011) Mono-and multivariable control and estimation: linear, quadratic and LMI methods. Springer Science & Business Media
Pourasghar M, Puig V, Ocampo-Martinez C (2019) Interval observer versus set-membership approaches for fault detection in uncertain systems using zonotopes. Int J Robust Nonlinear Control 29(10):2819–2843
Puig V (2010) Fault diagnosis and fault tolerant control using set-membership approaches: application to real case studies. Int J Appl Math Comput Sci 20(4):619–635
Puig V, Quevedo J, Escobet T, Stancu A (2003) Robust fault detection using linear interval observers. IFAC Proc Vol 36(5):579–584
Rego BS, Raffo GV, Scott JK, Raimondo DM (2020) Guaranteed methods based on constrained zonotopes for set-valued state estimation of nonlinear discrete-time systems. Automatica 111:108614
Ren C, Li X, Yang X, Ma S (2019) Extended state observer-based sliding mode control of an omnidirectional mobile robot with friction compensation. IEEE Trans Ind Electron 66(12):9480–9489
Rotondo D, Nejjari F, Puig V (2014) Robust state-feedback control of uncertain LPV systems: an LMI-based approach. J Franklin Inst 351(5):2781–2803
Shan Y, Zhu F (2022) Interval observer-based fault tolerant control strategy with fault estimation and compensation. Asian J Control 24(2):895–906
Trapiello C, Puig V (2022) A zonotopic-based watermarking design to detect replay attacks. IEEE/CAA J Autom Sin 9(11):1924–1938
Wang J, Lv X, Meng Z, Puig V (2023) An integrated design method for active fault diagnosis and control. Int J Robust Nonlinear Control 33(10):5583–5603
Wang X, Pin Tan C, Wang Y, Zhang Z (2021) Active fault tolerant control based on adaptive interval observer for uncertain systems with sensor faults. Int J Robust Nonlinear Control 31(8):2857–2881
Wang Z, Zhang L, Ra T, Shen Y et al (2024) Interval estimation for discrete-time Takagi-Sugeno fuzzy nonlinear systems with parameter uncertainties. IEEE Trans Fuzzy Syst
Xu F, Olaru S, Puig V, Ocampo-Martinez C, Niculescu SI (2017) Sensor-fault tolerance using robust MPC with set-based state estimation and active fault isolation. Int J Robust Nonlinear Control 27(8):1260–1283
Yang D, Wang Y, Chen Z (2020) Robust fault diagnosis and fault tolerant control for PEMFC system based on an augmented LPV observer. Int J Hydrogen Energy 45(24):13508–13522
Yang S, Xu F, Wang X, Liang B (2020) A novel online active fault diagnosis method based on invariant sets. IEEE Control Syst Lett 5(2):457–462
Yang X, Scott JK (2018) A comparison of zonotope order reduction techniques. Automatica 95:378–384
Zhang S, Puig V, Ifqir S (2023) Zonotopic set-membership state estimation for switched LPV systems. IFAC-PapersOnLine 56(2):9442–9447
Zhang W, Wang Z, Raïssi T, Su R (2023) An ellipsoid-based framework for fault estimation and remaining useful life prognosis. Int J Robust Nonlinear Control 33(12):7260–7281

Parameter Identification for Cooperative SOFC Models on the GPU

Ekaterina Auer, Andreas Rauh, and Lorenz Gillner

Abstract Over the last decade, using graphics processing units (GPUs) for computations has given a significant boost to varied scientific fields and even to our everyday lives. An emergent branch is modeling, parameter identification, simulation, and control of solid oxide fuel cells (SOFCs). SOFCs convert chemical energy into electricity with a high degree of efficiency and in a low-waste manner. Models for SOFC temperature are based on partial-differential equations, which are usually discretized with respect to space and time into algebraic equations. A disadvantage of this technique is the lack of flexibility, which is unsuitable for control or for guaranteed prevention of overheating that can accelerate stack degradation. In our past publications, it has been shown that it is possible to arrive at dynamic SOFC models consisting of a set of ordinary differential equations which are cooperative. In this contribution, we extend the GPU-based parameter identification technique we used for a reduced distributed heating system to cover cooperative SOFC models with the help of interval analysis to deal with uncertainty. Additionally, we show how the amount of data for controlled and measured system inputs and outputs can be reduced by using interval Bézier curves.

1 Introduction

A recent development in the area of high performance computing (HPC) is the use of graphics processing units (GPUs) for scientific computations. GPUs are specialized programmable devices developed originally for speeding up graphic rendering but

E. Auer (✉) · L. Gillner
University of Applied Sciences Wismar, Wismar, Germany
e-mail: ekaterina.auer@hs-wismar.de

L. Gillner
e-mail: lorenz.gillner@hs-wismar.de

A. Rauh
Carl von Ossietzky Universität Oldenburg, Oldenburg, Germany
e-mail: andreas.rauh@uni-oldenburg.de

T. N. Dinh et al. (eds.), *Set-Valued Approaches to Control and Estimation of Uncertain Systems*, Mathematical Engineering,
https://doi.org/10.1007/978-3-031-94239-6_9

used now also in science due to increased flexibility in their programming. Since HPC applications are usually understood as requiring massively parallel processing with relatively similar processes, GPUs represent the logical choice for a relatively low cost support hardware if a problem at hand maps to the GPU's so-called stream processing model. The employment of GPUs has helped to advance such varied fields as neural networks, bioinformatics, medical imaging, or cryptography.

One of the fields profiting from the use of the GPUs is control engineering, where this paradigm opens up possibilities which remained unexplored until now (Lee et al. 2022; Maceina and Manduchi 2017; Romero 2018). In particular, modeling, parameter identification, simulation, and control of solid oxide fuel cells (SOFCs) can benefit from cheap computing power (Auer et al. 2020). This application, on the one hand, requires involved (dynamic) computations and, on the other hand, exhibits high data parallelism in certain settings, the latter making it a classical choice for the GPU use.

SOFCs convert chemical energy into electricity with a high degree of efficiency and in a low-waste manner. Models for SOFC temperature are based on partial-differential equations, which are usually discretized with respect to space and time into algebraic equations (Huang et al. 2012). A disadvantage of this technique is the lack of flexibility: Only stationary states of SOFC systems can be simulated this way, which is not well-suited for control or for prevention of overheating that can accelerate stack degradation. Using the same finite volume method without discretization in time, it is possible to arrive at *dynamic* system models consisting of a set of ordinary differential equations (ODEs) which can be transformed into a cooperative form Ifqir et al. (2019). A challenge here is to identify a large number of parameters based on uncertain measured values from experiments obtained using a real SOFC system. A further difficulty is that such models might be sensitive wrt. parameter regions in which the property of cooperativity (Smith 1995) holds.

For a cooperative ODE system with uncertain but bounded parameters, two bracketing systems with crisp parameters can be defined to catch the bulk of uncertainty and, possibly, to reduce conservativeness for the obtained result regions. These bounding systems might be coupled with each other if lower and upper interval bounds for the system parameters to be identified appear simultaneously in an equation. Moreover, it is possible to devise several brute force approaches on the GPU to reduce the uncertainty in set-based simulation or to identify the parameters of such (cooperative) ODEs. The parameter search space (or parameter uncertainty) can be partitioned into a large number of small boxes, each of which can be processed with a separate GPU kernel. The considered ODE system is processed over these subintervals in parallel. In case of parameter identification, regions inconsistent with available measurements can be eliminated in this way. Finally, to avoid a prohibitively large number of system evaluations due to overly large search spaces, a set of additional simple consistency tests can be employed. They represent knowledge from physics such as non-negativity and strict monotonicity of heat capacities and reaction enthalpies that occur multiple times in the dynamic SOFC model (common subexpressions). Such constraints in combination with inequalities reflecting physically meaningful temporal variation rates of the measured SOFC stack temperature help to reduce the number of subintervals. This preprocessing stage is carried out prior to the parallelized evaluation of the system model.

In this paper, we extend the GPU-based experimental technique we applied for a reduced distributed heating system in Auer et al. (2020). Additionally, we show how the amount of data for controlled and measured system inputs and outputs can be reduced by using interval Bézier curves. For quantifying uncertainty, we rely on methods with result verification in addition to the property of cooperativity, in particular, on interval analysis (IA) (Moore et al. 2009).

This chapter is structured as follows. First, we overview the background, explain the proposed techniques and illustrate them using a well-known example of a two-compartment problem from Jaulin et al. (2001); Kieffer et al. (2011) in Sect. 2. Then, we describe the control-oriented model we use for the SOFC and apply the methods from the previous section for its simulation and parameter identification in Sect. 3. Conclusions and an outlook on our future work are in the last section.

2 Experimental GPU-Based Parameter Identification Techniques

In this paper, the models we consider have the form of an initial value problem (IVP) for ODEs

$$y' = f(t, y, p, u),\ \ y : \mathbb{R} \mapsto \mathbb{R}^{n_y},\ \ p \in \mathbb{R}^{n_p},\ \ u \in \mathbb{R}^{n_u},\ f : \mathbb{R}^{1+n_y+n_p+n_u} \mapsto \mathbb{R}^{n_y} \tag{1}$$

with $y(t_0) \in \mathbf{y}$ for a $t_0 \in \mathbb{R}$ and $p \in \mathbf{p}$, where bold face letters denote set-valued, uncertain characteristics. Here, y is the sought solution function depending on the time t, p are parameters, and u are control inputs.

We assume that the principles of IA are well-known to the readers, see Moore et al. (2009). In this paper, we consider all set-valued, uncertain characteristics to be (machine) intervals of the form $\mathbf{x} = \{x \in \mathbb{R} \mid \underline{x} \leq x \leq \overline{x}\}$, with $\underline{x} \leq \overline{x}$ being the lower and the upper bound of the interval $\mathbf{x}$. The extension to the vector case is performed in an element-wise manner. For machine intervals, $\underline{x}, \overline{x}$ belong to the set of floating point (FP) numbers defined according to the IEEE 754 standard, obtained from possibly real-valued exact bounds by directed rounding (cf. IEEE standard 1788–2015 for intervals (IEEE standard for interval arithmetic 2015)). Out of the whole set of methods from IA, we rely mainly on interval arithmetic in combination with the property of cooperativity in this paper. Interval arithmetic defines interval versions of operations such as addition along with elementary functions such as the sine. By replacing all FP operations by their interval counterparts, it is possible to obtain so-called (naive) interval extensions of any further composite, continuous functions. If uncertain quantities can be bounded by intervals, the uncertainty can be propagated through (static) systems in this way. Another advantage is that the result is verified. A disadvantage is the possibility of too conservative enclosures (e.g.,

$(-\infty, +\infty)$) that do not provide any meaningful information about the considered system. This can usually happen because of overestimation (Lohner 2001) and/or because the problem to be solved is ill-posed.

Although there are a lot of CPU-based implementations for IA and more involved set-based methods, for example, those for solving IVPs to ODEs, the adequate software for the GPU is still lacking. In this paper, we rely on an extended version of the library from Collange et al. (2012), which is available in CUDA, a proprietary, closed-source API for general-purpose programmable NVIDIA GPUs. At the moment, only an algorithm for obtaining rough enclosures of solutions to IVPs is available for the GPU (Auer et al. 2017). However, the provided sets are not tight enough to allow for meaningful parameter identification or even simulation; therefore, we rely on the property of cooperativity described in Sect. 2.1 in combination with FP-based IVP solvers provided by the BOOST library ODEINT and made available for the GPU through ODEINT's compatibility layer with the standard algorithms library THRUST (implemented using CUDA, cf. Ahnert et al. (2014)). Note that since the time discretization error introduced by ODEINT is negligible compared to parameter uncertainty present in the considered application, this procedure is not a major restriction.

The remainder of this section is structured as follows. In Sect. 2.1, the property of cooperativity is described briefly. In Sect. 2.2, general possibilities for parameter identification are pointed out. In Sect. 2.3, the proposed experiment-based identification technique is formulated for the GPU in different versions. In Sect. 2.4, there is a short overview of the corresponding implementation. In Sect. 2.5, we illustrate the introduced experimental techniques using the two-compartment problem.

2.1 Cooperativity

The property of cooperativity holds for the system in Eq. (1) if

$$\frac{\partial f_i}{\partial y_j} \geq 0 \text{ for all } i \neq j,\ i, j = 1 \ldots n_y. \tag{2}$$

Müller's theorem (Gennat and Tibken 2008; Meslem and Ramdani 2011) combined with the property in Eq. (2) and positivity of y_j results in the Smith's theorem (Smith 1995). It proves that the uncertainty in the system (1) can be quantified by working with two systems with crisp parameters instead of one system with uncertain parameters

$$y'_{\text{lb}} = \underline{f}(t, y_{\text{lb}}, p_{\text{b}}) \quad \text{and} \quad y'_{\text{ub}} = \overline{f}(t, y_{\text{ub}}, p_{\text{b}}) \tag{3}$$

(two bracketing systems). The notation p_{b} means that only bounds of the interval $\mathbf{p}$ are used (and not the values in between). The true result $y(t)$ lies in the interval $[\underline{y}_{\text{lb}}(t), \overline{y}_{\text{ub}}(t)]$ of the enclosures for $\mathbf{y}_{\text{lb}}(t)$, $\mathbf{y}_{\text{ub}}(t)$ from (3). Considering such enclosures in each point of time t_k of the discretization grid gives us the verified flow of the uncertain system (1). A good approximation to this verified solution are $y_{\mathrm{lb}}^{(k)}$, $y_{\mathrm{ub}}^{(k)}$ obtained by solving the systems in (3) using traditional FP arithmetic with sufficiently small discretization step sizes. This procedure still captures the bulk of output uncertainty. The idea in this chapter is therefore to combine the property of cooperativity (2) with parameter identification and to parallelize it on the GPU, with the additional use of data reduction. Due to the lack of verified solvers providing tight enclosures on the GPU, we rely on non-verified ones.

2.2 General Possibilities for Parameter Identification

A common technique to identify unknown parameters $p \in \mathbb{R}^{n_p}$ of a dynamic model of the form in Eq. (1) given a search space $\mathbf{p} \in \mathbb{IR}^{n_p}$ and measured data over a time grid $t_k \in \{T_b, T_b + 1, \ldots, T_e\}$ is based on the least squares approach

$$\Phi(p) = \sum_{k=T_b}^{T_e} \sum_{j=1}^{n_m} \left(y_j(t_k, p) - y_{j,\mathrm{m}}(t_k)\right)^2 \overset{p\in\mathbf{p}}{\longrightarrow} \min, \tag{4}$$

where $y_j(t_k, p)$ is the jth component of the solution to the model in Eq. (1); $y_{j,\mathrm{m}}(t_k)$ is the measured data for $y_j(t_k, p)$ at the time point $t_k \in [T_b, T_e]$; T_b, T_e are the start and end times of data recording. In some applications, not every component of the solution can be measured. Therefore, n_m denotes the number of the measured components, for which we suppose that they are consecutively numbered from 1 to n_m. There are different possibilities to tackle this problem, with varying degrees of verification associated with them.

The topic of a verification degree is not standardized yet. One possibility is described in Auer (2014): It is convenient to classify models implemented on computers into four classes wrt. the extent up to which we can be sure that their results are correct. The corresponding verification degrees, from C4 (lowest) to C1 (highest), can then be given as follows.

C4: The computer-based model (the program) uses FP or fixed point arithmetic.
C3: The model is meaningfully divided into its components; the numeric implementation uses at least the standardized IEEE 754 FP arithmetic.
C2: Relevant subsystems of the model are implemented using tools with result verification or providing reliable error limits; the convergence of numerical algorithms is substantiated by analytical solutions, computer-based proofs or fixed point theorems.
C1: The whole model is implemented using formal verification and algorithms with result verification/ real-number algorithms/ analytical solutions/ computer-based proofs; software and hardware comply with the IEEE standards 754 and 1788–2015; uncertainty is quantified for the entire computer-based model using verified or stochastic approaches (or both); sensitivity analysis is performed.

Additionally, we add a + to the degree if sensitivity analysis is performed for uncertain parameters or uncertainty is quantified using traditional (e.g., Monte Carlo) or verified methods. Then, the available options for parameter identification can each possess different verification degrees in dependence on the choices made for questions F0–F3:

F0	Is the formula in Eq. (4) used?	F0.a (no) Experimental/ANN/... F0.b (yes) Least squares
F1	How is $y(t_k, p)$ obtained?	F1.a A closed-form solution F1.b Approximation by an expression (e.g., using the Euler method) F1.c Numerical solution from a "black-box" solver
F2	What is the implementation technique (i.e., arithmetic)?	F2.a Fixed or floating point F2.b Interval F2.c Other verified
F3	How are the measured data represented?	F3.a As provided by sensors: files with decimal FP numbers F3.b With the help of any reliable means for data reduction

In this paper, we always rely on a black-box solver to obtain the solution to the IVP (F1.c). We consider options F0.a/b, F2.a,b and F3.b both for the simpler illustration in Sect. 2.5 and for the SOFC model in Sect. 3.3. For data reduction in F3.b, we use enclosures by Bézier curves as described in Sects. 2.5, 3.3. For the example in Sect. 2.5, we also consider F3.a, with further variants for its parameter identification implemented and compared in Gillner and Auer (2024). We specify the verification degree that can be attained according to the classes C4–C1 in the corresponding sections.

2.3 Experimental Parameter Identification

Subdividing the initial search space **p** into smaller boxes creates data parallelism, which is perfectly suitable for the GPU use. Although it is not really necessary to employ the GPU in this way for the illustration from Sect. 2.5, it makes computations feasible in the case of the SOFC application from Sect. 3. In this sense, all techniques presented in this section for parameter identification are brute-force approaches which would be impracticable using the CPU, serially or even parallelized. Likewise, the GPU can be employed for reducing the overestimation in IVPs with uncertainties directly (i.e., without any parameter identification) by corresponding parallelized simulations over the small boxes in an acceptable time as compared to the CPU implementation.

The idea of employing the GPU for global optimization with intervals is not new, cf. Stakhiv et al. (2012); Eriksen and Rasmussen (2013); Bagóczki and Bánhelyi (2017); Sanders (2022); Rafiei and Chevalier (2024). Some of the cited references also use data parallelism as their basis. However, these works consider a static problem (i.e., option F1.a), whereas our goal is to tackle a dynamic problem in Sect. 3 using a black-box IVP solver (option F1.c).

The basic technique we consider in this paper is shown in Algorithm 1. Steps 2 and 3 are carried out on the GPU. The returned list $\mathcal{L}$ of indices of suitable boxes can be used to build the hull of $\mathbf{p}_j$ for $j \in \mathcal{L}$ to obtain a reduced search space. The algorithm can then be reiterated with tighter search intervals and a lower bisection bound w. Another use is to compute the convex hull of enclosures for $\mathbf{y}(t_k, \mathbf{p}_j)$ for $j \in \mathcal{L}$ (again on the GPU). This new enclosure can be good enough without subsequent least squares identification.

Different exclusion strategies can be employed in Step 3. In Auer et al. (2020), we used

Criterion 1: Exclude $\mathbf{p}_j$ if $\exists k, i\colon \mathbf{y}_i^{(k)} \cap [y_{i,\mathrm{m}}(t_k) - \delta,\, y_{i,\mathrm{m}}(t_k) + \delta] = \emptyset$,

where $[y_{i,\mathrm{m}}(t_k) - \delta,\, y_{i,\mathrm{m}}(t_k) + \delta]$ were plausibility bounds on measured data $y_{i,\mathrm{m}}(t_k)$ with δ determined by the quality of sensors. Here, we also rely on

Criterion 2: Exclude $\mathbf{p}_j$ if $\exists k, i\colon \mathbf{y}_i^{(k)} \cap B_{y_{i,\mathrm{m}}}\left(a_{y_{i,\mathrm{m}}}(t_k), \Delta_{y_{i,\mathrm{m}}}(t_k)\right) = \emptyset$,

where $B_{y_{i,\mathrm{m}}}\left(a_{y_{i,\mathrm{m}}}(t_k), \Delta_{y_{i,\mathrm{m}}}(t_k)\right)$ is an enclosing Bézier curve for the data (cf. Sects. 2.5, 3.3).

Additionally, a SIVIA-like algorithm can be employed in Step 3, SIVIA being a well-known set-inversion based optimization approach (Jaulin and Walter 1993). Here, we need to consider the cost function in Eq. (4) that has a known (ideal) minimum of zero. The parallelized version shown in Algorithm 2 introduced in Gillner and Auer (2024) can be applied to it. In the algorithm, $\mathbf{z}$ stands for the enclosure of the optimum ($\mathbf{z} = [0, tol]$ for some tolerance bound tol in our case), $\mathcal{S}$ contains

Algorithm 1

Basic experimental parameter identification

Require: $\mathbf{p}$, w
Step 1: Bisect $\mathbf{p} = \bigcup_{j=1}^{l} \mathbf{p}_j$ until a predefined bound w on the width of $\mathbf{p}_j$
Step 2: Compute enclosures $\mathbf{y}_i^{(k)}$ of $\mathbf{y}_i(t_k, \mathbf{p}_j)$ for $k = T_b \dots T_e$ in parallel
Step 3: Exclude $\mathbf{p}_j$ according to a certain exclusion criterion
Return: List $\mathcal{L}$ containing (indices of) suitable boxes

indices of definite solution boxes, $\mathcal{N}$ of boxes that are definitely to be excluded, and $\mathcal{E}$ of boxes for which a decision cannot be made without a further subdivision. The same subdivision strategy as in Algorithm 1 can be used. Also here, Bézier curves, plausibility bounds, or just crisp values for $y_{i,\mathrm{m}}$ can be used while building $\Phi(\mathbf{p})$.

Algorithm 2

SIVIA-like identification

Require: Φ, $\mathbf{z}$, $\mathbf{p}$, w

Step 1: Bisect $\mathbf{p} = \bigcup_{j=1}^{l} \mathbf{p}_j$ until a predefined bound w on the width of $\mathbf{p}_j$

Step 2: Compute enclosures $\mathbf{y}_i^{(k)}$ of $\mathbf{y}_i(t_k, \mathbf{p}_j)$ and $\Phi(\mathbf{p}_j)$ in parallel

Step 3: Check in parallel for each $\mathbf{p}_j$: if $\Phi(\mathbf{p}_j) \subset \mathbf{z}$ then add j to list $\mathcal{S}$
else if $\Phi(\mathbf{p}_j) \cap \mathbf{z} = \emptyset$ then add j to list $\mathcal{N}$ else add j to list $\mathcal{E}$

Return: Index list $\mathcal{S}$ of solution boxes, index list $\mathcal{N}$ of excluded boxes,
index list $\mathcal{E}$ of undecided boxes

2.4 Implementation Overview

Algorithms 1 and 2 are implemented on the GPU using the CUDA library THRUST.[1] This library allows users to describe their computations in terms of high-level abstractions, avoiding the necessity to deal with low-level CUDA programming such as specifics of memory allocation. The implementation structure for both the illustration from Sect. 2.5 and the SOFC application from Sect. 3 is similar to the one we described in Auer et al. (2020) for the reduced distributed heating system. Compared to Auer et al. (2020), the main algorithmic extensions in this paper are the use of data reduction and the use of SIVIA-like identification with a least squares cost function.

THRUST does not only simplify CUDA implementations, but also provides an interface to make the collection of IVP solving methods ODEINT from BOOST available for the GPU execution. We use this functionality in combination with the property of cooperativity (cf. Sect. 2.1) in order to deal with parameter uncertainty for dynamic systems. That is why we need lower and upper bounds of input/output variables and data separately, as demonstrated for the lower bounds in the implementation overview below.

The first step of Algorithm 1 is carried out on the CPU at the moment. Step 2 of the Algorithm corresponds to defining the right hand side of the system from

[1] https://nvidia.github.io/cccl/thrust/.

```
struct cooperative_system
{
 struct cooperative_functor
 {
  template< class T >   __host__ __device__
  void operator()( T t ) const
  { //Define your system here...
    //unpack y, lower bound, from zip_iterator
    value_type ylb = thrust::get< 0 >( thrust::get< 0 >(t) );
    //define the right side for y, lower bound
    thrust::get< 1 >( t ) = ...;
  }
 };
 cooperative_system( size_t N , ...): m_N( N ) { }
 template< class State , class Deriv >
 void operator()(  const State &x , Deriv &dxdt , value_type t) const
 {
    thrust::for_each(data.begin(), data.end(), cooperative_functor() );
 }
 size_t m_N;
};
```

Fig. 1 Implementing the IVP system in THRUST

Eq. (1) in THRUST using a C++ structure called `cooperative_system`, cf. Fig. 1. THRUST offers a useful feature of a so-called `zip_iterator` to bundle all the data, input and output variables into one entity (represented by `data` in the Figure) which just needs to be unpacked for the use inside the `cooperative_functor()`. In `data`, vectors of length N (determined by the number l from the subdivision scheme and the GPU memory requirements) for the inputs, outputs, and data are collected, which first need to be defined and copied to the GPU. This can be done very easily via THRUST:

```
vector< double > plb_host( N ); //parameter, lower bound, CPU
thrust::device_vector< double > plb = plb_host; //GPU
```

Step 3 is implemented in THRUST using a C++ structure `step3_functor` within `my_observer`, cf. Fig. 2. If data reduction is employed, then a function implementing the Bézier curve as given by Eq. (6) needs to be defined and called from the operator `my_observer()` instead of defining a data vector. For implementing Algorithm 2, the same overall scheme can be used, extended by an implementation of a functor for Φ in the observer. Finally, a solver from BOOST ODEINT collection (e.g., the Runge-Kutta Dormand-Prince method with the constant step size `dt`) can be run for times between 0 and `t_end` on the GPU from the `main()` function, with the defined observer `obs` called at each step:

```
typedef runge_kutta_dopri5<...> stepper_type;
cooperative_system coop( N );
my_observer obs( N ) ;
integrate_const( make_controlled(...stepper_type() ) ,
         coop ,  y , 0.0 , t_end , dt, boost::ref(obs) );
```

The variable `y` contains the values of states at `t_end` for each subdivision box. Now, depending on the observer results, obtained by using the function `fill_res(...)` of the observer `obs`, the postprocessing is carried out on the CPU (computing the convex hull of admissible states or the reduced search space for parameters).

```
struct my_observer
{
 struct step3_functor
 {
  template< class T >   __host__ __device__
  void operator()( T t ) const
  {
    //Define Criterion 1 or 2 by modifying m_res
    //zip_iterator'ed in "data"
  }
 };
 my_observer( size_t N , ...): m_N( N ) { }
 template< class State , class Deriv >
 void operator()(  const State &x , value_type t) const
 {
  thrust::for_each(data.begin(), data.end(), step3_functor() );
 }
 template< class Res >
 void fill_res( Res &res ) //copy results
 {
  thrust::copy( m_res.begin() , m_res.end() , res.begin() );
 }
 size_t m_N;
 thrust::device_vector<bool> m_res;//store results
};
```

Fig. 2 Implementing box exclusion in THRUST

Table 1 Artificial crisp measurements $y_{2,\mathrm{m}}$ for y_2

t	1	2	3	4	5	6	7	8
$y_{m,2}$	0.051801	0.046753	0.040787	0.032543	0.024291	0.021511	0.020577	0.012213
t	9	10	11	12	13	14	15	16
$y_{m,2}$	0.008919	0.007634	0.004929	0.008158	0.006349	0.002386	0.005543	0.004355

2.5 *Illustration: Two-Compartment Model*

The two-compartment model described in Jaulin et al. (2001); Kieffer et al. (2011) has been a standard example in parameter optimization since many years. This dynamic system has the form

$$\begin{aligned} y_1' &= -(p_3 + p_1) \cdot y_1 + p_2 \cdot y_2 \\ y_2' &= p_3 \cdot y_1 - p_2 \cdot y_2 \end{aligned} \tag{5}$$

with the initial conditions $y_1(0) = 1$, $y_2(0) = 0$. We are interested in the values of p_1, p_2, p_3 given synthetic (measurement) data shown in Table 1 as well as in enclosures of y. The problem is formulated such that only the second component y_2 is measured. The data are generated in the way described in detail in Gillner and Auer (2024). The parameter values for the generation are $p_1 = 1.925403$, $p_2 = 0.232718$, $p_3 = 0.145076$, which provides the reference for the optimum. The linear ODEs in (5) can be solved analytically. However, to illustrate the methods we propose for the SOFC case where finding analytical solutions is not an option, we obtain enclosures of the solution using the property of cooperativity and the BOOST solver ODEINT (Option F1.c).

For Criterion 1, we use the (hypothetical) plausibility bounds

$$\mathbf{y}_{2,\mathrm{m}} = [y_{2,\mathrm{m}} - 0.007, y_{2,\mathrm{m}} + 0.007]$$

($\delta = 0.007$). For Criterion 2, we use the following Bézier curve of degree $n = 5$ enclosing the data $y_{2,\mathrm{m}}$:

$$y_{2,\mathrm{m}}(t_k) \in \sum_{i=0}^{n} a_{y_{2,\mathrm{m}},i}(t_k) \cdot b_{i,n}(t_k) + \Delta_{y_{2,\mathrm{m}}}(t_k) =: B_{y_{2,\mathrm{m}}}(a_{y_{2,\mathrm{m}}}(t_k), \Delta_{y_{2,\mathrm{m}}}(t_k)) \,, \tag{6}$$

where the basis Bernstein polynomials $b_{i,n}(t)$ are defined as

$$b_{i,n}(t) = \binom{n}{i} \left(\frac{t - t_b}{s}\right)^i \left(\frac{t_e - t}{s}\right)^{n-i} \,, \; i = 0, \ldots, n, \tag{7}$$

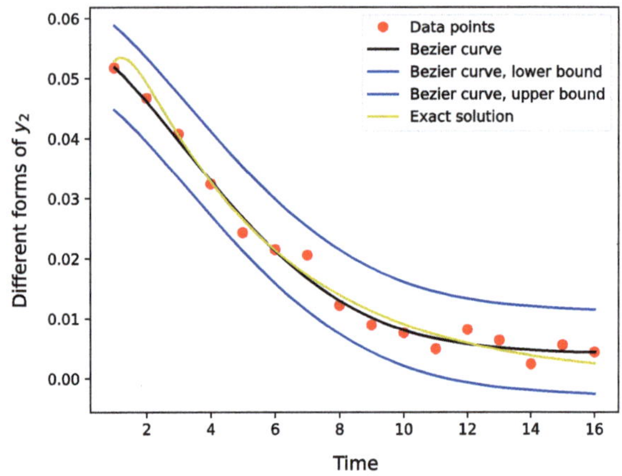

Fig. 3 Enclosure of data from Table 1 by a Bézier curve in Eq. (6)

with $s = t_e - t_b$ being the scaling factor. The (piecewise) constant coefficients $a_{y_{2,\mathrm{m}},i}(t)$ are crisp and $\Delta_{y_{2,\mathrm{m}}}(t)$ are intervals. For the considered example, we chose a single constant interval $\Delta_{y_{2,\mathrm{m}}} = [-\delta, \delta]$ that did not depend on time. The coefficients $a_{y_{2,\mathrm{m}},i}$ also do not depend on t and were determined by a PYTHON algorithm implementing the general approach to Bézier curves[2] as

$$\begin{aligned} a_{y_{2,\mathrm{m}},0} &= 0.004355 & a_{y_{2,\mathrm{m}},1} &= 0.0044165489511391876 \\ a_{y_{2,\mathrm{m}},2} &= 0.0072886805620881 & a_{y_{2,\mathrm{m}},3} &= 0.0031257843762404793 \\ a_{y_{2,\mathrm{m}},4} &= 0.0365115054020798 & a_{y_{2,\mathrm{m}},5} &= 0.051801. \end{aligned}$$

The shape of the corresponding curve is shown in Fig. 3. All computations were performed using

- a CPU system with two Intel Xeon Gold 5215 64 bit CPUs with 192 GB of main memory and
- a GPU system with an NVIDIA Quadro RTX 6000 GPU with 24 GB of memory,

running the operating system Ubuntu 22.04 with CUDA 12 and the NVIDIA driver 525.105 installed. Experiments on the GPU were written in CUDA C++. Kernel execution times were measured by wrapping the relevant parts of the program between two calls to `std::chrono::high_resolution_clock::now()` and calculating their difference at nanosecond resolution, averaged over 100 program calls. The execution time of entire programs was captured by the standard UNIX command `time`. Since the testing environment was accessed remotely, measurements of (peak) power were limited to software-based tools, which might have distorted the values somewhat. Although these measurements are considerably influenced by noise (Lang and Rünger 2013), they nonetheless provide an idea about the magnitude of this characteristic.

[2] see, for example, https://pomax.github.io/bezierinfo/.

Table 2 Test results for the parameter space $\mathbf{p} = [1, 2] \times [0.01, 1] \times [0.05, 2]$ and $w = 0.05$

	$\mathbf{p_1}^* \times \mathbf{p_2}^* \times \mathbf{p_3}^*$	Boxes	$\mathbf{y}_2(16)$	M_{wid}	t_{kern} (s)	t_{wall} (s)	P_{peak} (W)
I	$[1, 2]\times$ $[0.1028, 0.4741]\times$ $[0.0500, 0.2024]$	443	$[1.909 \cdot 10^{-4}, 1.849 \cdot 10^{-2}]$	$1.829 \cdot 10^{-2}$	0.0488	0.35	41.12
II	$[1, 2]\times$ $[0.0718, 0.5979]\times$ $[0.05, 0.2024]$	610	$[6.522 \cdot 10^{-5}, 1.971 \cdot 10^{-2}]$	$1.964 \cdot 10^{-2}$	0.0489	0.36	36.34
IIIa	$[1, 2]\times$ $[0.01, 1]\times$ $[0.05, 0.5680]$	8690	$[3.863 \cdot 10^{-8}, 6.685 \cdot 10^{-2}]$	$6.685 \cdot 10^{-2}$	0.0472	0.54	38.19
IIIb	$[1, 2]\times$ $[0.01, 1]\times$ $[0.05, 0.6594]$	10101	$[9.680 \cdot 10^{-9}, 7.005 \cdot 10^{-2}]$	$7.005 \cdot 10^{-2}$	0.0474	0.58	40.70
IIIc	$[1, 2]\times$ $[0.01, 1]\times$ $[0.05, 0.6594]$	10076	$[9.680 \cdot 10^{-9}, 7.170 \cdot 10^{-2}]$	$7.170 \cdot 10^{-2}$	0.0474	0.58	41.08

Note that the parameters for this example, although it looks simple, are actually not so easy to identify. The reasons are, first, that the data are limited (only 16 points for y_2). Second, optimal solutions are symmetric, which makes it hard to reduce the search space using the convex hull (cf. Kieffer et al. (2011)).

In Table 2, the results for parameter identification are shown. By $\mathbf{p}^*$, we denote the convex hull of all suitable $\mathbf{p}_j$; by $\mathbf{y}_2(16)$, the hull of all $y_2(t, \mathbf{p}_j)$ for $t = 16$ and $j \in \mathcal{L}$. For comparison, the exact value for y_2 at $T_e = 16$, obtained by evaluating the exact solution in FP arithmetic, is approximately $2.5010 \cdot 10^{-3}$ (rounded to 4 digits after the decimal point, similar to all values from the table). Intervals are rounded outwards, crisp values to the next decimal number. The value M_{wid} is defined as

$$M_{\text{wid}} = \max_{i=1,\dots,n_y} \text{wid}(\mathbf{y_i}(T_e))$$

with $n_y = 2$ in this example. We consider the following scenarios:

I: experimental identification using Criterion 1 in Step 3 (F0.a/F1.c/F2.a/F3.a)
II: experimental identification using Criterion 2 in Step 3 (F0.a/F1.c/F2.a/F3.b)
IIIa: identification using Algorithm 2 in Step 3, with Φ based on crisp data, cf. Table 1 (F0.b/F1.c/F2.ab/F3.a)
IIIb: identification using Algorithm 2 in Step 3, with Φ based on the plausibility bounds for the data (F0.b/F1.c/F2.ab/F3.a)
IIIc: identification using Algorithm 2 in Step 3, with Φ based on the Bézier curve enclosing the data (F0.b/F1.c/F2.ab/F3.b)

Table 3 Test results for the parameter space $\mathbf{p} = [1, 2] \times [0.01, 1] \times [0.05, 2]$, $w = 0.05$ and Scenario III for boxes from $\mathcal{S}$ only

	$\mathbf{p}_1{}^* \times \mathbf{p}_2{}^* \times \mathbf{p}_3{}^*$	Boxes	$\mathbf{y}_2(16)$	M_{wid}
a	$[1, 2] \times [0.07187, 1] \times [0.05, 0.4461]$	5650	$[8.752 \cdot 10^{-8}, 1.937 \cdot 10^{-2}]$	$19.36 \cdot 10^{-3}$
b	$[1, 2] \times [0.1337, 0.6907] \times [0.05, 0.2633]$	928	$[1.694 \cdot 10^{-5}, 9.210 \cdot 10^{-3}]$	$9.193 \cdot 10^{-3}$
c	$[0.01, 1] \times [0.1337, 0.6907] \times [0.05, 0.2633]$	963	$[1.694 \cdot 10^{-5}, 9.412 \cdot 10^{-3}]$	$9.395 \cdot 10^{-3}$

In III.a–c, the value $tol = 0.005$ is used and the hull is built over not only the solution set of boxes $\mathcal{S}$ but also the undecided regions $\mathcal{E}$ ($\mathcal{L} = \mathcal{S} \cup \mathcal{E}$). Whereas the verification degree of Scenarios I and II is C3+, it can be determined as C2 for III since the SIVIA-related subprocess is verified. Additionally, we show the results where $\mathcal{L} = \mathcal{S}$ for Scenario III in Table 3 (of degree C3+ again since we do not consider undecided regions). In the column *Boxes*, the number of boxes in the corresponding list(s) is shown. The initial number of boxes for $w = 0.05$ is 2^{16}. Kernel time t_{kern} represents the execution time of each method on the GPU, while t_{wall} is the total number of seconds elapsed (wall-clock time) for the entire program execution, including subdivision into boxes (Step 1 of the algorithms) and post-processing that are preformed entirely on the CPU at the moment. Subdivision takes most of the CPU time and is a part of the simulation that can be performed offline once for all simulations. The smaller the subdivision bound w, the slower the overall program. During each test, peaks in GPU power usage were recorded. The highest enduring peaks during kernel execution P_{peak} are listed in the last column. We do not provide t_{kern} or P_{peak} for results in Table 3 since all three lists $\mathcal{S}$, $\mathcal{N}$ and $\mathcal{E}$ are obtained in one GPU run so that only the post-processing varies.

Tables 2 and 3 demonstrate that good enclosures for y_2 at the end of the integration interval ($t = 16$) can be obtained in short times using the proposed methods on the GPU. The search space is also reduced. It can be observed that energy consumption is nearly the same for all scenarios. Employing data reduction via Bézier curves does not shorten execution times since the data set in Table 1 is too small and can be handled by the GPU memory efficiently (I compared to II). A slight worsening of the maximal enclosure width from I to II is expected and reflected by Table 2. Finally, we notice that using the global optimization for the least squares cost function leads to somewhat worse results for the considered example (Scenario III). However, a good enclosure of the true result $y_2(16)$ can be obtained by considering only definitely verified boxes from the list $\mathcal{S}$ (e.g., the last two rows in Table 3).

In Table 4, we show what happens if we do not perform any parameter identification for the model. We consider the same search space for the parameters and plug it directly into our solver for differential equations. If the intervals of the search

Table 4 Test results for simulation without parameter identification over the parameter space $\mathbf{p} = [1, 2] \times [0.01, 1] \times [0.05, 2]$ with different subdivision bounds w

w	Boxes	$\mathbf{y}_2(16)$	M_{wid}	t_{kern} (s)	$t_{\text{kern,a}}$ (s)	P_{peak} (W)
0.05	2^{16}	$[9.6805 \cdot 10^{-9}, 0.8767]$	0.8766	0.0460	0.0211	40.03
0.025	2^{19}	$[9.9831 \cdot 10^{-9}, 0.7438]$	0.7438	0.1807	0.1107	70.31
0.005	2^{25}	$[1.0220 \cdot 10^{-8}, 0.6578]$	0.6578	7.7827	6.2787	188.91

space are substituted for parameters as they are, the resulting maximal width M_{wid} of the enclosures for $\mathbf{y}$ is over four millions. Using the same subdivision strategy until the bound $w = 0.05$ as for parameter identification, this overestimation can be dramatically reduced to the value of the order of magnitude of one, with GPU computations requiring just under one tenth of a second. However, we notice, obviously, that this result is worse compared to the case where parameter identification is performed, independent of the used scenario (cf. Table 2). Not much more additional kernel time is needed for parameter identification, which confirms the statistics from the NVIDIA profiler `nvprof` purporting that the most work happens in evaluations of the right-hand side of the model. Subdivision into more boxes (2^{19} for $w = 0.025$ and 2^{25} for $w = 0.005$) reduces M_{wid} somewhat but arguably not enough to justify the increase in runtimes and power. This is only to be expected since the original uncertainty in the parameters is relatively large.

The simulations were performed by the Runge-Kutta Dormand-Prince method of order 5 with a constant stepper from the ODEINT collection. However, we can also use an adaptive stepper with the same method. This improves the GPU runtimes by approximately the factor of two in the best case as shown in Column 6 of Table 4 ($t_{\text{kern,a}}$). This result is obvious; however, we still draw attention to it since using a constant stepper cannot be avoided in Scenarios I, IIIa, IIIb for parameter identification. The obtained enclosures are the same for this example.

The general tendencies described by Tables 2 and 3 for parameter identification with subdivision until $w = 0.05$ can be observed for lower subdivision bounds w as shown in Table 5. Here, the number of boxes as well as the resulting enclosures for $y_2(16)$ and their widths obtained by considering just the solution set S in Scenario III are given in each second row. Again, using global optimization based on the least squares approach does not necessarily improve the quality of enclosures with decreasing w because of the overestimation. Kernel execution times are also somewhat higher for Scenario III since there is more work to be done and IA is used. Note that although SIVIA-like versions seem to perform worse than the experimental-based ones, their verification degree is higher. The enclosure width is slightly worsened by the use of data reduction via interval Bézier curves. The best enclosure of width $5.781 \cdot 10^{-3}$ is obtained by Scenario I with $w = 0.005$ in under 8 s; however, the enclosure for $w = 0.025$ and the same scenario is only slightly worse and is obtained approxi-

Table 5 Test results for parameter identification over the parameter space $\mathbf{p} = [1, 2] \times [0.01, 1] \times [0.05, 2]$ with two subdivision bounds w

w		Boxes	$\mathbf{y}_2(16)$	M_{wid}	t_{kern} (s)	P_{peak} (W)
0.025	I	1817	$[3.6643 \cdot 10^{-4}, 0.9324 \cdot 10^{-2}]$	$8.957 \cdot 10^{-3}$	0.1888	77.92
	II	2697	$[1.4710 \cdot 10^{-4}, 1.0770 \cdot 10^{-2}]$	$1.062 \cdot 10^{-2}$	0.1884	68.71
	IIIa	63767	$[5.7532 \cdot 10^{-8}, 4.3902 \cdot 10^{-2}]$	$4.390 \cdot 10^{-2}$	0.1912	64.65
	IIIa($\mathcal{S}$)	51614	$[8.4769 \cdot 10^{-8}, 2.2426 \cdot 10^{-2}]$	$2.243 \cdot 10^{-2}$		
	IIIb	76313	$[9.9831 \cdot 10^{-9}, 5.4666 \cdot 10^{-2}]$	$5.467 \cdot 10^{-2}$	0.1910	72.35
	IIIb($\mathcal{S}$)	13091	$[7.8397 \cdot 10^{-6}, 1.1959 \cdot 10^{-2}]$	$1.195 \cdot 10^{-2}$		
	IIIc	76139	$[9.9831 \cdot 10^{-9}, 5.4666 \cdot 10^{-2}]$	$5.467 \cdot 10^{-2}$	0.1911	69.51
	IIIc($\mathcal{S}$)	13502	$[6.9522 \cdot 10^{-6}, 1.2063 \cdot 10^{-2}]$	$1.206 \cdot 10^{-2}$		
0.005	I	55413	$[6.4951 \cdot 10^{-4}, 0.6431 \cdot 10^{-2}]$	$5.781 \cdot 10^{-3}$	7.9475	187.9
	II	89224	$[2.5187 \cdot 10^{-4}, 0.7276 \cdot 10^{-2}]$	$7.024 \cdot 10^{-3}$	7.9507	189.6
	IIIa	3796799	$[7.4911 \cdot 10^{-8}, 3.0065 \cdot 10^{-2}]$	$3.006 \cdot 10^{-2}$	8.1198	189.7
	IIIa($\mathcal{S}$)	3602485	$[8.2538 \cdot 10^{-8}, 2.5090 \cdot 10^{-2}]$	$2.509 \cdot 10^{-2}$		
	IIIb	4671268	$[1.0220 \cdot 10^{-5}, 3.9322 \cdot 10^{-2}]$	$3.932 \cdot 10^{-2}$	8.1200	190.6
	IIIb($\mathcal{S}$)	1161303	$[5.5356 \cdot 10^{-6}, 1.3583 \cdot 10^{-2}]$	$1.358 \cdot 10^{-2}$		
	IIIc	4660028	$[1.0220 \cdot 10^{-5}, 3.9322 \cdot 10^{-2}]$	$3.932 \cdot 10^{-2}$	8.1271	191.1
	IIIc($\mathcal{S}$)	1192218	$[4.9769 \cdot 10^{-6}, 1.3781 \cdot 10^{-2}]$	$1.378 \cdot 10^{-2}$		

mately 40 times faster. The increase in the quality of the enclosure from $w = 0.05$ to $w = 0.025$ is one order of magnitude, at the cost of the same increase in the runtime.

In the parameter identification tests described in Tables 2, 3 and 5, we relied on the test data with the frequency of $\Delta t = 1$ s (at the sampling frequency; cf. data in Table 1) and used a constant stepper from ODEINT for that. However, it is also possible to use an adaptive stepper for parameter identification if data is reduced by interval Bézier curves since we have an enclosure of measurements at any t in this case. This improves the runtimes approximately by the factor of two in scenarios II and IIIc (the ones relying on Bézier curves). The peak power remains nearly the same. For Scenarios II and IIIc, the enclosure width also remained approximately the same, although it seems to be possible in principle that the widths might become bigger since not the real data but its approximated description is used. The improved runtimes is the reason we used the adaptive stepper for the SOFC model in the next section.

As mentioned earlier, more scenarios are studied for the two-compartment problem in Gillner and Auer (2024), for example, the option to rely on the closed-form solution F1.a. For this case, a comparison of GPU techniques with the SIVIA and an interval global optimizer from OCTAVE on the CPU was carried out. The GPU-based SIVIA implementation slightly outperformed both of the CPU-based ones in terms of runtimes and power, even though the GPU use is not really necessary here. To compare the developed GPU-based methods with the CPU-based ones (serial/parallelized) for further benchmark problems from the angle of enclosure tightness, computing times, and energy consumption is a topic for our future research. However, to cover all possible versions would be unfeasible for the SOFC application from the next section. Here, we only plan to implement a parallelized CPU version hoping to increase the verification degree.

3 Parameter Identification for a Control-Oriented SOFC Model

The general goal of the GPU employment is to speed up computations, for example, for more reliability (a higher verfication degree) or for online use. Certainly, not every type of a control algorithm well-suited for the parallelization on the CPU is also efficient on the GPU (Maceina and Manduchi 2017). Parameter identification for control-oriented models, however, can be efficiently performed using the GPU, a claim we intend to substantiate further in this section. In the past, we worked successfully with different options for parameter identification out of the ones described in Sect. 2.2 on the GPU, for example, the choice of F0.b/F1.b/F2.b/F3.a for an SOFC model ("verified approximation" with the C2 verification degree, cf. Kiel et al. (2013)) or F0.a/F1.c/F2.a/F3.a ("experimental identification", C3+) for a simple distributed heating system with 8×2 equations and 3 parameters (Auer et al. 2020). For the latter, F2.a means exploiting the information about the uncertainty using the property of cooperativity as described in Sect. 2.1.

In this section, we apply our findings from Sect. 2 to a much more complex problem of a control-oriented SOFC model. First, we describe the model we use in detail in Sect. 3.1. In Sect. 3.2, we introduce possible consistency tests that can additionally be carried out (either in parallel or in a pre-processing step) and reflect the physical constraints present in the model. Results and their discussion are in Sect. 3.3.

3.1 SOFC Temperature Models: Assumptions and Properties

The usual assumption for modeling the SOFC temperature with the help of a finite-volume semi-discretization is that it is piecewise homogeneous inside the stack. In this way, it is possible to develop a system of ODEs for the temperature (instead of relying on PDEs) if the stack is subdivided into finite-volume elements as shown in Fig. 4. As explained in the introduction, the ability to represent time-varying hotspot locations can be retained by using the ODE-based model. As an additional simplifying assumption, specific heat capacities of the anode and cathode gases as well as the reaction enthalpy are represented as second-order polynomials in temperature.

The model described in this section has the general form shown in Eq. (1), but can be represented in the specific quasi-linear form

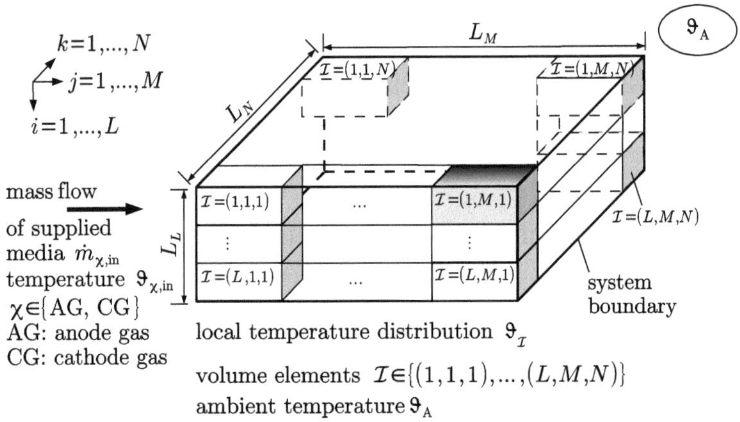

Fig. 4 Modeling scheme for the temperature distribution inside the SOFC stack

$$y' = A(t, y, p) \cdot y + B(t, y, p) \cdot u, \text{ where } A(t, y, p) \in \mathbb{R}^{n_y \times n_y},\ B(t, y, p) \in \mathbb{R}^{n_y \times n_u} \quad (8)$$

so that the bracketing systems bounding the flow according to the property of cooperativity can be given as

$$\underline{\mathbf{A}} \cdot v + \underline{\mathbf{B}} \cdot u = v' \leq y' \leq w' = \overline{\mathbf{A}} \cdot w + \overline{\mathbf{B}} \cdot u, \quad (9)$$

where $y \in [v, w]$, $\underline{\mathbf{A}} \leq A(t, y, p) \leq \overline{\mathbf{A}}$ and $0 \leq \underline{\mathbf{B}} \leq B(t, y, p) \leq \overline{\mathbf{B}}$, $B(t, y, p) \cdot u \geq 0$.

In this paper, we work with three volume elements ($L = 1$, $M = 3$, $N = 1$ in Fig. 4). The ambient temperature ϑ_A is supposed to be in the interval [295, 300] K. States y_i are temperatures ϑ_{1i1} for $i = 1, 2, 3$. In an experiment with the SOFC test rig (see Rauh and Auer (2022); Rauh et al. (2021) for details), the temperatures were measured with the frequency of 10 Hz from $T_b = 0$ s to $T_e = 34256.5$ s. Additionally, the corresponding values were recorded for mass flows of hydrogen, nitrogen and cathode gas $\dot{m}_{H2}, \dot{m}_{N2}, \dot{m}_{CG}$, the current I and the temperatures at the anode/cathode $\vartheta_{AG}/\vartheta_{CG}$. These values represent the model inputs u_j. The heating phase took place in the time interval between 0 and 29184.8 s during the experiment; the reaction phase between 29184.8 and 34256.5 s.

The parameters to be identified from the measured data for the SOFC system we consider in this paper are the heat conductivity $\lambda > 0$; the convective heat transfer $\alpha > 0$; three coefficients of each quadratic approximation to the heat capacities of hydrogen, nitrogen, water vapor and cathode gas (air); three coefficients of a quadratic approximation to the reaction enthalpy H_R; and the Ohmic cell resistance $R > 0$. That is, overall of 36 parameters need to be identified ($n_p = 18$ for each bracketing system). Note that the property of cooperativity holds only for physically meaningful parameters. This results in the model being quite sensitive to the chosen parameter search spaces. In principle, it is possible to obtain the partial derivatives of the right hand side of Eq. (1) using algorithmic differentiation and then check if Eq. (2) holds for the given parameter values $p \in \mathbf{p}$, that is, if the lower bound of the interval evaluation of $\frac{\partial f_i(t_k, \mathbf{y}, \mathbf{p})}{\partial y_j}$ is greater or equal to zero. This procedure corresponds to a special centered-form representation of the ODE system. For the case of the model representation in Eq. (9), the property of cooperativity can be verified by the sign pattern of the matrix $A(t, y, p)$ (Rauh et al. 2021). To implement a procedure for checking cooperativity programmatically is a topic for our future research. At the moment, we use the Nelder-Mead simplex method `fminsearch` from MATLAB for the crisp model without uncertainty to determine possible vicinity of the parameters.

As mentioned earlier, nine measurements are available, namely, for $\dot{m}_{H2}$, $\dot{m}_{N2}$, $\dot{m}_{CG}$, I, ϑ_{AG}, ϑ_{CG}, ϑ_{111}, ϑ_{121}, ϑ_{131} ($n_u = 6$ for inputs u and $n_y = 3$), obtained with the frequency of 10 Hz in the experiment that lasted over nine and a half hours. The file with measurements has therefore the size of 48 MiB which would need to be transferred to the GPU. The transfer itself takes approximately 4 milliseconds using our test system from Sect. 2.5 and can therefore be almost neglected since the runtimes of the SOFC application have the magnitude of seconds. However, the repeated access to the global GPU memory the kernels need then would probably influence the runtime negatively (Maceina and Manduchi 2017). To circumvent this, we enclose the data by nine interval functions using piecewise Bézier curves. Moreover, the corresponding kernel functions can be implemented much easier on the GPU in this way.

The enclosure function definitions using Bézier curves are visualized in Fig. 5 for the example of input temperatures at the anode and cathode ϑ_{AG} and ϑ_{CG}. For the anode temperature, four time intervals are necessary for the Bézier curve of degree $n = 5$ shown in the plot. In general, this number depends on the requested tightness of the enclosure curve, the interrelation being limited by numerical considerations. Using the same notations as in Sect. 2.5, we reproduce the crisp coefficients $a_{\vartheta_{\mathrm{AG}},i}(t)$ and enclosure intervals $\Delta_{\vartheta_{\mathrm{AG}}}(t)$ as a reference:

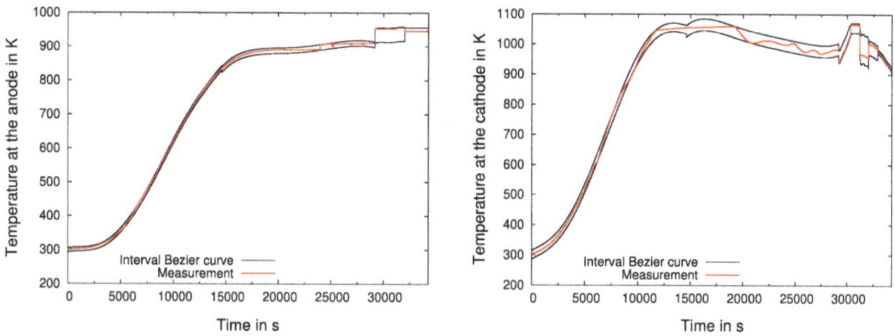

Fig. 5 Interval enclosures of the data for the anode (left) and cathode (right) input temperatures via piecewise interval Bézier curves

$$
\begin{aligned}
a(t) &= \begin{pmatrix} 300.073 & 319.191 & 224.691 & 619.272 & 729.644 & 838.250 \end{pmatrix} && \text{for } t \in [0, 14592.3] \\
&= \begin{pmatrix} 832.048 & 930.574 & 861.783 & 887.743 & 925.134 & 906.265 \end{pmatrix} && \text{for } t \in [14592.3, 29184.8] \\
&= \begin{pmatrix} 950.484 & 958.969 & 948.253 & 951.291 & 958.878 & 955.774 \end{pmatrix} && \text{for } t \in [29184.8, 32009.5] \\
&= \begin{pmatrix} 946.341 & 945.673 & 946.786 & 945.674 & 946.340 & 946.118 \end{pmatrix} && \text{for } t \in [32009.5, 34256.5]
\end{aligned} \tag{10}
$$

$$
\begin{aligned}
\Delta_{\vartheta_{\mathrm{AG}}}(t) &= [-7.43701, 4.49219] && \text{for } t \in [0, 14592.3] \\
&= [-9.21037, 6.95606] && \text{for } t \in [14592.3, 29184.8] \\
&= [-40.7487, 2.58995] && \text{for } t \in [29184.8, 32009.5] \\
&= [-0.189503, 9.81035] && \text{for } t \in [32009.5, 34256.5]
\end{aligned} \tag{11}
$$

Such enclosure functions can be fitted to the data using an arbitrary FP-based optimization algorithm inside the general approach to constructing Bézier curves, for example, the least squares fit implemented in PYTHON as we did in Sect. 2.5. For the bigger problem in this section, we used the Nelder-Mead simplex method `fminsearch` from MATLAB. Note that, if we use such enclosure functions for the data, the inputs u are not crisp-valued anymore, but rather belong to a certain interval $\mathbf{u}$ given by the enclosure function. That is, the model in (9) needs to be reformulated as

$$\underline{\mathbf{A}} \cdot v + \underline{\mathbf{B}} \cdot \underline{\mathbf{u}} = v' \leq y' \leq w' = \overline{\mathbf{A}} \cdot w + \overline{\mathbf{B}} \cdot \overline{\mathbf{u}}, \tag{12}$$

if data reduction is to be used. In this way, additional uncertainty is introduced into the model. However, the test from Sect. 2.5 demonstrates that this does not worsen the results too much if the maximal admissible enclosure widths are chosen to be sufficiently small. Note that these widths can be influenced by the choice of time intervals into which the overall horizon is subdivided.

3.2 Consistency Tests

For our specific application described in detail in Sect. 3.1, there is a number of consistency checks that parameters should comply to in order for the model to be feasible. The corresponding tests can be carried out before and/or after the subdivision in Step 1 of Algorithms 1 and 2, using the CPU or the GPU in parallel.

First, several of the quantities mentioned in Sect. 3.1 must be positive: the heat conductivity coefficient $\lambda > 0$, the heat transfer coefficient $\alpha > 0$, the heat capacities c of all gases (modeled as second-order polynomials in temperature ϑ)

$$c(\vartheta) \approx \gamma_{c,0} + \gamma_{c,1}\vartheta + \gamma_{c,2}\vartheta^2 > 0,$$

and the reaction enthalpy

$$H_R(\vartheta) \approx \gamma_{H_R,0} + \gamma_{H_R,1}\vartheta + \gamma_{H_R,2}\vartheta^2 > 0.$$

The requirement of positivity helps to eliminate those boxes for which the following inequalities that can be checked on a computer hold:

$$\begin{aligned}
&\overline{\gamma}_0 + \overline{\gamma}_1\underline{\vartheta} + \overline{\gamma}_2\underline{\vartheta}^2 < 0 && \text{if } \overline{\gamma}_1 < 0,\ \overline{\gamma}_2 < 0 \\
&\overline{\gamma}_0 + \overline{\gamma}_1\underline{\vartheta} + \overline{\gamma}_2\overline{\vartheta}^2 < 0 && \text{if } \overline{\gamma}_1 < 0,\ \overline{\gamma}_2 \geq 0 \\
&\overline{\gamma}_0 + \overline{\gamma}_1\overline{\vartheta} + \overline{\gamma}_2\underline{\vartheta}^2 < 0 && \text{if } \overline{\gamma}_1 \geq 0,\ \overline{\gamma}_2 < 0 \\
&\overline{\gamma}_0 + \overline{\gamma}_1\overline{\vartheta} + \overline{\gamma}_2\overline{\vartheta}^2 < 0 && \text{if } \overline{\gamma}_1 \geq 0,\ \overline{\gamma}_2 \geq 0
\end{aligned} \tag{13}$$

Here, γ_i means the corresponding coefficients of heat capacities of any of the gases or of the reaction enthalpy.

Second, the heat capacities of gases and the reaction enthalpy should be monotonically increasing, that is,

$$\frac{\partial}{\partial\vartheta}c(\vartheta) > 0,\ \frac{\partial}{\partial\vartheta}H_R(\vartheta) > 0.$$

This leads to the further conditions for eliminating the boxes:

$$\begin{aligned}
&\overline{\gamma}_{c_{CG},1} + 2\overline{\gamma}_{c_{CG},2}\underline{\vartheta} < 0 && \text{if } \overline{\gamma}_{c_{CG},2} < 0 \\
&\overline{\gamma}_{c_{CG},1} + 2\overline{\gamma}_{c_{CG},2}\overline{\vartheta} < 0 && \text{if } \overline{\gamma}_{c_{CG},2} \geq 0
\end{aligned} \tag{14}$$

For example, consider the heat capacity of the cathode gas with the approximation coefficients $\gamma_{c_{CG},i}$ in the initial search intervals $[-183, 183]$, $[-0.02, 0.02]$, $[-7.3 \cdot 10^{-5}, 7.4 \cdot 10^{-5}]$. After subdividing this search space into 100 intervals in each direction (overall 10^6 boxes) and applying the positivity and monotonicity tests, 334075 boxes remain, that is, 66% of the boxes are eliminated. The convex hull is $[-95, 183] \times [-0.02, 0.02] \times [-3.5 \cdot 10^{-5}, 7.4 \cdot 10^{-5}]$ for these parameters, which is reduced compared to the original one.

In addition to the simple consistency tests described above, more advanced checks are possible. For example, checks relying on stability-based subdivisions for λ or coordinate transformations for c can be devised. We can check the stability of bounding systems during the subdivision step of the general algorithm (cf. Algorithms 1 and 2) before identification. For example, a subdivision strategy based on open-loop stability analysis using the Gershgorin circle theorem can be implemented. Moreover, the use of contractors, for example, a forward-backward contractor applied to the subintervals for γ_i, might improve the results obtained above by the simple inequality checks. Working on these ideas and applying them in the SOFC context is a subject of our future research.

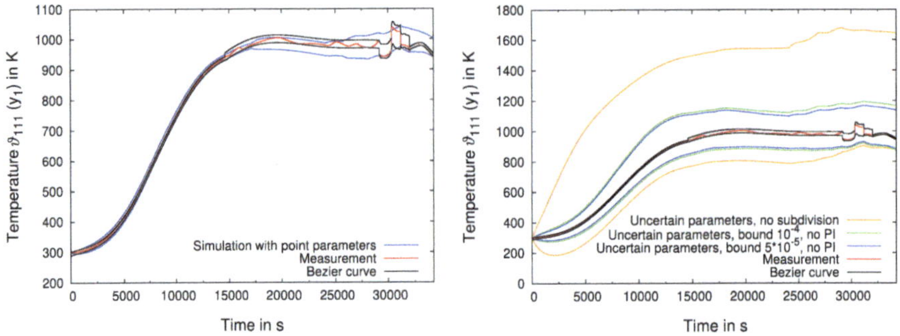

Fig. 6 Temperature ϑ_{111} for the crisp parameters identified in MATLAB (left) and for uncertain parameters on the GPU (right)

3.3 Results

In this section, we report the simulation and parameter identification results for the system described in Sect. 3.1 using methods from Sect. 2.3. The final model consists of 6 equations with with 36 parameters to identify depending on 6 (uncertain) control inputs. We use point initial conditions $\vartheta_{111} = 299.60666$ K, $\vartheta_{121} = 298.84343$ K, $\vartheta_{131} = 298.46319$ K and the same test environment as in Sect. 2.5.

First, we only simulate the behaviour of the SOFC temperatures by using Smith's theorem on the GPU, without any parameter identification but with data reduction for the model inputs (the mass flows of hydrogen, nitrogen and cathode gas, the current, and the temperatures at the anode/cathode). In Fig. 6, left, the results for ϑ_{111} are shown for crisp parameters identified in MATLAB. On the right in this figure, we show the temperature simulation for uncertain parameters (uncertainty of $\pm 10\%$ of the nominal values) without any subdivision of the initial parameter space (orange) as well as for subdividing until the bound of 10^{-4} (green) and until $5 \cdot 10^{-5}$ (blue). As a comparison, the measured values are shown in red and their Bézier enclosure in black.

In Table 6, the maximum width[3] M_{wid} of the simulated temperatures ϑ_{1i1}, $i = 1, 2, 3$, at the end of the integration interval $T_e = 34256.5$ s is shown for four subdivision cases (in columns titled PS for *pure simulation*). Additionally, we record the kernel time, the wall-clock time, the peak power and the approximate overall energy consumption. The peak power values are provided by software since we have only a remote access to the test system at the moment. The value E_{appr} is the sum of the power multiplied by time in 1/10 s increments. Because the measurement starts shortly before the GPU code begins and stops shortly after it ends, there is a slight overestimation in times in addition to usual inaccuracies, leading to a possible overestimation of energy consumption. We plan to develop a test rig setup and to

[3] The numbers for maximal temperature widths are rounded up to integers.

Table 6 Performance data for the pure simulation (PS) and with parameter identification (PI) for heating and reaction phases together ($t \in [0, 34256.5]$)

Bound w	—	10^{-4}			$5 \cdot 10^{-5}$			$3 \cdot 10^{-5}$		
	PS	PS	PI	PIS	PS	PI	PIS	PS	PI	PIS
M_{wid} (K)	912	354	346	346	299	297	297	278	275	275
Boxes	1	2^{14}	13982	13988	2^{16}	51464	51499	2^{18}	194201	194345
t_{kern} (s)	—	3.40	3.41	3.42	6.44	6.49	6.54	28.03	28.21	28.24
t_{wall} (s)	—	4.28	4.20	4.22	9.44	8.98	9.10	39.76	37.36	37.39
P_{peak} (W)	—	93.27	92.13	93.47	116.4	114.0	118.2	123.4	119.3	125.0
E_{appr} (W·s)	—	286.6	288.2	286.6	676.7	659.1	691.4	3208	3082	3275

compare the software-sided electric power values with actual hardware-based ones in the future.

For the number of subdivision boxes varying from 1 to 2^{18} (in the row *Boxes*), the simulations need at most 28 s on the GPU, achieving an uncertainty reduction of approximately 67% in the temperature. Note that the wall-clock time t_{wall} is high at the moment because it includes times for the subdivision, which is currently performed exclusively using the CPU. Implementing more efficient subdivision strategies is a subject for our future work. The characteristic t_{wall} is higher for pure simulation than for parameter identification since more boxes have to be post-processed on the CPU to obtain the enclosures for ϑ_{1i1} (or y_i in the notation from Eq. (1)). Employing the NVIDIA profiler `nvprof` for the simulation with the bound $w = 5 \cdot 10^{-5}$, we observe that most of the GPU time is used by calls to `for_each` routines containing the system (with `zip_iterator`). Memory transfers from the GPU to the host took approximately 2.15%, from the host to the GPU approximately 0.01% of the overall time. The results show that just the application of the Smith's theorem on the GPU allows us to deal with the considerable uncertainty which is present in the system. Moreover, the use of the GPU makes SOFC model simulations with reduced uncertainty possible in an online setting from the point of view of computing times. We did not test our methods with respect to continuous receiving of data.

In Table 6, the performance characteristics of tests including the parameter identification are shown for the heating and reaction phases together in columns titled *PI* (for parameter identification, the same setup as for Scenario II described in Sect. 2.5) and *PIS* (PI with SIVIA, Scenario IIIc), respectively. The GPU times for both parameter identification and pure simulation are quite similar in all cases in spite of the additional work in the former case. Through parameter identification, an improvement of the enclosure width of approximately 8 K can be achieved for $w = 10^{-4}$ in reasonable time of 3.4 seconds. Although other tested bounds reduce the width even

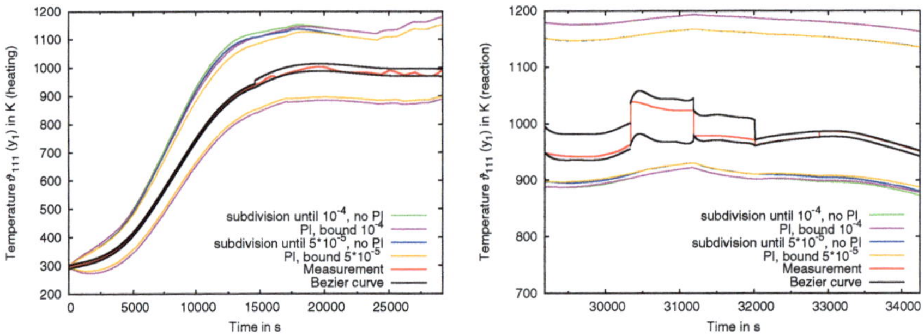

Fig. 7 Temperature ϑ_{111} after experimental parameter identification for the heating (left) and reaction stages (right)

further, the difference to pure simulation is not quite as dramatic, about 2 and 3 K for $w = 5 \cdot 10^{-5}$ and $w = 3 \cdot 10^{-5}$, respectively.

Note that, whereas the use of Bézier curves just introduced an additional uncertainty and did not show any significant improvement in power usage or runtimes for the simple example from Sect. 2.5, it seems almost unavoidable in the application from this section. Since the amount of data is much larger, repeated accesses to the global GPU memory are inefficient. They are needed in calls of `for_each` to compute the right hand side of the model (because of the control variables u) and for each call to the routines implementing Step 3 of Algorithms 1 and 2. Therefore, a compromise was required (as also described in Maceina and Manduchi (2017)), resulting in our use of Bézier enclosure curves.

In principle, it is possible to consider the heating and reaction phases separately, cf. Fig. 7. The reduction in enclosure widths due to parameter identification during the reaction stage is approximately 3 K; during the heating stage approximately 8 K at the moment. During the reaction stage, the main influencing parameters are three coefficients of a quadratic approximation to the reaction enthalpy H_{R}. That is, the process of parameter identification can be reduced to six parameters whereas the remaining parameters can be assigned the values obtained at the preceding heating stage. Further improvements are possible if an interval observer is used at runtime. For this, it is necessary to adapt CPU-based approaches from Rauh et al. (2020) to the GPU case. These are topics for our future work.

4 Conclusions

In this paper, we introduced GPU-based optimization techniques for parameter identification of dynamic systems based on data parallelization. First, we analysed the methods with the help of an illustrative example to identify best approaches and to assess the influence of data reduction via interval Bézier curves. The latter seemed unavoidable in the case of the practical application we considered next in this paper, namely, the control-oriented SOFC model. For this, the GPU simulation alone reduced the uncertainty in the temperature dramatically in an acceptable runtime; adding a parameter identification stage reduced the maximal width even further and did not add any considerable overhead. Overall, it seems to be possible to apply the proposed approaches in an online setting from the point of view of computing times.

Although we managed to simulate and identify parameters under uncertainty for a quite complex model of an SOFC stack with a good verification degree, there is still room for further improvements and developments. Many have been already mentioned at the appropriate places in the text of the paper. The main of them are:

- developing a test environment for hardware-based measuring of the energy consumption;
- implementing better subdivision strategies (also in the form of further consistency tests and contractors);
- considering heating and reaction stages separately for the SOFC model;
- using an interval observer (Rauh et al. 2020).

Moreover, an improvement in runtimes might be achieved by implementing the model without the use of the library THRUST. This is indicated by the warnings from the visual NVIDIA profiler related to bottlenecks in memory management which is carried out automatically by THRUST in our case. For this, a CUDA-based IVP solver is necessary. Finally, if this solver could be efficiently verified on the GPU, then it would be possible to realize a truly high verification degree both for simulation and for parameter identification.

References

IEEE standard for interval arithmetic (2015) IEEE Std 1788–2015:1–97. https://doi.org/10.1109/IEEESTD.2015.7140721

Ahnert K, Demidov D, Mulansky M (2014) Solving ordinary differential equations on GPUs, pp. 125–157. Springer, Berlin (2014). https://doi.org/10.1007/978-3-319-06548-9_7

Auer E (2014) Result verification and uncertainty management in engineering applications. Verlag Dr. Hut (2014), Habilitation monograph

Auer E, Kiel S, Rauh A (2017) Towards a verified ODE solver for GPU-based parameter identification. In: Bucher C, Ellingwood B, Frangopol D (eds) Safety, reliability, risk, resilience and sustainability of structures and infrastructure (Procedings of the 12th international conference on structural safety and reliability (ICOSSAR), pp 2039–2049

Auer E, Rauh A, Kersten J (2020) Experiments-based parameter identification on the GPU for cooperative systems. J Comput Appl Math 371:112657. https://doi.org/10.1016/j.cam.2019.112657

Bagóczki Z, Bánhelyi B (2017) A parallel interval arithmetic-based reliable computing method on a GPU. Acta Cybern 23(2):491–501. https://doi.org/10.14232/actacyb.23.2.2017.4

Collange C, Daumas M, Defour D (2012) Interval arithmetic in CUDA. In: Hwu WW (ed) GPU computing gems Jade edition, Chap. 9, pp 99–107. Elsevier. https://doi.org/10.1016/b978-0-12-385963-1.00009-5

Eriksen MB, Rasmussen S (2013) GPU accelerated parameter estimation by global optimization using interval analysis. https://projekter.aau.dk/projekter/files/77291483/report.pdf

Gennat M, Tibken B (2008) Computing guaranteed bounds for uncertain cooperative and monotone nonlinear systems. IFAC Proc Vol 41:4846–4851. https://doi.org/10.3182/20080706-5-KR-1001.00814

Gillner L, Auer E (2024) GPU-accelerated, interval-based parameter identification methods illustrated using the two-compartment problem. Acta Cybern 26(4):913–932. https://doi.org/10.14232/actacyb.306774

Huang B, Qi Y, Murshed A (2012) Dynamic modeling and predictive control in solid oxide fuel cells. Wiley. https://doi.org/10.1002/9781118501054

Ifqir S, Rauh A, Kersten J, Ichalal D, Ait-Oufroukh N, Mammar S (2019) Interval observer-based controller design for systems with state constraints: application to solid oxide fuel cells stacks. In: 2019 24th International conference on methods and models in automation and robotics (MMAR). pp. 372–377 (2019). https://doi.org/10.1109/MMAR.2019.8864718

Jaulin L, Kieffer M, Didrit O, Walter E (2001). Applied interval analysis. Springer, Berlin. https://doi.org/10.1007/978-1-4471-0249-6

Jaulin L, Walter E (1993) Guaranteed nonlinear parameter estimation from bounded-error data via interval analysis. Math Comput Simul 35(2):123–137. https://doi.org/10.1016/0378-4754(93)90008-I

Kieffer M, Markót MC, Schichl H, Walter E (2011) Verified global optimization for estimating the parameters of nonlinear models. In: Modeling, design, and simulation of systems with uncertainties. Springer Berlin Heidelberg. https://doi.org/10.1007/978-3-642-15956-5_7

Kiel S, Auer E, Rauh A (2013) Uses of GPU powered interval optimization for parameter identification in the context of SO fuel cells. In: Procedings of NOLCOS 2013, pp 558–563. https://doi.org/10.3182/20130904-3-FR-2041.00169

Lang J, Rünger G (2013) High-resolution power profiling of GPU functions using low-resolution measurement. In: Wolf F, Mohr B, Mey D (eds) Euro-Par 2013 parallel processing, pp 801–812. Springer Berlin Heidelberg. https://doi.org/10.1007/978-3-642-40047-6_80

Lee S, Lee H, Kim Y, Kim J, Choi W (2022) GPU-accelerated PD-IPM for real-time model predictive control in integrated missile guidance and control systems. Sensors 22(12). https://doi.org/10.3390/s22124512

Lohner R (2001) On the ubiquity of the wrapping effect in the computation of error bounds. In: Kulisch U, Lohner R, Facius A (eds) Perspectives on enclosure methods, pp 201–218. Springer, Berlin. https://doi.org/10.1007/978-3-7091-6282-8_12

Maceina TJ, Manduchi G (2017) Assessment of general purpose GPU systems in real-time control. IEEE Trans Nucl Sci 64(6):1455–1460. https://doi.org/10.1109/TNS.2017.2691061

Meslem N, Ramdani N (2011) Interval observer design based on nonlinear hybridization and practical stability analysis. Int J Adapt Control Signal Process 25(3):228–248. https://doi.org/10.1002/acs.1208

Moore RE, Kearfott RB, Cloud MJ (2009) Introduction to interval analysis. Society for Industrial and Applied Mathematics, Philadelphia. https://doi.org/10.1137/1.9780898717716

Rafiei SS, Chevalier S (2024) GPU-accelerated DCOPF using gradient-based optimization. arXiv:2406.13191

Rauh, A., Auer, E.: Interval extension of neural network models for the electrochemical behavior of high-temperature fuel cells. Frontiers in Control Engineering **3** (2022). https://doi.org/10.3389/fcteg.2022.785123

Rauh A, Kersten J, Frenkel W, Kruse N, Schmidt T (2021) Physically motivated structuring and optimization of neural networks for multi-physics modelling of solid oxide fuel cells. Math Comput Model Dyn Syst 27(1):586–614. https://doi.org/10.1080/13873954.2021.1990966

Rauh A, Kersten J, Aschemann H (2020) Interval and linear matrix inequality techniques for reliable control of linear continuous-time cooperative systems with applications to heat transfer. Int J Control 93(11):2771–2788. https://doi.org/10.1080/00207179.2019.1708966

Romero M (2018) Applications of GPU computing to control and simulate systems. In: XXXIX Jornadas de Automática

Sanders DP (2022) Global optimization and interval constraint programming using Julia: Symbolics, code generation, GPU. https://www.youtube.com/watch?v=iXJwq8-kcvE

Smith HL (1995) Monotone dynamical systems: An introduction to the theory of competitive and cooperative systems, Mathematical surveys and monographs, vol 41. https://doi.org/10.1090/surv/041

Stakhiv P, Strubutska I, Kozak Y (2012) Parallelization of calculations using GPU in optimization approach for macromodels construction. Electr Rev 88(3a)

Distributed Moving Horizon Estimation Over a Sensor Network with Nonlinear Measurements and Pre-estimation

Matthieu Borelle, Sylvain Bertrand, Cristina Stoica, Teodoro Alamo, and Eduardo F. Camacho

Abstract This chapter addresses the localization problem of a non-cooperative mobile target by a multi-sensor network. More precisely, a Distributed Moving Horizon Estimation (DMHE) approach is proposed to deal with the case of non-linear measurements, in presence of bounded uncertainties (e.g., unknown input of the target, measurement noises) and accounting for a priori available information on the problem (e.g., ground mobile target evolving in a known bounded area). A set-based representation is used to incorporate constraints in the DMHE optimization problem to improve the estimation accuracy. This chapter proposes a new DMHE algorithm with pre-estimation based on an Extended Kalman Filter to account for nonlinearities, while reducing the computation load. An example of target tracking in a delimited area by a sensor network illustrates the considered approaches in terms of computation time and estimation accuracy.

M. Borelle (✉) · S. Bertrand
Université Paris-Saclay, ONERA, Traitement de l'information et systèmes, Palaiseau, France
e-mail: matthieu.borelle@onera.fr; matthieu.borelle@l2s.centralesupelec.fr

S. Bertrand
e-mail: sylvain.bertrand@onera.fr

M. Borelle · C. Stoica
Université Paris-Saclay, CNRS, CentraleSupélec, Laboratoire des signaux et systèmes, Gif-sur-Yvette, France
e-mail: cristina.stoica@l2s.centralesupelec.fr

T. Alamo · E. F. Camacho
Department of Ingeniería de Sistemas y Automática, Universidad de Sevilla, Sevilla, Spain
e-mail: talamo@us.es

E. F. Camacho
e-mail: eduardo@esi.us.es

T. N. Dinh et al. (eds.), *Set-Valued Approaches to Control and Estimation of Uncertain Systems*, Mathematical Engineering,
https://doi.org/10.1007/978-3-031-94239-6_10

1 Introduction

Sensor Networks (SNs) are composed of connected nodes (possibly distributed across various locations) equipped with sensing, processing, and communication capabilities. Nowadays, SNs are extensively employed in various domains such as area monitoring, smart cities, power grids, where multiple sensors collaborate by communicating information over transmission channels to estimate the state of a complex system or environment to be monitored (He et al. 2020). Existing literature typically explores two main approaches to address low cost sensor networks challenges. The first approach involves centralized filtering methods, where a central node or unit gathers and processes the data (measurements) from the entire network in order to obtain a common/shared estimate. In contrast, the second approach adopts a Distributed State Estimation (DSE) strategy, eliminating the need for a central unit. Distributed approaches rely on the exchange of information among neighboring sensors of the network. It enables to enhance autonomy, fault-tolerance, scalability, and computational efficiency by sharing computational burden among the network nodes. A smaller communication range is reached using local information from neighbor nodes instead of communicating with all the nodes. The goal of distributed state estimation is to make each node inside a sensor network able to reconstruct the state of the system, with suitable accuracy and computational efficiency, despite the individual sensor limitations. The primary challenge lies in designing DSE algorithms that maintain a performance (i.e., accuracy of the estimates and stability of the estimation error dynamics) comparable to their centralized counterparts. Consensus protocols are commonly used to fuse information from different nodes and make all the estimates computed in a distributed way converge to the same value among all the network nodes. The consensus can either be carried out on estimates, measurements (Battistelli et al. 2015) or information (Battistelli et al. 2011; Mahmoud and Khalid 2013).

Generally, in model-based approaches, the system is not accurately known (e.g., non cooperative target localization in an area monitoring scenario), leading to modeling uncertainties (e.g., neglected dynamics, uncertainties on some parameters, perturbations). Additionally, the measurements provided by sensors are also affected by different sources of uncertainties, such as noise, bias, etc. Gaussian distributions are commonly used in the literature when modeling uncertainties as random variables. This is the case of most of the Kalman-based filtering approaches (Alriksson and Rantzer 2006; Madhavan et al. 2002; Olfati-Saber 2007). Although convenient in terms of computation, this modeling assumption could be inaccurate in practice, especially when dealing with bounded uncertainties. In this case, set-based representations are an alternative to account for the bounded support of the uncertainties, without any statistical distribution. In some state estimation problems, contextual information or a priori knowledge can be formulated as bounds, e.g. position/acceleration bounds for a target in a localization scenario. In this case, set-based representations are well suited to account for such information. Distributed set-membership estimation methods (Wang et al. 2017) have been developed to handle set-valued representations of uncertainties. Set representations can also be translated into constraints. Moving Horizon Estimation (MHE), the counterpart of Model Predictive Control (MPC), is a well known approach that efficiently handles constraints in a state estimation optimization problem.

Real world state estimation problems often deal with nonlinearities in the model dynamics or in the measurements, e.g., sensors providing distance or angle measurements. However, Kalman Filter-based methods can become suboptimal in presence of strong nonlinearities and even unstable in case of large initial estimation errors. Moreover, they cannot easily deal with constraints. This encourages the development of MHE approaches for linear (Rao et al. 2001) and nonlinear systems (Rao et al. 2003), as this method can account for constraints and nonlinearities in its formulation. MHE is also proved to be more robust to poor initialization of the estimates. The basic principle of MHE lies in a fixed-size estimation window that moves forward in time, aiming to minimize a cost function at each sample time, while satisfying constraints. This cost function reflects the estimation accuracy concerning measurements and prior knowledge of noise and system dynamics. Contrary to the full information estimation problem which processes all the measurements from the initial time, MHE adopts a receding horizon strategy for computational tractability. Indeed, only the most recent information is processed, while the past evolution of the system is condensed in the so-called *arrival cost* (Rao et al. 2001). The arrival cost is designed to approximate the full information problem within a finite horizon window (Rao et al. 2001, 2003). Its update is crucial to ensure the estimation error convergence. In the linear unconstrained centralized case, the arrival cost can be approximated using the Kalman Filter (KF) covariance update (Rao et al. 2001), however when the system is nonlinear or constrained, an algebraic update expression for the arrival cost rarely exists. A reasonable solution is to approximate the arrival cost for the constrained problem with the one of the unconstrained problem (Rao et al. 2001) and use an Extended Kalman Filter (EKF) covariance update in the nonlinear case (Rao et al. 2003).

Previous work already proposed Distributed Moving Horizon Estimation (DMHE) using neighborhood measurements, with consensus steps on the a priori state at the beginning of the horizon (Farina et al. 2010) and on the arrival cost (Battistelli 2018; Simonetto et al. 2011). In Battistelli (2018), the authors proposed a DMHE algorithm with multiple-step consensus to ensure the stability of the estimation error dynamics in all the sensor network nodes, under the assumptions of network connectivity and collective observability. The consensus steps enable to spread the information through the network in order to provide each sensor node the necessary information to estimate parts of the system state (which could be locally or even regionally non observable) from other nodes (Farina et al. 2012). Distributed MHE has also been implemented within the context of the set-membership paradigm for large-scale systems under unknown but bounded disturbances and noises (Segovia et al. 2022). To reduce the computation time and, thus, to empower a real-time implementation on low-cost processors, the work in Venturino et al. (2020) and Venturino et al. (2023) proposed a DMHE with a "pre-estimation" strategy based on a Luenberger observer. In Venturino et al. (2020), the state estimation is propagated over the horizon using the equation of an auxiliary Luenberger pre-estimator, instead of the state equation as in classic DMHE formulations. This pre-estimator serves to mitigate model errors without having to estimate the optimal process noise sequence over the horizon. The initial state of the horizon window hence becomes the only optimization variable. This DMHE with pre-estimation has shown to require smaller computation time in comparison to standard DMHE approaches, while preserving its robustness and accuracy benefits (Suwantong et al. 2014). The pre-estimation strategy has initially been introduced in centralized MHE for linear systems (Sui and Johansen 2014), and then extended to nonlinear systems in Suwantong et al. (2014), where in the nonlinear case the Luenberger observer is replaced by an EKF (Suwantong et al. 2014). However, to the best of the authors' knowledge, in the distributed case, this pre-estimation strategy has only been designed for systems with linear dynamics and linear measurements (Venturino et al. 2020). The nonlinear nature of measurements provided by many types of sensors justifies the interest of developing DMHE algorithms to handle both nonlinear measurements and low computation capabilities.

This chapter explores the design of robust state estimation under set-based uncertainties. It focuses on how state estimators can leverage the knowledge of set constraints related to state variables, unknown control inputs, and noise sources to enhance the accuracy of their estimates. In this context, the main contribution of this chapter consists in extending the DMHE approach with pre-estimation of Venturino et al. (2020) in order to handle nonlinear measurements. Following the results in Borelle et al. (2023), the Luenberger observer used in Venturino et al. (2020) for pre-estimation is replaced by a nonlinear Extended Kalman Filter pre-estimator. Another contribution of this chapter consists in applying the developed DMHE technique to the constrained distributed localization of a non-cooperative target by a sensor network.

This chapter is organized as follows. Section 2 introduces the problem of distributed state estimation of a non-cooperative target in a sensor network. Then, the next section briefly explores a Distributed Extended Kalman Filter with its consensus steps. The standard Distributed Moving Horizon Estimation adapted to the considered problem is presented in Sect. 4. Then, the proposed approach of Distributed Moving Horizon Estimation with EKF pre-estimation is detailed in Sect. 5. Finally, the validation of theoretical approaches is illustrated with comparative simulations results in Sect. 6, while concluding remarks are provided in Sect. 7.

Notation. Common notations are further used in this chapter. Classically, x_t stands for the state of the system (target) at time t, while $\hat{x}_t^i$ stands for the state estimate at time t by the sensor i. The tilde notation $\tilde{x}_t^i$ represents the state estimate at time t obtained after a consensus between neighboring sensor nodes of sensor i. A bar notation $(\bar{.})$ is used to refer to regional information.

2 Distributed State Estimation of a Non-Cooperative Target by a Sensor Network

This section formulates the Distributed State Estimation (DSE) problem, focusing on the case of the constrained localization of a non-cooperative target by a Sensor Network.

2.1 Problem Description

The main theoretical problem addressed in this chapter refers to the constrained distributed localization of a non-cooperative target by a network of n_s sensors $\mathcal{N}_s = \{i_1, i_2, \ldots, i_{n_s}\}$ with limited communication ranges and different sensing capabilities.

To solve the problem, at each sample time, each sensor shares relevant information within its neighborhood based on the communication network that will be defined in Sect. 2.4. This information, including the sensor own measurements, estimates, or covariance matrices, is used to perform efficient data fusion like weighted averaging or other consensus strategies, in order to make each sensor estimate the target state with the best possible accuracy.

Remark 2.1 To simplify the notations and the implementation, only one intruder (non-cooperative target) is considered in this chapter. Nevertheless, the extension to the localization problem of several targets is possible by enlarging the state to be estimated to all the targets states.

2.2 Considered Models

The dynamics of the target is described by the following discrete-time Linear Time Invariant (LTI) model

$$x_{t+1} = Ax_t + Bw_t, \tag{1}$$

with t the time index, $x \in \mathcal{X} \subset \mathbb{R}^{n_x}$ the state vector of the target, $w \in \mathcal{W} \subset \mathbb{R}^{n_w}$ the unknown input and/or process noise vector, A the evolution matrix and B the input matrix of corresponding dimensions. The vector w can be used to model some unknown inputs, to account for model errors, perturbations or process noise. For the estimation purpose its estimate covariance matrix is denoted by Q.

Remark 2.2 Linear dynamics are considered in (1) for the sake of simplicity, and driven by the target tracking problem for which kinematic linear models (e.g., nearly constant-speed or nearly constant-acceleration) are often used in the literature (Rong Li and Jilkov 2003), when the dynamics of the target is assumed to be unknown. With the DMHE algorithm with EKF pre-estimation proposed in this chapter, nonlinear dynamics can be accounted for in a straightforward way.

Each sensor $i \in \mathcal{N}_s$ is able to perform measurements on the target state. The following nonlinear measurement equations are considered

$$y_t^i = h^i(x_t) + v_t^i, \tag{2}$$

with t the time index, $y^i \in \mathcal{Y}^i \subset \mathbb{R}^{n_y}$ the measurement vector and $v^i \in \mathcal{V}^i \subset \mathbb{R}^{n_y}$ the measurement noise of the sensor i. The measurement noise is supposed to be a zero mean noise with a covariance matrix denoted by R^i. Notice that the measurement function h^i depends on the sensor i, referring to the sensing capabilities.

2.3 Representation of Uncertainties and Constraints

Different sources of uncertainties are considered in (1) and (2), such as the unknown input of the target or the measurement noises of the sensors. Within the context of set-theoretic methods, uncertainties are assumed to belong to a bounded set, unlike classic stochastic filtering approaches usually involving Gaussian distributions.

Using set-theoretic methods allows us to encompass more realistic cases, e.g., noises with bounded supports. Therefore, $\mathcal{W}$ and $\mathcal{V}^i$ are assumed to be bounded compact sets. This set-based representation enables to account in the state estimation process for some possible a priori knowledge on the uncertainties bounds, if available.

A priori information on the problem can also be considered via a set representation. This includes constraints on the state of the target, e.g., speed limitation, position constrained in a restricted area (e.g., within road boundaries), or constraints on the target input (e.g., limited acceleration capability). To this end, state and process noise constraint sets $\mathcal{X}$ and $\mathcal{W}$ are assumed to be known bounded compact sets. Thus, the set $\mathcal{Y}^i$ can be used to represent constraints on the measurements provided by the sensors (e.g., limited range or field of view). This chapter aims to determine how state estimators can take advantage of the knowledge of these constraints to get more accurate estimates.

2.4 Communication Network Topology

The sensors of the network are equipped with diverse communication capabilities to exchange information. In the considered distributed state estimation scheme, the communication range is supposed to be limited. Therefore, each sensor is assumed to be able to broadcast information to its neighbors, i.e., to other sensors within the communication range. Denote by $\mathcal{N}_s^i$ the neighborhood of sensor i. No relay of information is considered between the sensors of the network, leading to a one-step communication (also known as "one-hop") scheme. Hence, the communication network can be described by an undirected connected graph $\mathcal{G} = (\mathcal{N}_s, \mathcal{E})$, where $\mathcal{E} \subseteq \mathcal{N}_s \times \mathcal{N}_s$ is the set of all edges (communication links between sensor nodes). Therefore, the pair $(i, j) \in \mathcal{E}$ exists if and only if sensor j can receive information from sensor i.

In this chapter it is assumed that there are no communication failures and that the communication network is time-invariant.

Some useful notations for the sensor network are further defined. Regional information (represented by a bar notation $(\bar{\cdot})$) refers to information exchanged within the one step neighborhood $\mathcal{N}_s^i$ of sensor i. For instance, the regional measurements $\bar{y}^i$ indicates the concatenation in a vector of the measurements of sensor i and of its one-step neighbors. Matrix $\bar{R}^i$ stands for the block diagonal matrix composed of R^i and the measurement noise covariance matrices of the sensors belonging to the neighborhood of sensor i.

2.5 *Observability and Distributed State Estimation*

The notion of observability is paramount in state estimation, it refers to the ability to reconstruct or infer the complete internal state of a dynamical system based on its inputs and outputs/measurements. In the case of linear systems, the observability property of a system boils down to verify a simple rank condition (e.g., computing the rank of the observability matrix). Whereas for a nonlinear system it is more complicated. In Hermann and Krener (1977), the notion of *local weak observability* of a non-linear system was introduced and an observability rank condition was derived, consisting in iteratively computing the rank of the span of the measurement function Lie derivative along the flow of the system dynamics (Martinelli 2019). Notice that this property is even more difficult to verify in the presence of unknown inputs (Martinelli 2019).

In a distributed state estimation scheme, when considering a state consensus phase (and thus exchanges of information between the sensors), the local/individual observability of each sensor is not necessary to ensure a stability of each estimation error as long as *collective observability* (or even detectability) holds (Farina et al. 2010). To satisfy this property, it is also necessary to assume a (strongly) connected graph, a deterministic (noise-free) and a disturbance-free framework.

The main steps of the different distributed state estimation strategies at a time index t can be summarized as follows: (i) each sensor $i \in \mathcal{N}_s$ performs a new measurement y_t^i, then sends it to its one-step neighbors and receives the new measurements from its one-step neighbors[1] (nodes in $\mathcal{N}_s^i$) collected in the regional measurements $\bar{y}_t^i$, (ii)

[1] Sensor i exchanges information with its neighbors, for instance to conduct a consensus on a priori state and/or on past/novel information (see Battistelli et al. (2015)).

based on this information, each sensor $i \in \mathcal{N}_s$ computes a local estimate of the target state at the current time, denoted by $\hat{x}_t^i$.

Remark 2.3 As the target is non-cooperative (antagonistic) with the sensors in $\mathcal{N}_s$, it is assumed that the sensors estimators do not have access to the target input w_t to reconstruct its state x_t. Thus, w_t can be seen as an unknown input in the estimation scheme.

3 Distributed Extended Kalman Filtering

This section provides key elements on the implementation of the Extended Kalman Filter in a distributed state estimation framework. Even though the distributed Extended Kalman filter is not suitable to handle constraints, it is used for comparison with the proposed methods based on distributed moving horizon estimation. Furthermore, the L-step consensus on information (Battistelli and Chisci 2014; Battistelli et al. 2015) is introduced and is used to fuse the information from neighbor sensors in the Distributed Extended Kalman Filter implementation.

3.1 Distributed Extended Kalman Filter (DEKF)

The extended Kalman filter is the well known extension of the classic Kalman filter, designed to handle nonlinearities in the system and measurement models. The standard Kalman filter is optimal for linear systems with Gaussian noise, but when dealing with nonlinearities, the EKF can only provide an approximation of the optimal solution. The basic idea of the extended Kalman filter is to linearize the nonlinear system and measurement equations by using a first-order Taylor series expansion around the current estimate of the state. Implicitly, it is assumed that the nonlinearities are "sufficiently small" to be effectively well approximated by linearizations. Beyond this principle of linearisation, the EKF and Kalman filter implementations are similar. Indeed, the Extended Kalman assumes in its formulation that the errors in the system and measurement models can be modeled by the process and measurement noises, which are assumed to be Gaussian. We refer the interested reader to Welch and Bishop (2006) for derivations of EKF formulas and more explanations about its different steps, which are adapted here to the considered problem formulated in Sect. 2 for a system with linear dynamics.

Denote the state estimate of sensor $i \in \mathcal{N}_s$ at time t by $\hat{x}_t^i$ and its covariance by Π_t^i. The considered DEKF approach is an adaptation of the Consensus on Information (CI) algorithm in Battistelli et al. (2015), rewritten in the standard Kalman filter form (equivalent to the information filter form) and with the neighbors sensors measurements incorporated at once directly in the state update (7). Therefore, the DEKF is computed as described below

1. State prediction

$$\hat{x}^i_{t+1|t} = A\tilde{x}^i_t, \tag{3}$$

where $\hat{x}^i_{t+1|t}$ is the predicted state at time $t+1$ based on the measurements at time t and on the dynamical model, while $\tilde{x}^i_t$ is used to simplify the notation $\tilde{x}^i_{t|t}$, representing the consensus estimated state at time t,

2. Linearization of regional measurements

$$\bar{C}^i_t = \left.\frac{\partial \bar{h}^i(x_t)}{\partial x_t}\right|_{\hat{x}^i_{t+1|t}}, \tag{4}$$

3. Error covariance prediction[2]

$$\Pi^i_{t+1|t} = A\tilde{\Pi}^i_t A^\top + BQB^\top, \tag{5}$$

with $\tilde{\Pi}^i_t$ the consensus covariance of the estimated state error at time t,

4. Kalman gain update

$$K^i_t = \Pi^i_{t+1|t}(\bar{C}^i_t)^\top \left(\bar{C}^i_t \Pi^i_{t+1|t}(\bar{C}^i_t)^\top + \bar{R}^i\right)^{-1}, \tag{6}$$

5. State update (with regional measurements)

$$\hat{x}^i_{t+1} = \hat{x}^i_{t+1|t} + K^i_t\left(\bar{y}^i_{t+1} - \bar{h}^i\left(\hat{x}^i_{t+1|t}\right)\right), \tag{7}$$

6. Error covariance update

$$\Pi^i_{t+1} = \left(I_{n_x} - K^i_t\bar{C}^i_t\right)\Pi^i_{t+1|t}. \tag{8}$$

In the state prediction step (3) it is important to highlight that only the known dynamical evolution model is propagated, excluding the presumed unknown input w_t. In the linearization step (4) and in the state correction (7), it is worth noticing that to take advantage of the distributed implementation, each sensor considers its regional measurement (not solely its own measurements), which allows to improve the state estimation performed by sensors with weak local observability. The state and the covariance obtained in (7) and (8), respectively, are used as initial conditions of a consensus scheme, detailed in the next section, allowing to broadcast information into the sensors network and to obtain a more accurate target state estimate. This consensus part on the state estimate and its covariance is developed in the next subsection.

[2] In (5) and (6), it is assumed that w_k and v_k are independent and zero mean uncorrelated noises.

3.2 Information Fusion/Consensus Step

After completing the DEKF state and covariance updating steps (7)–(8), a further information fusion or consensus step can be completed among neighboring sensors. This step aims to enhance information dissemination within the network and to improve the estimation performance. In the literature, various consensus step approaches have been implemented on distributed (extended) Kalman filters to enhance the estimation performance in distributed systems. In particular, the covariance intersection, consensus on information, consensus on measurement, hybrid consensus on measurements and information are summarized in Battistelli et al. (2011, 2015). The multiple consensus steps allow for broader diffusion of the information in the sensor network in spite of the communication range limitation. In addition, it enables sensors with uninformative or unreliable measurements to obtain a better state estimation. This subsection presents a review of the L-step consensus on information (Battistelli and Chisci 2014; Battistelli et al. 2015) which is a generalization to multiple consensus steps of the covariance intersection (Julier and Uhlmann 2001), consisting in a regional weighted average consensus on the information (matrix-vector) pair, defined in (9) and (10).

At step $l = 0$, the information matrix $P^i_{t,0}$ and the information vector $\xi^i_{t,0}$ of sensor i are initialized at time t with the state and covariance estimate

$$P^i_{t,0} = \left(\Pi^i_t\right)^{-1}, \tag{9}$$
$$\xi^i_{t,0} = P^i_{t,0}\hat{x}^i_t, \tag{10}$$

with Π^i_t the covariance of the estimation error of the initial state of the receding horizon (see Farina et al. (2010)). Each step $l \in \{1, \dots, L\}$ of the information consensus consists in computing

$$P^i_{t,l} = \sum_{j \in N^i_s} \kappa_{i,j} P^j_{t,l-1}, \tag{11}$$
$$\xi^i_{t,l} = \sum_{j \in N^i_s} \kappa_{i,j} \xi^j_{t,l-1}, \tag{12}$$

with the consensus weights $\kappa_{i,j}$ defined as a convex combination further detailed in this subsection. This computation corresponds to a weighted average of both the information matrix and the information vector among neighbors.

After L-steps of consensus, the consensus covariance of the estimation error $\tilde{\Pi}^i_t$ and the consensus estimated state $\tilde{x}^i_t$ at time t can finally be retrieved

$$\tilde{\Pi}^i_t = (P^i_{t,L})^{-1}, \tag{13}$$
$$\tilde{x}^i_t = (P^i_{t,L})^{-1}\xi^i_{t,L}. \tag{14}$$

This consensus steps allow to broadcast information deeper into the network and permits to take into account the confidence of each sensor on each component of the system state to be estimated. It also allows convergence of the estimates even in the case of weak local observability (Battistelli and Chisci 2014; Julier and Uhlmann 2001).

3.3 Consensus Weights

The weights $\kappa_{i,j} \geq 0$ used in the consensus steps (11) and (12) can be defined as the components of a stochastic matrix $\mathbf{K}$ associated to the graph $\mathcal{G}$. Several techniques to compute the convex combination of the consensus weights $\kappa_{i,j}$ with $(i, j) \in \mathcal{N}_s \times \mathcal{N}_s$ have been studied in the literature. We recall the condition for this consensus weights to form a convex combination for a neighborhood of a sensor

$$\sum_{j \in \mathcal{N}_s^i} \kappa_{i,j} = 1, \ \forall i \in \mathcal{N}_s. \tag{15}$$

The simplest way to tackle (15) is to make no assumption on the different sensors in the neighborhood and to give the same confidence to all the sensors $\forall (i, j) \in \mathcal{N}_s \times \mathcal{N}_s$, considering uniform weights, which leads to

$$\kappa_{i,j} = \begin{cases} \frac{1}{\text{card}(\mathcal{N}_s^i)}, & \text{if } j \in \mathcal{N}_s^i, \\ 0 & \text{otherwise}, \end{cases} \tag{16}$$

where card$(\cdot)$ denotes the cardinality of a given set.

One can also take advantage of some observability properties of the neighborhood sensors in the choice of the consensus weights. The observability rank-based weights tuning technique of (Venturino et al. 2022, 2021a, b) can be extended to the nonlinear case (Borelle et al. 2023) by making use of the linearization (4). Consider a sensor i. Its regional (Kalman) observability matrix is given by

$$\bar{\mathcal{O}}_t^i = \text{col}\left(\bar{C}_t^i, \bar{C}_t^i A, \ldots, \bar{C}_t^i A^{n_x}\right), \tag{17}$$

where the matrix $\bar{C}_t^i$ is defined in (4) and col$(\cdot)$ concatenates the considered sequence of vectors in a single column vector. The observability matrix is of full rank if and only if the pair $(A, \bar{C}^i)$ is observable, i.e., rank$(\bar{\mathcal{O}}_t^i) = n_x$.

For the sake of simplicity, the following notation is further used

$$\rho_{\mathcal{O}}^i = \text{rank}(\bar{\mathcal{O}}_t^i). \tag{18}$$

This information could be used as the reliability of sensor i when choosing the weights $\kappa_{i,j}$, which must be averaged among the neighbors $j \in \mathcal{N}_s^i$

$$\kappa_{i,j} = \frac{\rho_O^j}{\sum_{j \in N_s^i} \rho_O^j}. \tag{19}$$

This method relies only on *regional* information (measurements) available to each sensor to compute its own component $\kappa_{i,j}$ of $\mathbf{K}$. Thus, it is suitable for a distributed scheme. The interest of this approach is that it enhances the accuracy of the estimates by means of exploiting observability properties of the neighborhoods.

4 Distributed Moving Horizon Estimation (DMHE)

Extending, the results in Battistelli (2018) to nonlinear measurements, this section addresses the distributed moving horizon estimation approach for the considered problem.

4.1 DMHE Principles and Formulation

The development of moving horizon estimation methods together with their distributed versions (DMHE) is encouraged by their ability to handle constraints and deal with high nonlinearities and robustness with respect to poor initial estimates. MHE is an optimal state observer approach (dual to MPC for the control framework) computing a state estimate by minimizing a cost function (with possible nonlinearities and constraints) involving a plant model and a sequence of past measurements over a finite horizon window (see Fig. 1) that moves forward at each sample time.

The presence of nonlinearities in the considered constraints and system dynamics could render the optimization problem non-convex and thus difficult to solve.

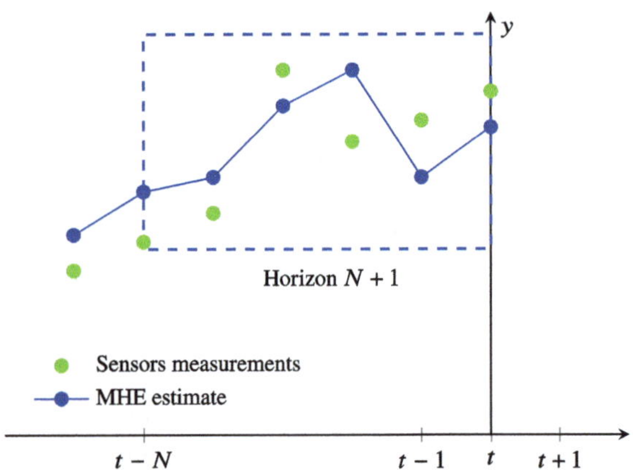

Fig. 1 Illustration of the receding horizon paradigm used in DMHE

Moreover, the complexity of the optimization problem grows with the size of the estimation/horizon window $N + 1$ (i.e., the number of considered measurements).

In DMHE, at time t, each sensor $i \in \mathcal{N}_s$ determines (based on regional information) its sequence of estimates of the target state $\hat{x}^i_{t-N|t}, \ldots, \hat{x}^i_{t|t}$ by solving the constrained optimization problem (20)–(24)

$$\min_{\hat{x}^i_{t-N|t},\ldots,\hat{x}^i_{t|t}} J^i_N(\cdot) \tag{20}$$

$$\text{s.t. } \hat{x}^i_{k+1|t} - A\hat{x}^i_{k|t} \in B\mathcal{W}, \qquad \forall k = t-N,\ldots,t-1. \tag{21}$$

$$\hat{x}^i_{k|t} \in \mathcal{X}, \qquad \forall k = t-N,\ldots,t, \tag{22}$$

$$\bar{y}^i_{k+1} - \bar{h}^i\left(\hat{x}^i_{k+1|t}\right) \in \bar{\mathcal{V}}^i, \qquad \forall k = t-N,\ldots,t-1, \tag{23}$$

$$\bar{y}^i_{k+1} \in \bar{\mathcal{Y}}^i, \qquad \forall k = t-N,\ldots,t-1, \tag{24}$$

where $\bar{y}^i_{k+1}$ and $\bar{h}^i(\hat{x}^i_{k+1|t})$ are the regional measurements and the predicted estimated regional measurements, respectively, and $\hat{x}^i_{k|t}$, $\forall k \in [t-N,\ldots,t]$ refers to an estimated state at time k based on the measurements up to time t. The constraints (21)–(24) regroup the bounds on the prior knowledge on the estimated state and measurements, with the corresponding noise bounds.

The standard cost function used in DMHE typically takes the form of a sum of a quadratic penalty cost $\Gamma^i_t(\cdot)$ and two squared norm difference terms, both summed over the estimation window, with the horizon length $N + 1$

$$J^i_N(\cdot) = \sum_{k=t-N}^{t} \left\|\bar{y}^i_k - \bar{h}^i\left(\hat{x}^i_{k|t}\right)\right\|^2_{(\bar{R}^i)^{-1}} + \sum_{k=t-N}^{t-1} \|\hat{x}^i_{k+1|t} - A\hat{x}^i_{k|t}\|^2_{(BQB^\top)^{-1}} + \Gamma^i_t(\cdot). \tag{25}$$

The first difference term in (25) represents a weighted norm of the error between the actual regional measurements $\bar{y}^i_k$ and the predicted regional measurements $\bar{h}^i(\hat{x}^i_{k|t})$, obtained from the sequence of estimated states. These errors are weighted by the inverse of the definite positive matrix $\bar{R}^i$ which is the regional measurement noise covariance matrix.

The second term of the cost function, which is the weighted norm of the difference between the estimated state $\hat{x}^i_{k+1|t}$ and a priori estimated state obtained using the dynamic model knowledge $A\hat{x}^i_{k|t}$, reflects the process noise possibly accounting for model mismatch/perturbation.

The last term of the cost function $\Gamma^i_t(\cdot)$, known in the literature as *initial penalty function* or *arrival cost* (Rao et al. 2001), is crucial to ensure the estimation convergence. This term aims at summarizing information about past measurements and prior knowledge on the state, before the current estimation window. An algebraic expression does not exist in the case of activated constraints and nonlinearities (Rao et al. 2003). Therefore, the arrival-cost $\Gamma^i_t(.)$ is commonly approximated by a squared error weighted norm between the predicted state at the beginning of the estimation window $\hat{x}^i_{t-N|t}$ and a prior estimated state $\tilde{x}^i_{t-N}$, symbolizing past knowledge

$$\Gamma^i_t(.) = \|\hat{x}^i_{t-N|t} - \tilde{x}^i_{t-N}\|^2_{(\tilde{\Pi}^i_{t-N})^{-1}}, \tag{26}$$

with $\tilde{\Pi}^i_{t-N}$ the estimated covariance error of this prior estimation. Similar to the DMHE scheme proposed in Battistelli (2018), a consensus is performed on the arrival cost through a regional average, further detailed in Sect. 4.2.

The state estimate is deduced from the last element of the solution to the optimization problem, i.e., $\hat{x}^i_t = \hat{x}^i_{t|t}$ (20) at time t. Based on the receding horizon principle, at the next sample time $t+1$, the horizon is shifted one step forward for the formulation of the new optimization problem. The second element $\hat{x}^i_{t-N+1|t}$ of the previously computed solution is retained for the initialization of the consensus step at $t+1$, as described in the next section.

4.2 Consensus Step for Arrival Cost in DMHE

The consensus step on the arrival cost (originated from Battistelli (2018)) permits to carry out L times a weighted average of all the arrival costs in a neighborhood. The multiple steps consensus allows broadcasting the information among the sensor network in a broad way despite the communication range limits, which is particularly relevant in the case of regional non observability

$$\forall i \in \mathcal{N}_s, \forall l \in [0, L-1], \quad \Gamma^i_{t,l+1}(\cdot) = \sum_{j \in \mathcal{N}^i_s} \kappa_{i,j} \Gamma^j_{t,l}(\cdot). \tag{27}$$

An equivalent result can be obtained by performing a L-step information consensus with the information matrix and the vector pair corresponding to the prior estimate. Notice that, for the DEKF, such a consensus scheme is considered, but on the current state estimate $\hat{x}_t$ (see Sect. 3.2). At time t, the initialization of the L-step consensus procedure considered for the DMHE is

$$P^i_{t-N,0} = (\Pi^i_{t-N})^{-1}, \tag{28}$$

$$\xi^i_{t-N,0} = P^i_{t-N,0} \hat{x}^i_{t-N|t-1}. \tag{29}$$

The weight averages among neighborhood are performed as in Sect. 3.2 with equations (11)–(12). After L steps of consensus, the following quantities are finally used to define the “fused” arrival cost (26)

$$\tilde{\Pi}^i_{t-N} = (P^i_{t-N,L})^{-1}, \tag{30}$$

$$\tilde{x}^i_{t-N} = (P^i_{t-N,L})^{-1} \xi^i_{t-N,L}. \tag{31}$$

Remark 4.1 Notice that $\tilde{\Pi}^i_{t-N}$ and $\tilde{x}^i_{t-N}$ correspond to the error covariance matrix Π^i_{t-N} and to the prior state estimation $\hat{x}^i_{t-N|t-1}$ obtained at time $t-1$, after proceeding the L-step information consensus, respectively.

The consensus phase is necessary to guarantee the convergence of the state estimates to the state of the observed system, even in lack of regional observability (Farina et al. 2010).

The positive definite matrix Π^i_{t-N+1} is then obtained from the matrix $\tilde{\Pi}^i_{t-N}$ using the discrete-time Riccati equation associated to an extended Kalman filter (as in Rao et al. (2003) for the centralized case).

$$\Pi^i_{t-N+1} = A\tilde{\Pi}^i_{t-N}A^\top + BQB^\top - A\tilde{\Pi}^i_{t-N}(\bar{C}^i_{t-N})^\top \cdot \left(\bar{C}^i_{t-N}\tilde{\Pi}^i_{t-N}(\bar{C}^i_{t-N})^\top + \bar{R}^i\right)^{-1} \cdot \bar{C}^i_{t-N}\tilde{\Pi}^i_{t-N}A^\top, \quad (32)$$

The regional measurement matrix $\bar{C}^i_{t-N+1}$ is obtained via the linearization of the regional measurement function around $\hat{x}^i_{t-N+1}$

$$\bar{C}^i_{t-N+1} = \left.\frac{\partial \bar{h}^i(x_t)}{\partial x_t}\right|_{\hat{x}^i_{t-N+1}}. \quad (33)$$

Notice that the update of the covariance matrix Π^i_{t-N} from Rao et al. (2003) has been extended here to the distributed case.

5 Distributed Moving Horizon Estimation with Pre-estimation

This section formulates the proposed Distributed moving horizon estimation approach with pre-estimation and observability rank-based weights to address the distributed state estimation problem with nonlinear measurements.

5.1 DMHE with Pre-estimation: Optimization Problem

At each sample time t, given an estimation horizon length $N \geqslant 1$, each sensor $i \in \mathcal{N}_s$ determines (based on the information received from its neighbors in $\mathcal{N}^i_s$) its estimate $\hat{x}^i_{t-N|t}$ of the state x_{t-N} by solving the following constrained optimization problem

$$\hat{x}^i_{t-N|t} = \arg\min_{\hat{x}^i_{t-N|t}} J^i_{N,\text{DMHE-pre}}(\cdot) \tag{34}$$

$$\text{s.t. } \hat{x}^i_{k+1|t} = A\hat{x}^i_{k|t} + K^i_k \left(\bar{y}^i_{k+1} - \bar{h}^i\left(A\hat{x}^i_{k|t}\right)\right), \tag{35}$$

$$\hat{x}^i_{k|t} \in \mathcal{X}, \quad \forall k = t-N, \ldots, t, \tag{36}$$

$$\bar{y}^i_{k+1} - \bar{h}^i\left(\hat{x}^i_{k+1|t}\right) \in \bar{\mathcal{V}}^i, \quad \forall k = t-N, \ldots, t-1, \tag{37}$$

$$\bar{y}^i_{k+1} \in \bar{\mathcal{Y}}^i, \quad \forall k = t-N, \ldots, t-1. \tag{38}$$

In the proposed formulation, the equation of an observer is used as pre-estimator in (35), replacing the constraint (21) of the original DMHE formulation which accounted for the dynamics of the system. The matrix gain K^i_k is computed as the Kalman gain of an EKF as detailed in Sect. 5.2. Compared to the classic DMHE of Battistelli (2018); Farina et al. (2010), the sole optimization parameter in the proposed DMHE with pre-estimation (34) is the estimated state at the beginning of the horizon. Therefore, the number of decision variables is decreased by the use of the auxiliary observer leading to a reduced computation time (Venturino et al. 2021b). The sequence of state estimates $\hat{x}^i_{t-N+1|t}, \ldots, \hat{x}^i_{t|t}$ is obtained from the optimal solution $\hat{x}^i_{t-N|t}$, using the pre-estimator expression (35). The objective function in (34) is defined as

$$J^i_{N,\text{DMHE-pre}}(\cdot) = \sum_{k=t-N}^{t} \left\| \bar{y}^i_k - \bar{h}^i\left(\hat{x}^i_{k|t}\right) \right\|^2_{\left(\bar{R}^i\right)^{-1}} + \Gamma^i_t(\cdot). \tag{39}$$

In comparison with the cost function (25) of the standard DMHE, in (39) the mismatch/perturbation term has been removed, as the unknown input (equivalently the sequence of state estimates over the horizon) is not directly estimated, but indirectly accounted for by the correction term of the pre-estimation auxiliary observer (35). The initial penalty term $\Gamma^i_t(\cdot)$ in (39) is the same as the one described earlier in (26). The consensus scheme retained is also the same as the one used in the standard DMHE and described in Sect. 4.2.

5.2 EKF as Pre-estimation Observer

At sample time t, the optimal Kalman gain K^i_k from the DMHE with pre-estimation optimization problem (35) is computed for $k \in \{t-N, \ldots, t-1\}$ using an EKF observer update, as in (6). First, the pre-estimation error covariance matrix is initialized at the beginning of the estimation window with

$$\Pi^i_{pre,t-N|t} = \tilde{\Pi}^i_{t-N}. \tag{40}$$

This equation represents a predicted (a priori) estimation error covariance. To lighten the notation, the reference to the current instant t is further omitted and this matrix is denoted by $\Pi^i_{pre,t-N}$. Then, for $k \in \{t-N, \dots, t-1\}$, the update of the predicted estimation error covariance and the update of the Kalman gain are respectively performed combining the same equations as in DEKF (5) and (8):

$$K^i_k = \Pi^i_{pre,k}(\bar{C}^i_k)^\top \left(\bar{C}^i_k \Pi^i_{pre,k}(\bar{C}^i_k)^\top + \bar{R}^i\right)^{-1}, \tag{41}$$

$$\Pi^i_{pre,k+1} = A\left(I_{n_x} - K^i_k \bar{C}^i_k\right) \Pi^i_{pre,k} A^\top + BQB^\top. \tag{42}$$

5.3 Overview of the Algorithm

The proposed algorithm can be summarized by the pseudo-code given in Algorithm 1. Its main steps are also described below.

At the initialization step, at time $t = 0$, each sensor/node $i \in \mathcal{N}_s$ stores its a priori initial estimate $\hat{x}^i_0$ of x_0 and associated covariance matrix Π^i_0. Then, the next step depends on the number of available past measurements over the estimation horizon window and thus on time t.

- For the case $1 \le t \le N+1$, then the number of available past measurements is smaller than the estimation horizon length N. The considered estimation horizon length N_w is therefore set to $N_w = t-1$. For the initialization of the consensus step at time t, each node $i \in \mathcal{N}_s$ performs the following steps:
  - set $\Pi^i_{1|t} = \Pi^i_{1|t-1}$, the initial state error covariance is initialized with the previous value,
  - set $\tilde{x}^i_1$ from $\hat{x}^i_{1|t-1}$, the initial a priori state is updated with the last computed solution of the optimization problem.
- For the case $t > N+1$, then the moving horizon estimation window has already reached its full size $N_w = N$. For the initialization of the consensus step at time t, each node $i \in \mathcal{N}_s$ performs the following steps:
  - compute Π^i_{t-N} from $\tilde{\Pi}^i_{t-N-1}$ according to (32),
  - set $\hat{x}^i_{t-N|t-1} = \hat{x}^i_{t-1-(N-1)|t-1}$, using the second element of the previously computed optimization problem solutions.

Each node computes the rank of it observability matrix, sends ρ^i_O to its neighbors $j \in \mathcal{N}^i_s$ and receive ρ^j_O from them in order to compute the consensus weights $\kappa_{i,j}$ (17), (18) and (19). Afterwards, each node performs the consensus step with initialization (28) and (29). The state covariance of the pre-estimator $\Pi^i_{pre,t-N_w}$ is initialized with (40). Then, each node solves the DMHE-pre optimization problem (34) with horizon size N_w in order to obtain $\hat{x}^i_{t-N_w|t}$. In this problem, the initial penalty $\Gamma^i_t(.)$ considered is

$$\Gamma^i_t(.) = \left\| \hat{x}^i_{t-N_w|t} - \tilde{x}^i_{t-N_w} \right\|^2_{(\tilde{\Pi}^i_{t-N_w})^{-1}} .$$

Next, the pre-estimator is run to obtain the state estimate sequence over the horizon $[\hat{x}^i_{t-N_w|t}, \hat{x}^i_{t-N_w+1|t}, \ldots, \hat{x}^i_{t|t}]$. Finally, each node stores the state estimate sequence and set its current estimate $\hat{x}^i_t = \hat{x}^i_{t|t}$.

Algorithm 1 DMHE with EKF pre-estimation procedure

Initialization: $\forall i \in \mathcal{N}_s$, at the first time step $t = 0$
  initialize $(\hat{x}^i_0, \Pi^i_0)$
  predict $(\hat{x}^i_{1|0}, \Pi^i_{1|0})$ using (3)–(5)
Online: $\forall i \in \mathcal{N}_s, \forall t > 0$
  exchange measurements with neighbors $j \in \mathcal{N}^i_s$, to form $\bar{y}^i_t$
  if $1 \leqslant t \leqslant N + 1$ **then**
    set the horizon length $N_w = t - 1$
  else
    set the horizon length $N_w = N$
    correct and **predict** the covariance $\tilde{\Pi}^i_{t-N_w-1}$ to obtain $\Pi^i_{t-N_w}$ using (32)
  end if
  compute $O^i_{t-N_w}$ and ρ^i_O according to (17) and (18)
  send ρ^i_O to neighbors $j \in \mathcal{N}^i_s$ and receive ρ^j_O from them
  compute the $\kappa_{i,j}$ components according to (19)
  perform L steps of the consensus algorithm (11)–(12) with the initialization $(\hat{x}^i_{t-N_w|t-1}, \Pi^i_{t-N_w})$ (9)–(10) to get $(\tilde{x}^i_{t-N_w}, \tilde{\Pi}^i_{t-N_w})$ (13)–(14)
  initialize $\Pi^i_{pre,t-N_w}$ with (40)
  solve the local optimization problem of DMHE with EKF pre-estimation, minimizing $J^i_{N_w,\text{DMHE-pre}}(t)$ as in (34) with (39) subject to the constraints (35)–(36)
  store the solution $[\hat{x}^i_{t-N_w|t}, \hat{x}^i_{t-N_w+1|t}, \ldots, \hat{x}^i_{t|t}]$ and **set** the estimate $\hat{x}^i_t = \hat{x}^i_{t|t}$.

6 Simulation Scenario and Numerical Results

6.1 Scenario and Target Dynamics

The considered application scenario consists in a network of $n_s = 14$ cooperative static sensors to estimate the state of a non-cooperative moving target (e.g., an intruder ground vehicle) evolving in a bounded area, e.g., limited section of a road. The road is defined by a rectangle of width $w_r = 4$m and length $l_r = 50$m (see Fig. 2), with the left-bottom vertex located at the position (10m, 10m).

The state vector of the target is composed of its 2D position and velocity components in a common inertial reference frame $x = \begin{bmatrix} p_x \; p_y \; v_x \; v_y \end{bmatrix}^\top$. In this reference

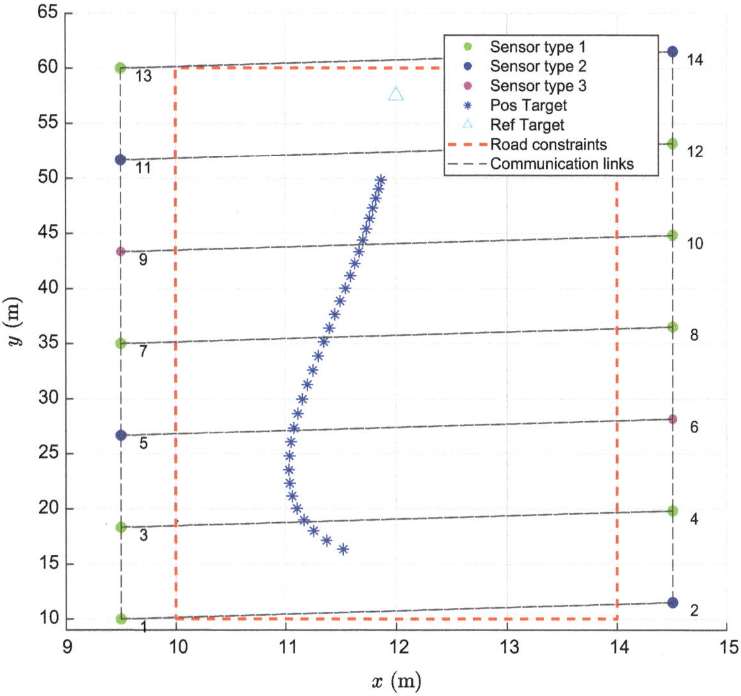

Fig. 2 Example of unknown target trajectory to be estimated by the sensor network

frame, the position vector of each sensor $i \in \mathcal{N}_s$ denoted by $x_s^i = \left[p_{s,x}^i \; p_{s,y}^i\right]^\top$ is assumed to be time-invariant. The translational dynamics of the target are considered to be modelled as in (1) with

$$A = \begin{bmatrix} 1 & 0 & \Delta t & 0 \\ 0 & 1 & 0 & \Delta t \\ 0 & 0 & 1 & 0 \\ 0 & 0 & 0 & 1 \end{bmatrix}, \quad B = \begin{bmatrix} \frac{\Delta t^2}{2} & 0 \\ 0 & \frac{\Delta t^2}{2} \\ \Delta t & 0 \\ 0 & \Delta t \end{bmatrix}, \tag{43}$$

with the sample time $\Delta t = 0.2$s. The input vector corresponds to the acceleration of the target $u = \left[a_x \; a_y\right]^\top$. The non-cooperative target is initially located in an area at the beginning of a rectilinear road and a controller is set up to allow the target to reach a reference point at the end of the road (without crossing the road boundaries), and generate its trajectory. A saturation of the obtained control signal is implemented as $[-3 \;\; -3]^\top \leq u \leq [3 \;\; 3]^\top$, to comply with limited acceleration capabilities assumed for the mobile target. For estimation purposes, this control signal limitation (formulated as constraints in the DMHE optimization problem) corresponds to assuming the type of target to be monitored.

A perturbation is added to this control signal u, in terms of an additive noise with the covariance matrix chosen as $Q_p = \text{diag}(q_p, q_p)$, with $q_p = 0.05^2$. The tuning of the controller is not the scope of this chapter and thus is not further detailed.

Monte-Carlo simulations have been run to generate several trajectories of the target from different initial conditions. To avoid crossing the road boundaries, the target initial state $x_0 \in \mathbb{R}^4$ is randomly drawn from a uniform distribution within the range $[11 \;\; 11 \;\; -1 \;\; 0.5]^\top \leq x_0 \leq [13 \;\; 21 \;\; 1 \;\; 4.5]^\top$.

6.2 Sensors Measurements and Noise Modelling

Three different types of sensors are considered in the network, with different sensing capabilities.

A sensor i of the first type (*Type* 1) produces noisy measurements of the distance to the target d_t^i and of the line of sight angle α_t^i of the target with respect to the sensor as in Battistelli and Chisci (2014). In this case, the measurement function is given by

$$h_1^i(x_t) = \begin{bmatrix} h_2^i(x_t) \\ h_3^i(x_t) \end{bmatrix} = \begin{bmatrix} d_t^i \\ \alpha_t^i \end{bmatrix} = \begin{bmatrix} \sqrt{(p_x - p_{s,x}^i)^2 + (p_y - p_{s,y}^i)^2} \\ \text{atan2}\left(p_y - p_{s,y}^i, p_x - p_{s,x}^i\right) \end{bmatrix}, \tag{44}$$

with $\text{atan2}(\Delta y, \Delta x)$ the 4-quadrant inverse tangent function of $\Delta y / \Delta x$.

The local measurement vector in (2) for a *Type 1* sensor corresponds to

$$y_t^i = h_1^i(x_t) + v_1^i,$$

with the measurement noise vector $v_1^i = \left[v_2^i \; v_3^i\right]^\top$ of covariance matrix

$$R_1 = \text{diag}(r_2, r_3) = \begin{bmatrix} 0.6^2 & 0 \\ 0 & 0.08^2 \end{bmatrix}. \tag{45}$$

The other two types of sensors are shown in Table 1.

Sensors of *Type 2* and *Type 3* produce measurements only on the distance and only on the line of sight angle, respectively. It is worth noticing that each *Type 2* or *Type 3* sensor is unable by itself to ensure observability of the entire target state vector, but

Table 1 Types and measurement models of the sensors

Type of sensor i	Measurement function	Measurement noise covariance
1	$h_1^i(x_t)$	R_1
2	$h_2^i(x_t)$	r_2
3	$h_3^i(x_t)$	r_3

collective observability holds for the sensor network. The chosen locations of the sensors (represented in Fig. 2 by a dot with the index i of each sensor) enable covering the complete road section. The color of the dot indicates the type of the sensor which has been chosen arbitrarily for the simulated case study: Type 1 in green, Type 2 in blue, and Type 3 in magenta. The communication links between the sensor network are depicted by gray dashed segments. An example of generated trajectory for the target is plotted with blue asterisks. The light blue triangle represents the reference point to be reached by the target (which is unknown by the sensor network). The road boundaries taken as constraints are represented by the red dashed rectangle.

Remark 6.1 Within this configuration, the communication network is undirected and the connectivity assumption is well respected.

Measurement noises are assumed to be bounded and following a beta distribution. This assumption of bounded support is more realistic for this application than the infinite support of Gaussian distributions. The beta distribution used in simulations has been designed to be symmetric and to ensure that the measurement noise v_1^i, with covariance R_1, satisfies the following bounds $\left[-4\sqrt{r_2} \quad -4\sqrt{r_3}\right]^\top \leq v_1^i \leq \left[4\sqrt{r_2} \ 4\sqrt{r_3}\right]^\top$.

6.3 *Results Analysis*

The performance of the proposed DMHE approach with EKF pre-estimation (denoted by DMHE-pre-EKF) is assessed by comparing its estimation accuracy and computation time as in Borelle et al. (2023) with

- a similar nonlinear extension of the DMHE algorithm of Battistelli (2018) (without pre-estimation), presented in Sect. 4, denoted by DMHE-1,
- the distributed EKF with consensus on information from Battistelli et al. (2015), introduced in Sect. 3, denoted by DEKF-CI.

We consider a scenario where the initial state of the target is poorly known, i.e. possible large error between x_0 and the initial estimate $\hat{x}_0^i$ given a priori to the observers. Moreover, it is assumed that the observers have too much confidence in these poor initial estimates, resulting in an initial estimation error covariance matrix Π_0^i that underestimates the true covariance Π_0 of the initial estimation error.

For simulations, the same values $\hat{x}_0^i$ and Π_0^i are given to all the observers. They are denoted by $\hat{x}_0$ and $\hat{\Pi}_0$, i.e., $\hat{x}_0^i = \hat{x}_0$, $\Pi_0^i = \hat{\Pi}_0$, $\forall i \in \mathcal{N}_s$. For randomly generated initial positions x_0, initial estimates $\hat{x}_0$ are chosen following a normal law centered on x_0 with covariance matrix $\Pi_0 = diag(30, 60, 4, 4)$. The following covariance matrix is provided to the observers $\hat{\Pi}_0 = diag(4, 4, 0.5, 0.5)$, which underestimates the true covariance.

Notice that the values chosen for $\hat{\Pi}_0$ could result in initial estimates outside the position constraints given by the road shape. This is intended to illustrate how the constraints can be efficiently used to improve the estimation performance of the observers, especially at the beginning of the scenario, correcting the poor initial estimate.

The input of the target is not known by the estimators (non cooperative target). Therefore, the covariance matrix $Q = Q_u + Q_p = \mathrm{diag}(3, 3) + \mathrm{diag}(0.05^2, 0.05^2)$ on w is tuned to encompass the possible variation of the unknown input and perturbation acting on the target. The following bounds are considered $[-3.2\ \ -3.2]^\top \leq w \leq [3.2\ 3.2]^\top$ on w for the DMHE estimator.

A Monte Carlo simulation of 30 runs has been carried out with different initial conditions for the target, different realizations of measurements noise and perturbation noise added to the target input.

Several metrics are considered for comparison: the Root Mean Square Error (RMSE), the RMSE averaged over the last eight time instants of the simulation (denoted by RMSE final values), and the computation time τ. These metrics are evaluated as averages over all the sensors of the network and over all the Monte Carlo runs. The time evolution of the RMSE and of the computation time are presented in Figs. 3 and 4 . Their average values over all the time instants are presented in Table 2 (for the entire state vector, and for the position sub-vector only), as well as the RMSE averaged over the final time instants, and the mean computation time. For the

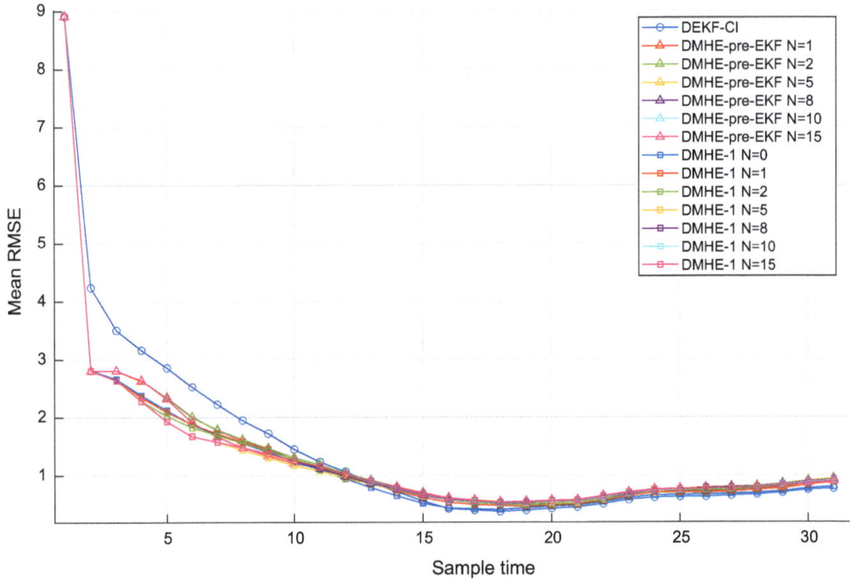

Fig. 3 Time evolution of the RMSE for the tested estimators (average value over all the sensors and all the Monte Carlo runs)

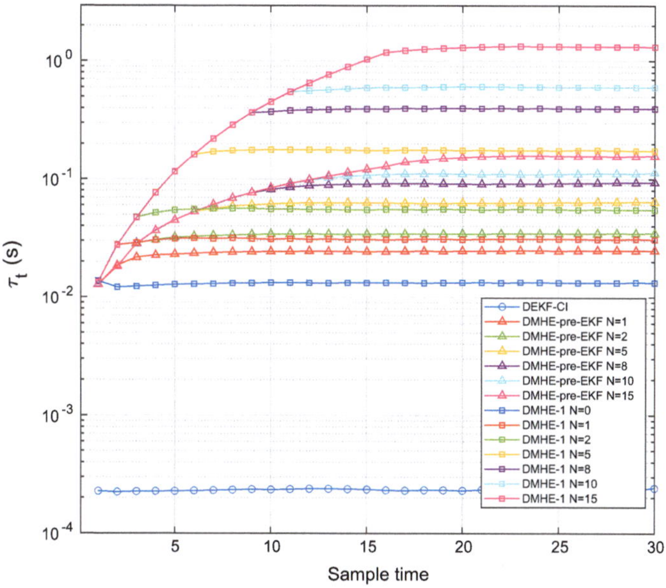

Fig. 4 Time evolution of the computation time τ_t for the tested estimators (average value over all the sensors and all the Monte Carlo runs)

sake of readability, we have only displayed on Figs. 3 and 4 the values corresponding to the observers of Table 2 with the following horizon size $N \in \{0, 1, 2, 5, 8, 10, 15\}$.

The simulations were carried out by a PC Linux Ubuntu 20.04.1 equipped with an Intel Core i7-10875H CPU. To solve the constrained nonlinear optimization problem of MHE we use the `fmincon` function of Matlab with the SQP solver (standing for Sequential Quadratic Programming).

Due to the complex and possibly non-convex nature of the optimization problem of DMHE, it is worth mentioning the interest of giving a warmstart solution to the solver in order to make it converge faster.

As it can be noticed in Table 2 and in Fig. 5 , the RMSE for the distributed Moving Horizon Estimation with pre-estimation are slightly larger than the ones of DMHE without pre-estimation. This is an expected behavior since the pre-estimation strategy compensates model mismatch by the correction term of equation (35) instead of estimating the process noise sequence w_k over the estimation horizon, as in "classical" DMHE without pre-estimation. This feature has previously been observed in the linear case in Venturino et al. (2020).

Table 2 does not show a decreasing trend of the RMSE with the horizon length N for the DMHE-based observers, as it can usually be the case in centralized MHE strategies (Müller 2017; Suwantong et al. 2014), where the more information is gathered the more accurate is the estimate, getting close to the one provided by the so-called full information estimator (Al-Matouq and Vincent 2014). This non-monotonic relationship between the RMSE and N can be observed in the graphs

Table 2 RMSE, final RMSE, and mean computation time (CT) for the tested estimators (average values over all the sensors, Monte Carlo runs, and time instants for RMSE and computation time)

Observer	Horizon N	RMSE	Final RMSE	Position RMSE	Mean CT (s)
DEKF-CI	–	1.4837	0.6680	0.8077	0.00023
DMHEpreEKF	0	1.2946	0.6986	0.6796	0.01320
DMHEpreEKF	1	1.3842	0.7750	0.6932	0.02366
DMHEpreEKF	2	1.4026	0.8045	0.7152	0.03252
DMHEpreEKF	3	1.3987	0.8153	0.7269	0.04083
DMHEpreEKF	4	1.3902	0.8176	0.7325	0.04874
DMHEpreEKF	5	1.3838	0.8162	0.7348	0.05626
DMHEpreEKF	8	1.3787	0.8070	0.7353	0.07655
DMHEpreEKF	10	1.3806	0.8031	0.7347	0.08783
DMHEpreEKF	15	1.3840	0.8004	0.7339	0.10965
DMHE-1	0	1.2946	0.6986	0.6796	0.01313
DMHE-1	1	1.3325	0.7486	0.6761	0.03018
DMHE-1	2	1.3452	0.7820	0.6997	0.05247
DMHE-1	3	1.3352	0.7955	0.7123	0.08055
DMHE-1	4	1.3276	0.8006	0.7194	0.11450
DMHE-1	5	1.3260	0.8029	0.7236	0.15519
DMHE-1	8	1.3360	0.8027	0.7308	0.31850
DMHE-1	10	1.3406	0.8002	0.7327	0.45269
DMHE-1	15	1.3436	0.7960	0.7320	0.83589

of Fig. 5. It has also been observed in the literature, e.g., Li et al. (2023); Sui and Johansen (2014); Venturino et al. (2021a) for DMHE and DMHE with pre-estimation strategies.

This can be explained by the accumulation of uncertainties related to unknown inputs along the estimation horizon. In the considered case, these unknown inputs are indeed not necessary well estimated. Due to the matrix B shape, the unknown input has more influence on the velocity part of the target state, which is not well represented in the measurement function. Thus, it makes more difficult to reconstruct the unknown input from the measurements information. This is also visible through the fact that, for $N = 0$, better RMSE are obtained. It is worth noting that the DMHE-1 optimization problem with an horizon $N = 0$ is exactly equivalent to the DMHE-pre-EKF formulation, which explains the same RMSE results as in Table 2.

It can be noticed that the constraints related to the road boundaries taken into account in the DMHE-1 and DMHE-pre-EKF methods allow to reduce faster the estimation error (RMSE), especially at the beginning of the simulations (see DMHE-1 and DMHE-pre-EKF curves in Fig. 3). This helps these two types of estimators to quickly compensate for the poor initialization, which is of practical interest for problems such as area monitoring. Figure 4 illustrates an increased computation

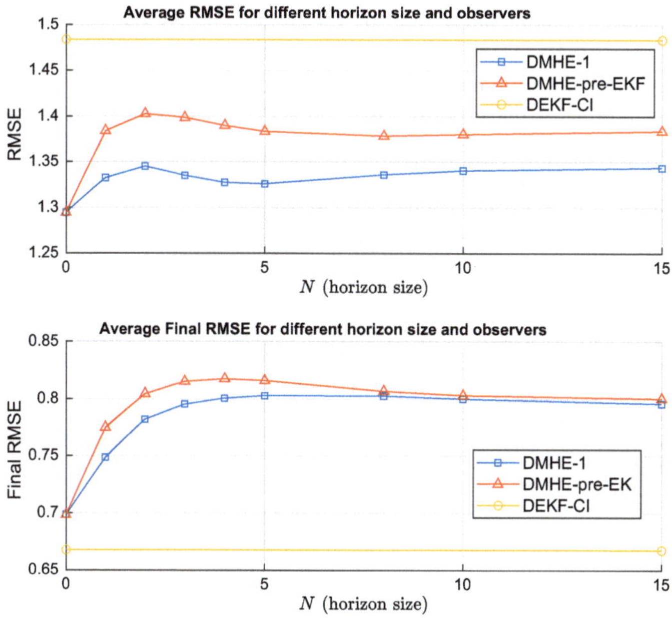

Fig. 5 RMSE and final RMSE obtained for the tested estimators with different horizons (average value over all the sensors and all the Monte Carlo runs)

time for DMHE-1 and DMHE-pre-EKF during the growing horizon phase. Indeed, during this phase, the optimization problem complexity increases in term of number of optimization variables (for DMHE-1) and constraints (for both DMHE-1 and DMHE-pre-EKF). Notably, the computation time increases more for DMHE-1 than for DMHE-pre-EKF. For a given horizon length, the DMHE-pre-EKF allows to significantly reduce the computation time compared to the DMHE-1, due to the reduction of the number of optimization variables, while keeping an estimation error sensibly close to the one obtained by the DMHE-1.

7 Conclusions

This chapter introduces a DMHE algorithm incorporating an EKF-based pre-estimation technique, tackling the issue of constrained localization of a non-cooperative target by a sensor network. The algorithm takes into consideration constraints on the target state, unknown inputs, measurement and process noises, as well as nonlinear measurements performed by the sensors and exchanges of information among them through a communication network. Comparisons are made between the proposed approach and various existing distributed observers, including classic DEKF that does not handle constraints.

The simulation results show that the distributed state estimation approach that can handle constraints (DMHE and DMHE with EKF pre-estimation) outperforms the DEKF (which do not handle constraints). Moreover, the simulation results also confirm the interest of the proposed method (DMHE with EKF pre-estimation) in handling constraints, and keeping a reduced computation load compared to standard DMHE approaches, while preserving a good estimation accuracy.

Future work will consider experiments of the proposed approach on a real hardware setup, as well as further performance comparisons with other distributed state estimation algorithms.

Acknowledgements The authors acknowledge support from the STIC doctoral school / Université Paris-Saclay, ONERA, Agence de l'innovation de défense (Grant 2022-65-0011-AID _ONERA), Agence Nationale de Recherche (ANR)-France (Grant ANR-21-CE48-0003), and the European Research Council under the advanced grant OCONTSOLAR Grant agreement ID: 789051. T. Alamo acknowledges support from grant PID2022-142946NA-I00 funded by MICIU/AEI/ 10.13039/501100011033 and by ERDF/EU.

References

Al-Matouq A, Vincent T (2014) Multiple window moving horizon estimation. Automatica 53

Alriksson P, Rantzer A (2006) Distributed Kalman filtering using weighted averaging. In: 17th International symposium on mathematical theory of networks and systems

Battistelli G (2018) Distributed moving-horizon estimation with arrival-cost consensus. IEEE Trans Autom Control 64(8):3316–3323

Battistelli G, Chisci L (2014) Kullback-Leibler average, consensus on probability densities, and distributed state estimation with guaranteed stability. Automatica 50(3):707–718

Battistelli G, Chisci L, Morrocchi S, Papi F (2011) An information-theoretic approach to distributed state estimation. IFAC Proc Vol 44(1):12477–12482

Battistelli G, Chisci L, Mugnai G, Farina A, Graziano A (2015) Consensus-based linear and nonlinear filtering. IEEE Trans Autom Control 60(5):1410–1415

Battistelli G, Chisci L (2014) Stability of consensus extended Kalman filtering for distributed state estimation. IFAC Proc Vol 47(3):5520–5525. 19th IFAC World Congress

Borelle M, Bertrand S, Stoica C, Alamo T, Camacho E (2023) Cooperative localization of an UAV fleet using distributed MHE with EKF pre-estimation and nonlinear measurements. 27th International conference on system theory. Control and computing. Timisoara, Romania, pp 143–148

Farina M, Ferrari-Trecate G, Scattolini R (2010) Distributed moving horizon estimation for linear constrained systems. IEEE Trans Autom Control 55(11):2462–2475

Farina M, Ferrari-Trecate G, Scattolini R (2012) Distributed moving horizon estimation for nonlinear constrained systems. Int J Robust Nonlinear Control 22(2)

He S, Shin HS, Xu S, Tsourdos A (2020) Distributed estimation over a low-cost sensor network: a review of state-of-the-art. Inf Fusion 54:21–43

Hermann R, Krener A (1977) Nonlinear controllability and observability. IEEE Trans Autom Control 22(5):728–740

Julier S, Uhlmann J (2001) General decentralized data fusion with covariance intersection (CI). Handbook of multisensor data fusion, Theory and practice

Li X, Bo S, Qin Y, Yin X (2023) Iterative distributed moving horizon estimation of linear systems with penalties on both system disturbances and noise. Chem Eng Res Des 194:878–893

Madhavan R, Fregene K, Parker L (2002) Distributed heterogeneous outdoor multi-robot localization. In: Proceedings 2002 IEEE international conference on robotics and automation, vol 1, pp 374–381

Mahmoud MS, Khalid HM (2013) Distributed Kalman filtering: a bibliographic review. IET Control Theory Appl 7(4):483–501

Martinelli A (2019) Nonlinear unknown input observability: extension of the observability rank condition. IEEE Trans Autom Control 64(1):222–237

Müller MA (2017) Nonlinear moving horizon estimation in the presence of bounded disturbances. Automatica 79:306–314

Olfati-Saber R (2007) Distributed Kalman filtering for sensor networks. In: 46th IEEE conference on decision and control, pp 5492–5498

Rao CV, Rawlings JB, Lee JH (2001) Constrained linear state estimation—a moving horizon approach. Automatica 37(10):1619–1628

Rao CV, Rawlings JB, Mayne D (2003) Constrained state estimation for nonlinear discrete-time systems: stability and moving horizon approximations. IEEE Trans Autom Control 48:246–258

Rong Li X, Jilkov V (2003) Survey of maneuvering target tracking. Part I. Dynamic models. IEEE Trans Aerosp Electron Syst 39(4):1333–1364

Segovia P, Puig V, Duviella E (2022) Set-membership-based distributed moving horizon estimation of large-scale systems. ISA Trans 128:402–413

Simonetto A, Balzaretti D, Keviczky T (2011) A distributed moving horizon estimator for mobile robot localization problems. IFAC Proc Vol 44(1):8902–8907

Sui D, Johansen TA (2014) Linear constrained moving horizon estimator with pre-estimating observer. Syst Control Lett 67:40–45

Suwantong R, Bertrand S, Dumur D, Beauvois D (2014) Moving horizon estimation with pre-estimation (MHE-PE) for 3D space debris tracking during atmospheric re-entry. In: 53rd IEEE conference on decision and control, pp 254–259

Suwantong R, Bertrand S, Dumur D, Beauvois D (2014) Stability of a nonlinear moving horizon estimator with pre-estimation. In: 2014 American control conference, pp 5688–5693

Venturino A, Stoica Maniu C, Bertrand S, Alamo T, Camacho EF (2023) Multi-vehicle localization by distributed MHE over a sensor network with sporadic measurements: further developments and experimental results. Control Eng Pract 132:105410

Venturino A, Bertrand S, Stoica Maniu C, Alamo T, Camacho EF (2020) Distributed moving horizon estimation with pre-estimating observer. In: 24th International conference on system theory, control and computing, pp 174–179

Venturino A, Bertrand S, Stoica Maniu C, Alamo T, Camacho EF (2022) Multi-vehicle system localization by distributed moving horizon estimation over a sensor network with sporadic measurements. In: 2022 IEEE conference on control technology and applications, pp 595–600

Venturino A, Bertrand S, Stoica Maniu C, Alamo T, Camacho EF (2021a) A new ℓ-step neighbourhood distributed moving horizon estimator. In: 60th IEEE conference on decision and control, pp 508–513

Venturino A, Stoica Maniu C, Bertrand S, Alamo T, Camacho EF (2021b) Distributed moving horizon state estimation for sensor networks with low computation capabilities. Syst Theory Control Comput J 1(1):81–87

Wang Y, Alamo T, Puig V, Cembrano G (2017) Distributed zonotopic set-membership state estimation based on optimization methods with partial projection. IFAC-PapersOnLine 50(1):4039–4044. 20th IFAC World Congress

Welch G, Bishop G (2006) An introduction to the Kalman filter. Proc SIGGRAPH Course 8

MIX
Papier aus verantwortungsvollen Quellen
Paper from responsible sources
FSC® C105338

If you have any concerns about our products,
you can contact us on
ProductSafety@springernature.com

In case Publisher is established outside the EU,
the EU authorized representative is:
Springer Nature Customer Service Center GmbH
Europaplatz 3, 69115 Heidelberg, Germany

Printed by Libri Plureos GmbH
in Hamburg, Germany